Electronic Devices and Circuits

Electronic Devices and Circuits

S Rama Reddy

Alpha Science International Ltd.
Pangbourne, U.K.

S Rama Reddy
Dean
Jerusalem College of Engineering
Chennai, India

Alpha Science International Ltd.
P.O. Box 4067, Pangbourne RG8 8UT, UK

ISBN 1-84265-124-2

Printed in India.

Preface

Electronic Devices and Circuits is a core course for degree students in Electrical, Electronics, Instrumentation and Control Engineering. This book is written to meet the needs of Under Graduate Students. It will also be useful for Polytechnic, AMIE and AMIETE Students.

This book deals with fundamentals of Electronic Devices and Circuits. The subject is presented in a lucid and easy manner.

Chapter 1 deals with with Electron Ballastics. Chapter 2 describes semi conductor materials. Chapter 3 describes P-N junction. Chapter 4 deals with Bipolar Junction Transistor. Chapter 5 describes various devices. Chapter 6 deals with rectifiers. Chapter 7 describes biasing methods of BJT. Chapter 8 discusses analysis of amplifiers. Chapter 9 deals with Power Amplifiers. Chapter 10 deals with negative feedback amplifiers and oscillators. Chapter 11 deals with tuned amplifiers and multivibrators. Chapter 12 deals with differential amplifiers. Chapter 13 deals with operational amplifiers.

The author wishes to express his deep sense of gratitude to CEO, Jerusalem College for their keen interest and encouragement in bringing out this book.

Special thanks are due to Mr. R. Kumar, Mr. Trinath, Ms. Darly, Mr. R. Rajesh and Mr. Beuton for their help. The author owes his gratitude to his colleagues for their help.

Constructive criticism for the improvement of this book will be highly appreciated.

Dr. S. RAMA REDDY MIE, FIETE
Dean
Jerusalem College of Enggineering, Chennai

Contents

7. Biasing Methods of transistor

1

Electron Ballastics

1.1 Introduction

In this chapter, details of Cathore ray oscilloscope, types of deflection and their comparison are discussed.

1.2 Cathode Ray Tube

The most important element of the CRO system is the cathode ray tube (CRT) shown in Fig. 1.1. The CRT mainly consists of three sections: Electron gun, Deflection plates and Screen.

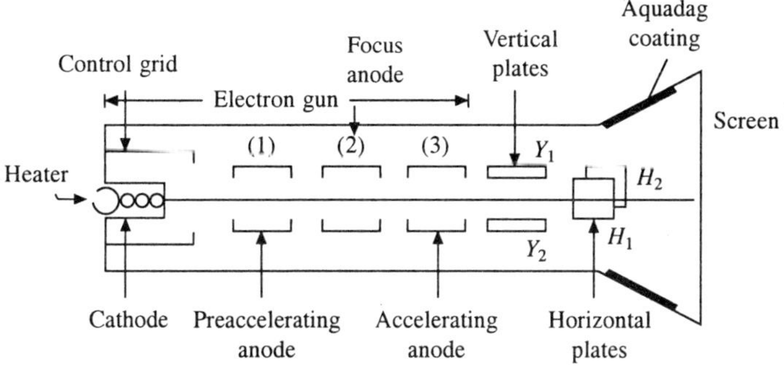

Fig. 1.1 Cathode ray tube.

Electron Gun: An indirectly heated cathode is used to generate electrons. The heat energy required for thermionic emission is supplied by the heater. A control grid with a small aperture is provided to control the electrons emitted. When the electron enters the electric field it experiences a force and it deflects. 3-anodes are used to provide electrostatic deflection. The potential value of anodes (1) and (3) are higher than that of anode (2). The electric field and equipotential lines act as double concave lens for the electron beam. The focus of the beam can be controlled by controlling the voltage of the accelerating anodes (1) and (3).

Deflection Plates: Vertical plates Y_1 and Y_2 are provided. The unknown voltage to be measured is applied to the vertical plates. These plates control the vertical motion of the beam. A saw tooth voltage is applied to the horizontal plates H_1 and H_2. This voltage forces the beam to move horizontally from left to right. The resulting motion of the beam depends on the horizontal and vertical forces.

Screen: The screen is provided with fluorescent coating. When the electron beam hits the screen a dot appears on the screen as the energy of the electron beam is converted into light energy. Electrostatic and Magnetic deflection systems are discussed in the mext section.

1.3 Electrostatic Deflection

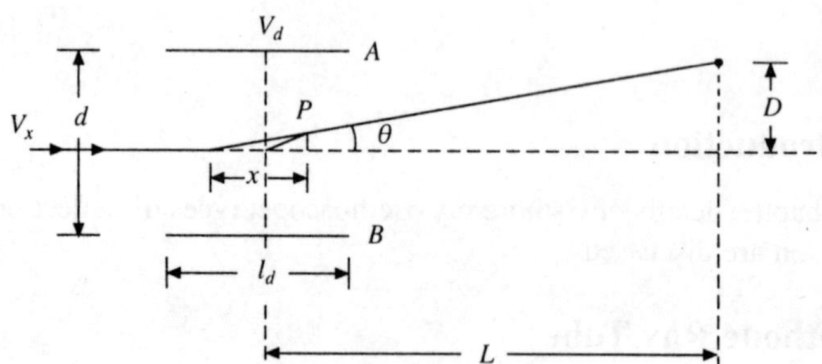

Fig. 1.2 Electrostatic deflection.

Let V_a be the voltage of pre-acceleration anode
V_d the voltage applied to the vertical plates
V_x the velocity of the electron before entering the *y*-plates
d the distance between the vertical plates A and B
l_d the length of the deflection plates
L the distance between the centre of *y*-plate to screen
D the deflection along the positive *y*-direction

When the electron resides in the cathode, it has some potential energy. By thermionic emission the electron comes out. This is accelerated by the field of accelerating anodes. When the electron enters into the field of vertical plates, it experiences a force and deflects. The beam appears as a dot along the positive *y*-direction. This is a single dot on the screen since no voltage is applied to the *x*-plates.

Potential is defined as the work done (W.D.) per charge

$$V_a = \frac{\text{W.D.}}{e}$$

where *e* is the charge of an electron.

From above, we have $$\text{W.D.} = V_a \times e$$

$$\text{Work Done} = \text{Potential Energy (P.E.)}$$

$$\text{P.E.} = V_a e$$

When the electron gets deflected it losses potential energy and gains *kinetic energy*. These two energy are equal.

$$\text{P.E.} = \text{K.E.}$$

$$V_a e = (1/2)\, m V_x^2$$

Therefore, $$V_x^2 = \frac{2 V_a e}{m} \tag{1.1}$$

When the electron enters the field of y-plates a force in the y-direction is also experienced.

$$\text{Electric field } [E] = V_d/d$$

Consider a point P on the trajectory. From the Newton's 2nd law, we know that

$$F = ma; \qquad E = F/e; \qquad F = E \cdot e;$$

Substitute $E = V_d/d$

$$F = \frac{V_d}{d}e; \qquad \frac{V_d e}{d} = ma \tag{1.1b}$$

$$a = \frac{V_d e}{d \cdot m} \tag{1.2}$$

From dynamics we know that

$$Y = ut + 1/2\ at^2$$

Just before entering y-plates the velocity component along the y-direction is zero since V_x has only x-component, $u = 0$.

Therefore, $$y = (1/2)\ at^2$$

Substitute (1.2); hence $$y = \frac{1}{2}\frac{V_d e}{d \cdot m}t^2 \tag{1.3}$$

We know that $v_x = \frac{x}{t}$ or $t = \frac{x}{v_x}$; $t^2 = \frac{x^2}{v_x^2}$

Substitute this in (1.3)

Therefore, $$y = \frac{1}{2}\frac{V_d e}{dm}\frac{x^2}{v_x^2} = \frac{1}{2}\frac{V_d e x^2}{dm v_x^2}$$

$$\frac{dy}{dx} = \frac{1}{2}\frac{V_d e}{dm}\frac{2x}{v_x^2} = \frac{V_d e x}{dm v_x^2}$$

Substituting $v_x^2 = \frac{2V_a e}{m}$ here

$$\frac{dy}{dx} = \frac{V_d}{dm}\frac{ex}{2V_a e/m} = \frac{V_d x}{2dV_a}$$

When $x = l_d$, $\tan\theta = D/L$, $y = D$. Hence

$$\frac{D}{L} = \frac{dy}{dx} = \frac{V_d l_d}{2dV_a}$$

When $x = l_d$ the entire length of the plates are considered or entire electric field is considered.

$$D = \frac{L l_d V_d}{2d\,V_a}$$

From the above equation, it can be seen that deflection is directly proportional to the voltage applied to y-plates.

Deflection sensitivity (S) is defined as the ratio of deflection to the voltage applied between the *y*-plates.

$$S = \frac{D}{V_d}$$

$$S = \frac{Ll_d}{2dV_a} \tag{1.4}$$

Deflection factor *G* is defined as inverse of sensitivity

$$G = \frac{1}{S}; \quad G = \frac{2dV_a}{Ll_d} \tag{1.5}$$

1.4 Magnetic Focusing

Cathode ray tube using magnetic deflection is shown in Fig. 1.3.

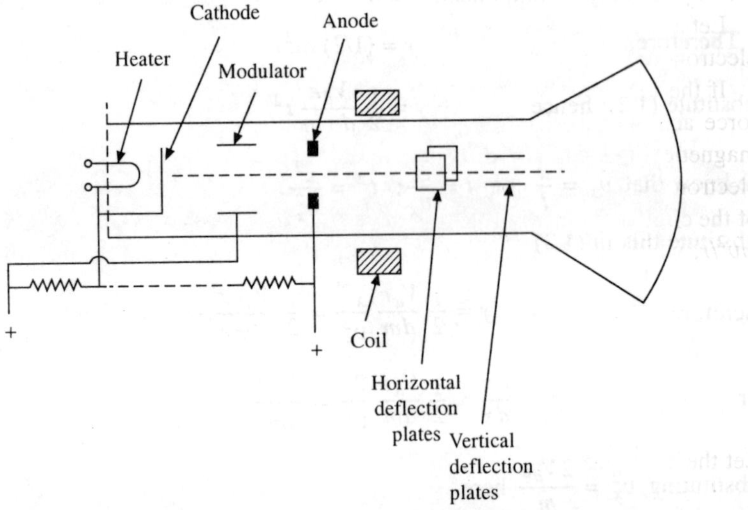

Fig. 1.3 CRT using magnetic deflection system.

An indirectly heated cathode is used to generate electrons. The electrons are produced using the principle of thermionic emission. Modulator is similar to the control grid. Modulator is made negative with respect to cathode. This prevents the electrons trying to move away from the main stream since electrons are repelled by the negative polarity of the modulator. Anode is made positive with respect to cathode, therefore electrons are accelerated towards the anode. The electrons flow through the aperture of the anode. A circular current carrying coil produces a magnetic field in a direction perpendicular to the plane of the coil. According to Lorentz, when an electron beam enters in the magnetic field it experiences a force. This force can be controlled by controlling the current through the coil. The beam is used for *x*-*y* plotting. The beam is controlled by controlling the voltages applied to the *x* and *y* plates. Horizontal motion is controlled by controlling the voltage applied to the x-plates and the vertical motion is controlled by controlling the voltage applied to the y-plates.

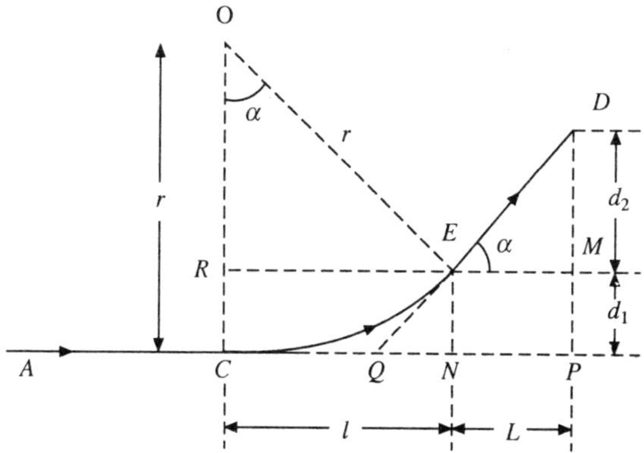

Fig. 1.4 Magnetic deflection.

Let a uniform magnetic field B act on the electron beam AC emitted from the electron gun over a length l of its path.

If the magnetic field is perpendicular to the plane of the paper, a magnetic force acts on the electron along the plane of the paper at right angles to the magnetic field and the direction of motion of the electron. As a result, the electron beam moves along a circular arc CE in the magnetic field. The radius r of the circle is found by equating the magnetic force Bev to the centripetal force mv^2/r, where is v the velocity of an electron of mass m and charge e. Hence

$$\frac{mv^2}{r} = Bev$$

or

$$r = \frac{mv}{Be} \tag{1.6}$$

Let the accelerating potential be V. Then

$$\frac{1}{2}mv^2 = eV$$

or

$$v = \sqrt{\frac{2eV}{m}} \tag{1.7}$$

Using Equation (1.7) in Equation (1.6), we get

$$r = \frac{1}{B}\sqrt{\frac{2mV}{e}} \tag{1.8}$$

Upon emergence from the magnetic field at the point E, the electron beam travels along the straight line ED and strikes the screen at the point D. The straight line ED is tangent to the circular arc CE at the point E.

At the point of exit from the magnetic field, the deflection of the electron is

$$d_1 = PM = NE = CR = OC - OR = r(1 - \cos\alpha) \tag{1.9}$$

where $\alpha = \angle COE \cdot AS\ \angle DEM = \angle COE = \alpha$, we obtain

$$d_2 = L \tan \alpha \tag{1.10}$$

Here $L = EM = NP$. In practice, the angle α is small enough so that only the terms up to α^2 are important. Therefore,

$$\tan \alpha = \alpha = \frac{l}{r} \tag{1.11}$$

and
$$\cos \alpha = 1 - \frac{\alpha^2}{2} = 1 - \frac{1}{2}\left(\frac{l}{r}\right)^2 \tag{1.12}$$

Substituting for tan α and cos α from equation (1.11) and (1.12) in equations (1.10) and (1.9) respectively, we have

$$d_2 = \frac{Ll}{r} \tag{1.13}$$

and
$$d_1 = \frac{l^2}{2r} \tag{1.14}$$

The total spot deflection on the screen is

$$d = d_1 + d_2 = \frac{l^2}{2r} + L\frac{l}{r} = \frac{l}{r}\left[L + \frac{l}{2}\right] \tag{1.15}$$

The distance $L + \frac{1}{2}$ equals QP, the distance of the screen from the centre of the magnetic field region. Substituting r from equation 1.8 into equation 1.15, gives

$$d = Bl\sqrt{\frac{e}{2Vm}}\left[L + \frac{l}{2}\right] \tag{1.16}$$

By definition, the magnetic deflection sensitivity is

$$S_m = \frac{d}{B} = l\sqrt{\frac{e}{2V_m}}\left[L + \frac{l}{2}\right] \tag{1.17}$$

If l and L are in meter, e in coloumb, V in volt, and mass in kg in equation (1.17), then S_m is expressed in *m*/tesla.

1.5 Comparison of Magnetic and Electrostatic Deflection Systems

1. The power required by magnetic deflection system is more than the power required by electrostatic system.
2. The length of the circuit using electrostatic deflection is much higher than that of magnetic system.
3. Electrostatic system is used in CRO. Magnetic system is used in TV picture tubes.

1.6 CRO System

Block diagram of CRO system is shown in Fig. 1.5. The intensity can be varied by varying the negative voltage applied to the control grid. Focus can be varied by varying the potential values of accelerating anodes. V_m is the unknown voltage to be measured using CRO. The vertical attenuator can reduce the voltage level if the input voltage is higher. If the input signal to be measured is very weak then it is amplified using the vertical amplifier before it is applied to the vertical plates.

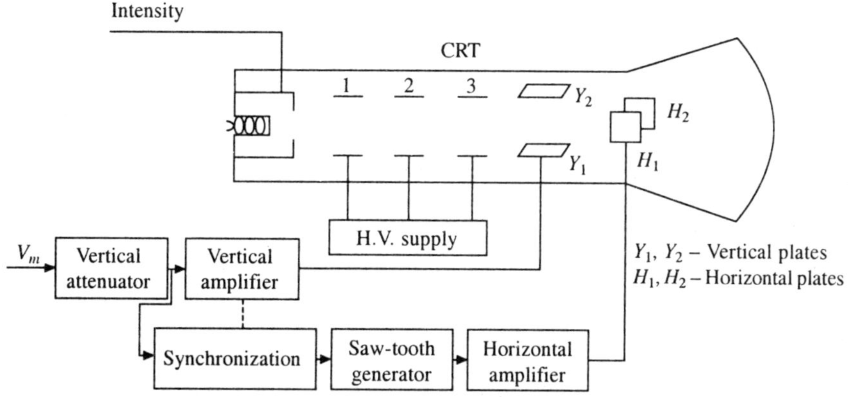

Fig. 1.5 CRO system.

The saw-tooth voltage or time base voltage applied to the horizontal plates has to be synchronized with the input voltage to obtain a steady wave form. The horizontal amplifier amplifies the saw-tooth voltage. Synchronization circuit has a zero crossing detector (ZCD). This ZCD gives the zero instants required for synchronization.

1.6.1 Saw-tooth Generator

Saw-tooth generator circuit, equivalent circuit and waveforms are shown in Fig. 1.6(a), (b) and (c). Rectified D.C is applied to the UJT relaxation oscillator circuit. This ensures that zero instant of the input voltage and zero instant of saw tooth voltage coincide. The capacitor charges through the high resistance R when the UJT is OFF. It slowly charges in t_1. Beyond t_1 the capacitor voltage is higher than intrinsic potential. The diode is on and the capacitor quickly discharges through the low resistance R_2. Beyond t_2 the capacitor once again gets charged slowly through the resistance R. Charging is very slow due to the high resistance value of R. Discharging is very fast due to the low resistance value of R_2. The voltage across the capacitor is a saw-tooth voltage. The frequency of this voltage can be varied by varying the resistance R.

1.6.2 Front View of CRO

Front view of a CRO is shown in Fig. 1.7. Intensity and focus of the beam can be varied by varying the upper two knobs. The frequency of the saw-tooth generator can be varied by varying the time base selector switch. The time period for one cycle is obtained as the product of milli sec/division and number of divisions per cycle. Frequency = 1/time period. Thus the frequency of unknown

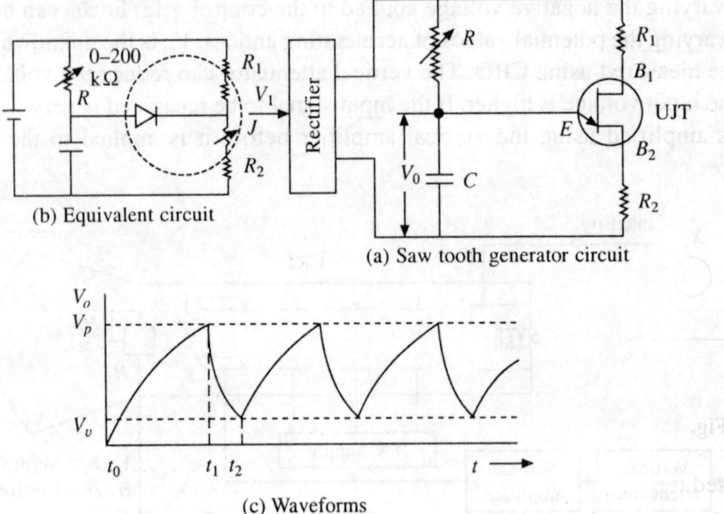

Fig. 1.6 Saw-tooth generator.

signal is obtained by using the frequency of the saw-tooth waveform. Volt/Division knob attenuates the input voltage to be measured. Unknown D.C. voltage is equal to the product of Volt/Div selected and number of Y-divisions. By using X-position the beam can be moved to the left or to right. By using Y-position the beam can be moved up-or down. The unknown voltage is applied across the terminals Y_1 and Y_2. Another signal can also be applied between X_1 and X_2. If an external voltage is applied to X_1 and X_2, the internal saw-tooth generator is switched off.

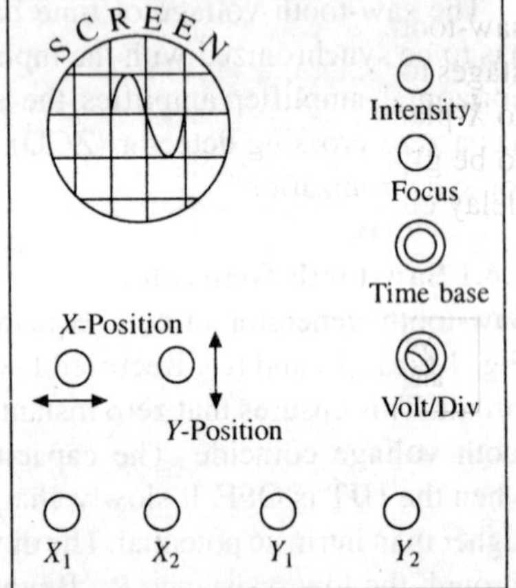

Fig. 1.7 Front view of CRO.

1.6.3 Measurements with CRO

1. Measurement of Voltage

Voltage measurement circuit is shown in Fig. 1.8. If no voltage is applied to the X-plates, the following method is used to obtain the unknown voltage.

If unknown D.C. voltage is applied between the vertical plates a dot appears on the screen. The height of the dot is proportional to the unknown voltage. If A.C voltage is applied between Y_1 and Y_2, the dot moves up and down at a faster rate and a vertical line can be observed on the screen. The total height on the screen represents the peak to peak value. RMS value = Peak value/ $\sqrt{2}$.

2. Measurement of Current

Current measurement circuit is shown in Fig. 1.9. The current to be measured is

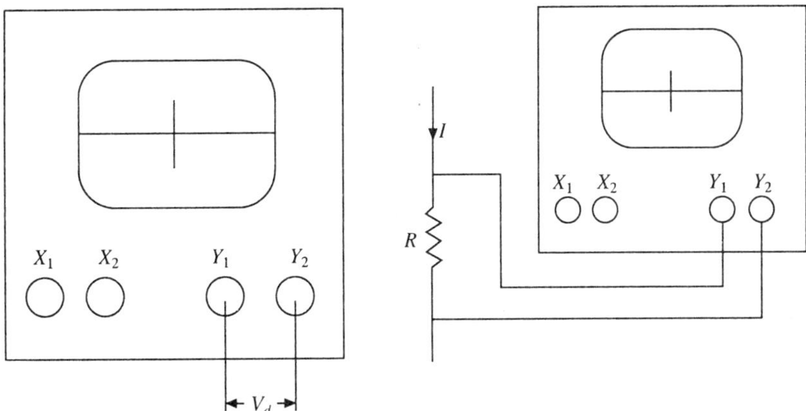

Fig. 1.8 Measurement of voltage. **Fig. 1.9** Measurement of current.

passed through the known resistance R. The voltage across R is applied to the Y plates. Unknown current is the ratio of measured voltage to the known resistance.

1.6.4 Delay Line

Delay line circuit is shown in Fig. 1.10. The horizontal plates receive the voltage using 3-stages. The 1st stage is synchronization circuit which synchronizes the saw-tooth voltage to the unknown voltage. Time base generator produces the saw-tooth voltage. This is amplified by using a horizontal amplifier. Both the stages together introduce a delay of about 90 nano-seconds. The voltages applied to X-plates and Y-plates should be in phase. An equal amount of time delay has to be given to the voltage to be applied to Y-plates. This is obtained by using a delay circuit. Delay circuit consists of several lumped R, L and C.

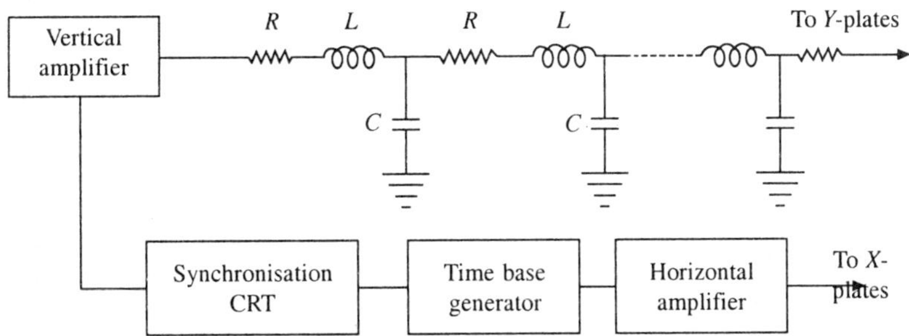

Fig. 1.10 Delay line circuit.

1.6.5 Z-Modulation

The waveforms for Z-modulation are shown in Fig. 1.11. An ideal saw-tooth wave has zero fall time. This will produce a perfect beam on the screen. A practical saw-tooth wave has a finite fall time. During this fall time, a small line appears on the screen. This can be avoided by applying large negative voltage to the control grid so that the beam can be suppressed during the short period. This is called Z-modulation.

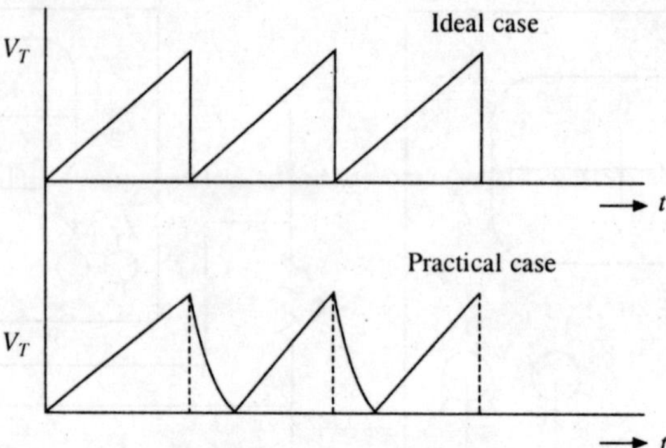

Fig. 1.11 Z-modulation.

1.6.6 Dual Trace CRO

Block diagram of dual trace CRO is shown in Fig. 1.12. A dual trace CRO consists of single CRT. Two channels are provided so that two wave forms can be observed simultaneously. Two voltage signals having common ground have to be applied at channel 1 and channel 2. A single focus control is provided. Also there is only one control knob for intensity. Amplitudes of wave forms can be independently controlled by varying the respective attenuators.

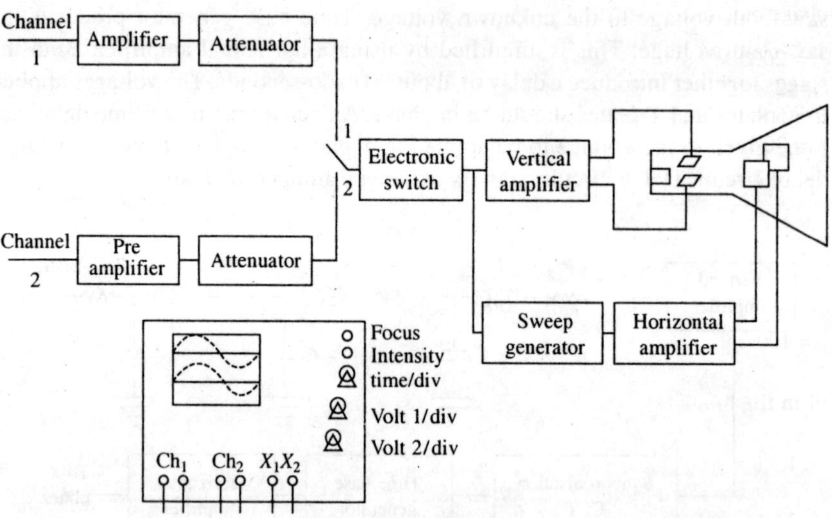

Fig. 1.12 Dual trace CRO.

Preamplifiers used are common drain FET amplifier or source follower circuits. This type of amplifier has very high input impedance, unity voltage gain and large current gain. A fast acting electronic switch connects alternately channel 1 and channel 2 to vertical amplifier circuit.

Two wave forms are observed on the screen since the electronic switch circuit operates at very high frequency. Saw-tooth voltage of the time base generator is amplified using horizontal amplifier before it is applied to the *X*-Plates.

Worked Examples

Ex. 1.1. An electrically deflected circuit has a final anode voltage of 2000 volts and parallel deflecting plates of 1.5 cm long and 5 mm apart. If the screen is 500 mm from the center of the deflecting plates. Find (a) beam speed (b) deflection sensitivity (c) deflection factor.

$V_a = 2000$ volt
$l_d = 1.5 \text{ cm} = 1.5 \times 10^{-2}\text{m}$
$d = 5 \times 10^{-3}$ m
$L = 500 \times 10^{-3} \text{ m} = 0.5$ m

$$V_x = \sqrt{\frac{2eV_a}{m}} = \sqrt{\frac{2 \times 1.602 \times 10^{-19} \times 2000}{9.1 \times 10^{-31}}}$$

$V_x = 0.59 \times 10^{-6}$ m/s

$$S = \frac{Lld}{2dV_a} = \frac{0.5 \times 1.5 \times 10^{-2}}{2 \times 5 \times 10^{-3} \times 2000}$$

$S = 0.375$ mm/V
$G = 1/s = 1/0.375 = 2.66$ V/mm

Ex. 1.2. A circuit has an anode voltage of 2000 volts and parallel deflecting plates 2 cm long and 5 mm apart. The screen is 30 cm from the center of the plates. Find the input voltage required to deflect the beam through 3 cm.

$V_a = 2000$ V
$l_d = 2 \times 10^{-2}\text{m} = 0.02$ m
$d = 5 \times 10^{-3}$ m
$L = 30 \times 10^{-2}\text{m} = 0.3$ m

$$D = \frac{Ll_d V_d}{2\,dV_a}$$

$$V_d = \frac{2dV_a D}{Ll_d} = \frac{2 \times 5 \times \overline{10}^{3} \times 2000 \times 3 \times 10^{-2}}{0.3 \times 0.02}$$

$V_d = 100$ volts

Fill in the blanks

1. *Electron gun* assembly produces the sharply focused beam of electrons which are accelerated to high velocity. Explain.
2. In a CRT, phosphor absorb *kinetic energy* of bombarding electron.
3. Dual trace CRO has *one* number of electron gun.
4. Internally generated ramp voltage in CRO is called *time base voltage*.
5. *Deflection sensitivity* of a CRT is defined as deflection per unit deflection voltage.
6. Screen of CRO is coated with *phosphorus* coating materials.
7. Time base is connected to *horizontal plates* of CRO.
8. Time delay and phase can be measured using *CRO*.

2

Semiconductor Basics

2.1 Introduction

This chapter deals with the fundamentals of intrinsic semiconductor and extrinsic semiconductor. Hall effect and light defendant resistor (LDR) are also discussed here.

2.2 Intrinsic Semiconductor

An intrinsic semiconductor is a pure semiconductor. Silicon and germanium are the semiconductor materials. To form a stable covalent bond, 8 valence electrons are required. The silicon atom at the centre has 4 valence electrons as shown in Fig. 2.1(a). It shares 4 electrons from the neighbour atoms to form the covalent bond. At absolute zero temperature, no energy is supplied to the crystal. All the electrons are engaged in forming the covalent bond and no free electrons are available. Hence there is no conduction. Thus the semiconductor acts as insulator at 0°K. The conduction band is empty, as no conduction electrons are available. When thermal energy is supplied to the semiconductor, some of the covalent bonds are broken due to the energy supplied. These electrons jump from the valence band (VB) to the conduction band (CB).

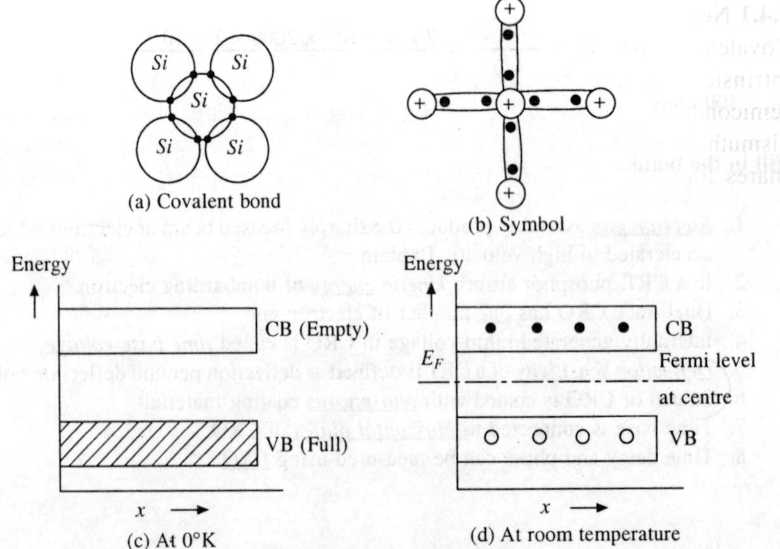

Fig. 2.1 Energy band diagram.

2.3 Drift Current

A semiconductor kept in the field of a capacitor is shown in Fig. 2.2. The direction of electric field is away from the positive charge. The energy supplied by the electric field can break a covalent bond at *E* and a vacancy is created at *E*. To fill this vacancy, the electron at *D* will jump and occupy the position *E*. Similarly, to fill the vacancy at *D* an electron from *C* will jump and occupy the position *D*. Thus there is a drift of electrons along the path *ABCDE*. The electrons move in a direction opposite to the direction of the applied electric field. The electrons will move towards the positively charged plate as the positive charge attracts the electrons. The current resulting due to the drift of electrons is known as the drift current. From the point form of ohm's law, $J = \sigma E$, where *J* is drift current density, a is the conductivity and *E* the electric field intensity. This equation indicates that the direction of *J* and *E* is same.

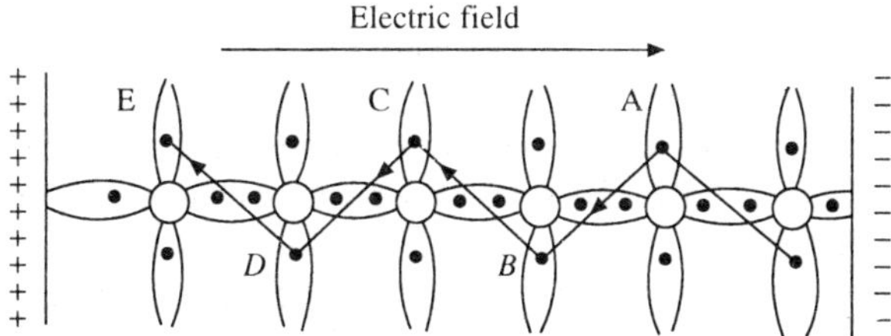

Fig. 2.2 Drift current.

2.4 Extrinsic Semiconductor

A doped semiconductor is called an extrinsic semiconductor. These semiconductors are classified as N-type and *P*-type.

2.4.1 Negative Type or *N*-Type

Covalent bond and energy band diagrams are shown in Fig. 2.3. When the intrinsic semiconductor is doped with pentavalent impurity, negative type semiconductor is formed. The pentavalent impurities are antimony, arsenic and bismuth. The pentavalent atom at the centre has 5 valence electrons. This atom shares four electrons from the neighbour atoms. For the formation of stable

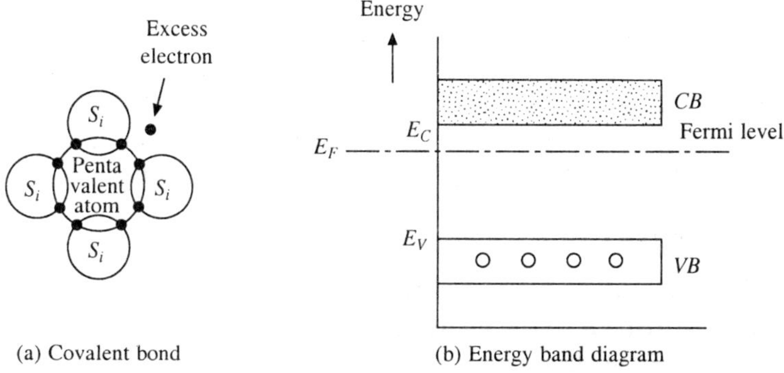

Fig. 2.3 Negative type or N-type.

covalent bond, only 8 electrons need to rotate in the valance orbit. Thus one excess electron is produced by each impurity atom. Several impurity atoms donate several electrons. Since the impurity atoms donate electrons, they are known as donors. A few covalent bonds are broken at room temperature due to the thermal energy supplied. The vacancies are shown as holes in the valence band. The majority carriers are electrons and the minority carriers are holes.

Fermi level corresponds to the centre of gravity of the electrons and holes. In the case of intrinsic semiconductor, the number of electrons are equal to the number of holes. The fermi level lies midway between the valence band and conduction band. In the N-type semiconductor the fermi level is lifted towards the conduction band as the conduction electrons are the majority carriers.

2.4.2 Positive Type or P-Type

When the intrinsic semiconductor is doped with trivalent atoms, positive type semiconductor is formed. The trivalent atoms are induium, gallium, boron and aluminium. The trivalent atom at the centre has 3 valence electrons. This atom shares 4 electrons from the neighbour atoms. 8 Electrons are required to form the valence orbit. In other words the trivalent atom can accept one electron. This vacancy is known as a hole. The holes have positive charge. Millions of impurity atoms can accept million of electrons. Hence they are called as acceptors. The majority carriers are holes and the minority carriers are electrons. The covalent bond and energy band diagram are shown in Fig. 2.4.

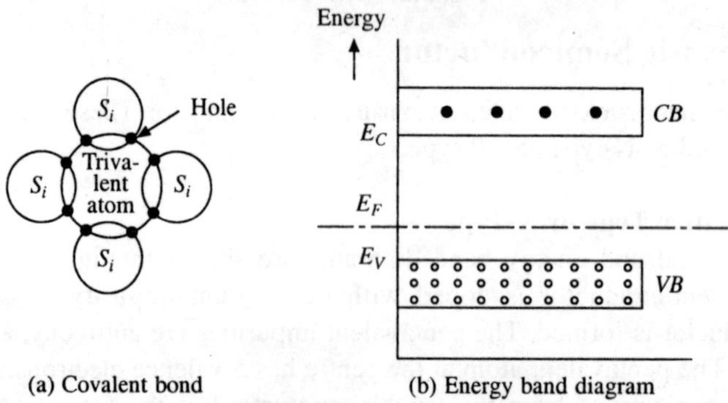

(a) Covalent bond (b) Energy band diagram

Fig. 2.4 Positive type or P-type.

The electrons jumped from the valence band to the conduction band due to the thermal energy are represented in the conduction band. The valence orbit of each impurity atom has one hole. Thus the holes in the valence orbits of the impurity atoms are represented in the valence band. The fermi level is shifted down as the majority carriers are the holes in the valence orbits.

2.5 Conductivity of Semiconductors

Let σ be the conductivity, μ_p the mobility of holes,
ρ the resistivity, μ_n the mobility of electrons,
A the area, E the electric field intensity,
n_i the number of electrons/m^3 and v the velocity.

Consider a semiconductor kept in the electric field. As electric field increases, the velocity of the electrons also increases. Therefore, velocity is proportional to E.

$$v \propto E$$

$$v = \mu E \tag{2.1}$$

$$\mu = \frac{v}{E}$$

Mobility μ is defined as average velocity per unit electric field. Consider a volume dV with area A and length dx. Let e be the charge of an electron. Let the charge in the volume dV be dQ.

Therefore, $dQ = n_i \times$ Volume $\times$ Charge of electron

$$dQ = n_i A\, dx\, e$$

Dividing both sides by dt

$$\frac{dQ}{dt} = n_i A \frac{dx}{dt} e$$

Substituting $\quad \frac{dx}{dt} = v$

$$I = n_i A v e \qquad \frac{I}{A} = n_i v e$$

Current density $\quad J = \frac{I}{A}$

Therefore,

$$J = n_i\, v e \tag{2.2}$$

Substituting (2.1) in (2.2), we get

$$J = n_i \mu E\, e$$

$$J = n_i\, \mu e\, E \tag{2.3}$$

By Ohm's law,

$$I = \frac{V}{R}$$

$$I = \frac{V}{\left(\frac{\rho l}{A}\right)} = \frac{VA}{\rho l}; \qquad \frac{I}{A} = \frac{V}{l}\frac{1}{\rho}$$

$$\frac{I}{A} = \frac{E}{\rho} = \sigma E \qquad \because \rho = \frac{1}{\sigma}$$

$$J = \frac{I}{A} = \frac{E}{\rho} = \sigma E$$

$$J = \sigma E \tag{2.4}$$

on comparing (2.3) and (2.4)

$$\sigma = n_i\, \mu e$$

In an intrinsic semiconductor the conductivity is due to electrons and holes. Total conductivity is the sum of electron conductivity and hole conductivity.

$$\sigma_T = \sigma_e + \sigma_h = n_i \mu_n e + n_i \mu_p e = n_i\, e\, [\mu_n + \mu_p]$$

$$\sigma_T = n_i e\, [\mu_n + \mu_p]$$

where σ_T is total conductivity of instrinsic semiconductor.

2.5.1 Conductivity of Extrinsic Semiconductor

In *N*-type, majority carriers are electrons and minority carriers are holes. The conductivity due to holes is neglected.

$$\sigma_T = \sigma_e + \sigma_h = \sigma_e + 0$$

$$\sigma_T = \sigma_e$$

$$\sigma_T = n \mu_n e \qquad (2.5)$$

where n = electrons or the number of impurity atoms/m^3.

In *P*-type the majority carriers are holes and minority carries are electrons.

$$\sigma = p \mu_p e \qquad (2.6)$$

where p is the number of holes/m^3.

Worked Problems

Ex. 2.1 Show that the resistivity of intrinsic germanium at 300°K is 45 ohm cm. If a donor impurity is added to the extent of 1 atom per 10^8 germanium atoms, prove that the resistivity of the intrinsic germanium drops to 3.7 ohm-cm. Assume the temperature to be at 300°K.

μ_n = 3800 cm^2/V sec
μ_p = 1800 cm^2/V sec
$n_i = 2.5 \times 10^{13}$ electrons/cm^3
No. of germanium atoms/cm^3 = 4.41×10^{22}

(i) Intrinsic or pure germanium

$$\sigma_T = n_i e(\mu_n + \mu_p) = 2.5 \times 10^{13} \times 1.6 \times 10^{-19}\,(3800 + 1800)$$

$$= 2.5 \times 1.6 \times 10^{-6}\,(5600) = 0.0224$$

$$\rho = \frac{1}{\sigma} = \frac{1}{0.0224} = 44.6 \text{ ohm-cm.}$$

(ii) For extrinsic semiconductor

$$\text{No. of electron } n = \frac{\text{No. of germanium atoms/cm}^3}{\text{Impurity concentration}}$$

$$= \frac{4.41 \times 10^{22}}{1 \times 10^8} = 4.41 \times 10^{14}$$

$$\sigma = n\, \mu_n\, e = 4.41 \times 10^{14} \times 3800 \times 1.6 \times 10^{-19} = 0.268128$$

$$\rho = \frac{1}{\sigma} = 3.7 \text{ ohm-cm}$$

2.6 Hall Effect

When a current carrying semiconductor is kept in the magnetic field, a voltage is induced in the hall element. This effect is called *Hall effect.*

In the Fig. 2.5 current flows along positive Y-direction. Magnetic field is applied along positive X-direction. Hall voltage is induced along the positive Z-direction in the case of N-type semiconductors. In the case of P-type, voltage is induced along negative Z-direction.

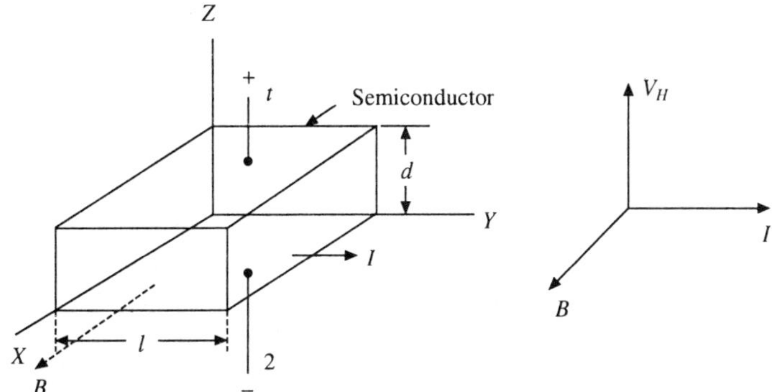

Fig. 2.5 Hall effect.

It can also be stated that electric field is produced along positive Z-direction in the case of N-type since,

$$E = \frac{V}{d}$$

The electrons in the N-block experience a force since they are kept on the electric field

$$F = Eq \tag{2.7}$$

According to Lorentz's Lorentz law, when a current carrying block is kept in the magnetic field it experiences a force

$$F = BIl \tag{2.8}$$

Under equilibrium these two forces must balance each other.

$$Eq = BIl$$

$$\frac{V_H}{d} q = BIl$$

$$\frac{V_H}{d} q = B\frac{q}{t} l$$

$$V_H = B\left(\frac{dl}{t}\right) \tag{2.9}$$

$$V_H = KB, \quad \text{where } K = \frac{dl}{t}$$

$$V_H \propto B$$

Hall voltage is directly proportional to the magnitude flux density.

Hall sensors are used for the measurement of flux density. They are very compact.

2.7 Light-Dependent Resistor (LDR)

A light-dependent resistor is a semiconductor device whose resistance varies with the amount of light energy imparted to it. Resistance of such a device is very high (of the order of several kilo-ohms or even mega ohms) when kept in total darkness. Resistance of the same device becomes very low (of the order of a few ohms only) when kept in a well illuminated area. As the amount of light intensity increases, the resistance of the device decreases. Resistance of this device is indirectly proportional to the amount of light falling on its surface. LDRs are also known as photo resistors. It has been observed that lower the initial conductivity of a semiconductor, higher is its photosensitive property. That is why photo resistors are prepared from lightly doped semiconductors.

Photo resistors are generally prepared either by coating a layer of powdered photosensitive material or by depositing a semiconductor film on an insulating base. Silicon is usually preferred for this purpose. The semiconductor film deposited on an insulating substance constitutes a unit which is placed in either a metallic or a plastic case whose top surface is covered with glass. The glass cover of the compact unit allows light rays to reach the semiconductor coated surface and accordingly the resistance changes. They are usually made in the form of cadmium sulphide discs provided with two tinned copper connecting leads. Rating and performance of the device, shown in Fig. 2.6 are characterized by the value of current flowing through the device at a given voltage and the amount of light flux. The rating and control capabilities of the device will depend upon the amount of semiconductor film deposited on the insulating substance. More the amount more will be the control capabilities. In other words, the cross-sectional area of the surface to be exposed to the light is one of the factors related to its rating. An LDR is specified according to the diameter of its surface. Because of very large difference in the value of its resistance between the (i) illumination state and (ii) dark state, the LDR behaves like a switch. In the presence of light it offers a very little (almost negligible) resistance in the circuit giving the ON state, whereas in darkness it offers a very high resistance which causes almost no flow of current thus resulting in the OFF state of the switch. Hence, the device acts like an automatic switch whose ON and OFF states are dependent on the total illumination and dark conditions respectively.

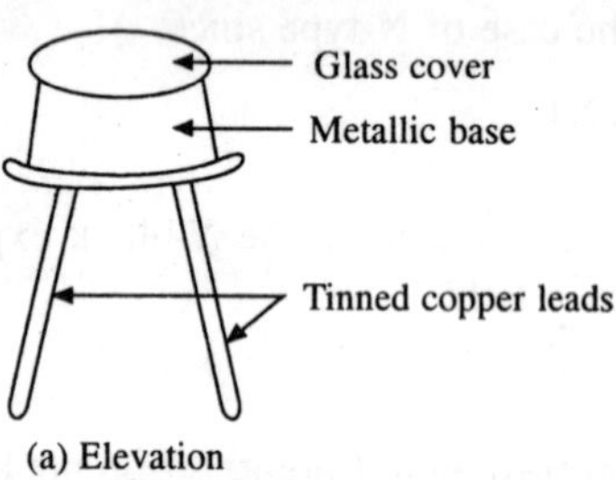

Fig. 2.6 Construction of an LDR.

Light-dependent resistors have a wide range of applications. Automatic street lighting, OFF at dark circuit, burglar alarm etc., are some of its important applications.

Worked Problems

Ex. 2.2 A sample of *N*-type semiconductor has a Hall co-efficient of 160 cm^3/col. If resistivity is 0.16 Ω-cm, estimate the electron mobility in the sample.

$$\sigma = \frac{1}{\rho} = \frac{1}{0.16}$$

$$\mu_n = k_H\,\sigma$$

$$\mu_n = 160 \cdot \frac{1}{0.16} = 1000\ \text{cm}^2/V\text{-Sec}$$

Ex. 2.3 A current of 50 A is passed through a semiconductor strip which is subjected to a flux density of 1.2 wb/m^2. Flux density is perpendicular to *I*. Hall volt is 100 V, thickness is 0.5 mm. Calculate number of electrons per m^3.

$$V_H = \frac{BI}{nqt}$$

$$n = \frac{B \cdot I}{V_H \cdot q \cdot t} = \frac{1.2 \times 50}{100 \times 1.6 \times 10^{-19} \times 0.5 \times 10^{-3}} = 0.75 \times 10^{22}$$

number of electrons/m^3 $\quad n = .75 \times 10^{22}$ electrons/m^3

$$n = 7.5 \times 10^{21}\ \text{m}^{-3}$$

Ex. 2.4 An electron at rest is accelerated to a potential difference of 100 V. Calculate final kinetic energy in Joules. What is the final velocity?

$$\text{Potential } V = \frac{\text{Work done}}{\text{Charge}} = \frac{\text{Kinetic energy}}{\text{Charge}}$$

once charge starts moving work done is converted into kinetic energy

$$V = \frac{\text{Kinetic energy}}{e}$$

$$\text{K.E.} = Ve = 100 \times 1.6 \times 10^{-19}$$

$$\text{K.E.} = 1.6 \times 10^{-17}\ \text{Joules}$$

$$\text{K.E.} = 1/2\ mV^2$$

$$1.6 \times 10^{-17} = \frac{1}{2} \times 9.11 \times 10^{-31} \times v^2$$

$$v^2 = \frac{1.6 \times 10^{-17} \times 2}{9.11 \times 10^{-31}}$$

(∵ mass of electron $e = 9.11 \times 10^{-31}$)

$$v = 5.9 \times 10^6\ \text{m/s}$$

Short Questions and Answers

Q. 1. Distinguish between majority and minority carriers in doped semiconductor.

Ans. In P-type, majority carriers are holes and minority carriers are electrons.
In N-type, majority carriers are electrons and minority carriers are holes.

Q. 2. Define mobility of charge carriers.

Ans. It is the ratio of average drift velocity to the electric field. $\mu = \frac{v}{E}$. It is the case with which charge carriers drift in the material.

Q. 3. How does the impurity when added to the pure semiconductor change its conductivity?

Ans. With doping conductivity increases.

Q. 4. Define cut in voltage of $P - N$ diode.

$$V_k = 0.3\ V \text{ for germanium}$$

$$V_k = 0.7\ V \text{ for silicon}$$

Q. 5. What is forbidden energy gap of germanium?

Ans. for Ge 0.72 eV

3

P-N Junction Diode

3.1 Introduction

This chapter deals with *P-N* diode, zener diode, tunnel diode, varactor diode, *P-N* diode, *LED* and liner doubler circuit.

3.1.1 *P-N* Junction

Unbiased *P-N* junction diode and its waveforms are shown in Fig. 3.1. The holes are represented by positive sign and the impurity atoms are represented by negative ions in the case of *P*-type semiconductor. The electrons are represented by negative sign and the impurity atoms are represented as positive ions in the case of *N*-type semiconductor.

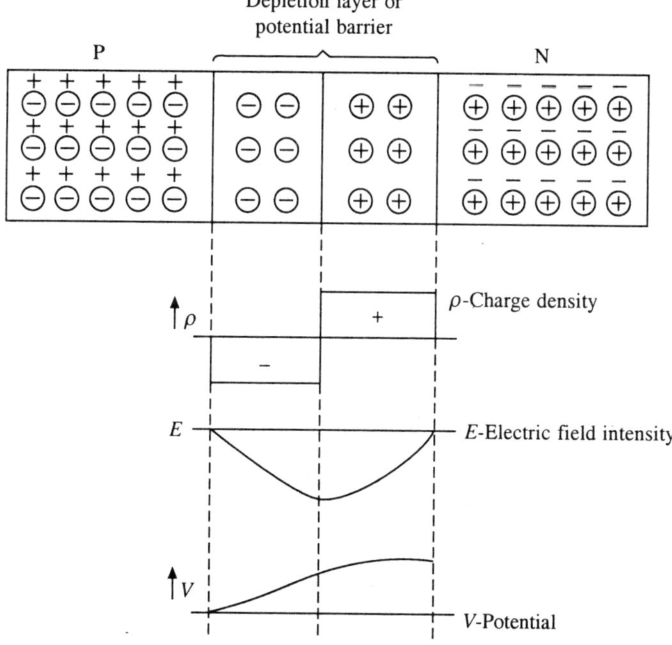

Fig. 3.1 *P-N* junction.

When the *P-N* junction is formed, the conduction electrons in the *N*-region will diffuse (penetrate or enter) into the *P*-region. The holes in the *P*-region will diffuse into the *N*-region. The electrons will fall into the holes. This process is known as recombination. The average time for which the electron travels before it recombines, is known as *life time*. A depletion layer is formed at the junction. The recombination stops after sometime. The conduction electrons on the *N*-side are repelled by the negative ions on the *P*-side. The holes on the *P*-side are repelled by the positive ions on the *N*-side. Thus a restraining force is set up at the juction and this prevents further recombination. The charge density on the left side of the potential barrier is negative and it is positive on the right side of the potential barrier as indicated in the charge density curve. The electric field intensity is maximum at the centre and the magnitude of electric field intensity decreases on either side. Potential is the work done in moving a unit positive charge from the left to the right side of the depletion layer. The work done has to be increased as the charge is moved to the right side.

3.1.2 Depletion Layer

When the *P-N* junction is formed, the electrons and holes move towards the junction. The electrons fall into the holes. At the junction a charge free region is formed. The positive ions and the negative ions are separated by a small distance. These are nothing but the dipoles. The direction of the electric field is given by the force experienced by the unit positive charge kept in the electric field. The direction is away from the positive charge. An electric field at this layer implies that there is a potential due to the separation of the charges. This potential is known as the barrier potential. This is equal to 0.3 V for germanium and 0.7 V for silicon. The region near the junction is called depletion layer since it is depleted (not contained) with free changes.

We know that $$E = \frac{dv}{dx} \tag{3.1}$$

differentiate with respect to x

$$\frac{dE}{dx} = \frac{-d^2V}{dx^2}$$

Substituting $E = D/\varepsilon$, we get

$$\frac{1}{\varepsilon}\frac{dD}{dx} = \frac{-d^2V}{dx^2}$$

Substituting $D = Q/A$

$$\frac{1}{A\varepsilon}\frac{dQ}{dx} = \frac{-d^2V}{dx^2}$$

Substitute $$dQ = \rho\, dV = \rho\, Adx$$

$$\frac{\rho\, Adx}{A\varepsilon\, dx} = \frac{-d^2V}{dx^2}$$

$$-\frac{\rho}{\varepsilon} = \frac{-d^2v}{dx^2} \tag{3.2}$$

The solution for (3.2) gives potential curve. Differentiating the solution of equation (3.2), we can get the equation for electric field.

3.2 Diffusion Current

A semiconductor is doped non-uniformly. Consider an imaginary axis Y-Y'. On the left side of this axis, the impurity concentration is more. On the right side of axis $Y Y'$, the impurity concentration is less. It is found that a current is flowing from the heavily doped region to the lightly doped region without applying any external voltage. This current is similar to the flow of gas from high pressure region to low pressure region. This current is proportional to the concentration gradient.

$$J_2 = -K\frac{dp}{dx}$$

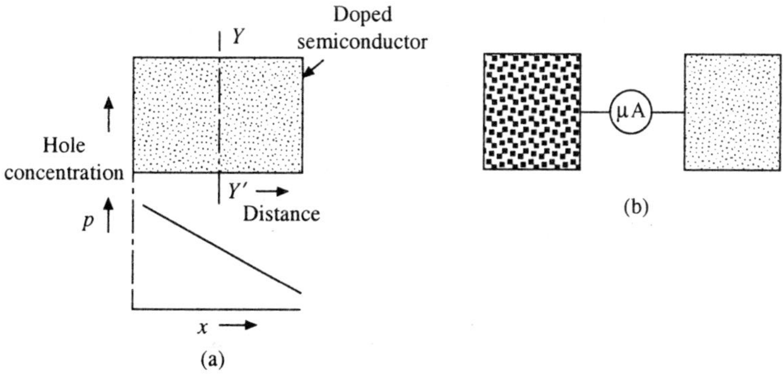

Doped semiconductor

Y

Hole concentration

Y' →

Distance

p

x →

(a)

μA

(b)

Fig. 3.2 Doped Semiconductor

The current due to the drift of electrons is defined as drift current. The total current in a semiconductor is the sum of drift current and diffusion current.

$$J = J_1 + J_2;\ J = \sigma E + \left(-K\frac{dp}{dx}\right)$$

3.3 Biasing of *P-N* Junction

3.3.1 Forward Biasing

When the positive terminal of the battery is connected to P and the negative terminal of battery is connected to *N*, the *P-N* junction is said to be forward biased. The direction of the applied electric field is opposite to the direction of the electric field of the depletion layer. Stream of electrons start from the negative terminal of the battery and they enter into the *N*-region. Through the N-region, they travel as conduction electrons. They cross the junction and then the conduction electrons fall into the holes. They become valence electrons. As valence electrons they travel through *P*-region. These electrons are attracted by the positive plate of the battery. Thus a continuous flow of electrons exist in the circuit. The junction offers minimum resistance and hence the current is maximum. The width of the depletion layer is less.

Large potential due to open circuited *P-N* junction is shown in Fig. 3.3(a). The potential gets reduced after closing the switch as shown in Fig. 3.3(b).

3.3.2 Reverse Biasing

Before closing the switch the potential graph is as shown in Fig. 3.4(a) and after closing, the potential increases as shown in Fig 3.4(b). When the positive terminal of the battery is connected to *N* and negative terminal of the battery is connected to *P*, the *P-N* junction is said to be reverse biased. The direction of the applied electric field is similar to the direction of the electric field of the potential barrier. The holes in the *P*-region are attracted by the negative plate and the electrons in the *N*-region are attracted by the positive plate. Thus the electrons and holes move away from the junction. The region containing positive and negative ions will be wider as the holes and electrons are moving away. In other words the width of the depletion layer will be increased. This width grows until the barrier potential is equal to the voltage applied. Thus the junction offers very high resistance when it is reverse biased. However, a small current due to the minority carriers is present in the *P*- and *N*-regions. These minority carriers are

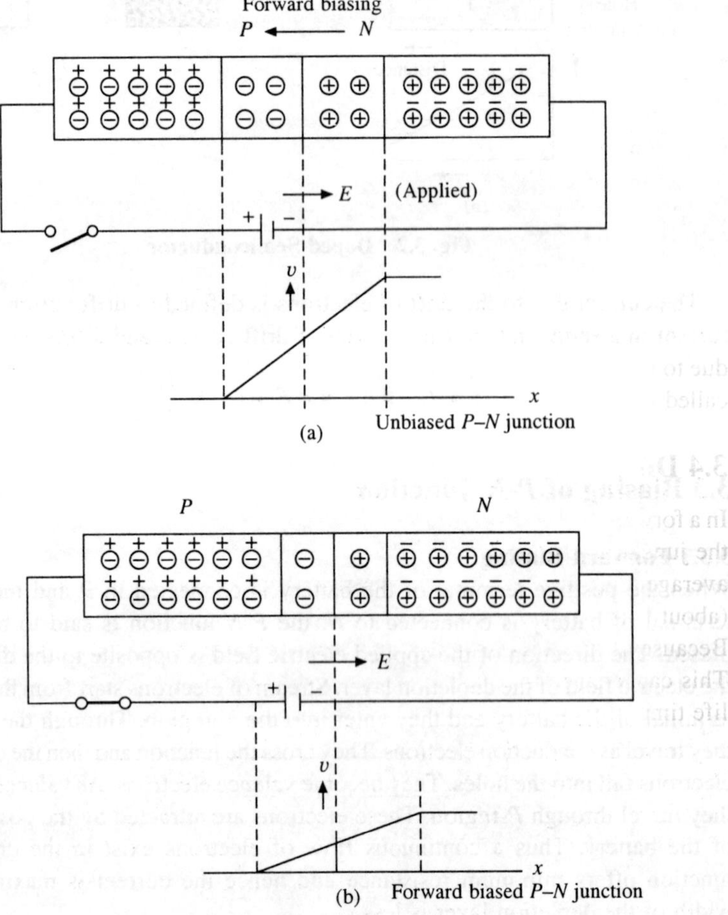

Fig. 3.3 Forword biasing.

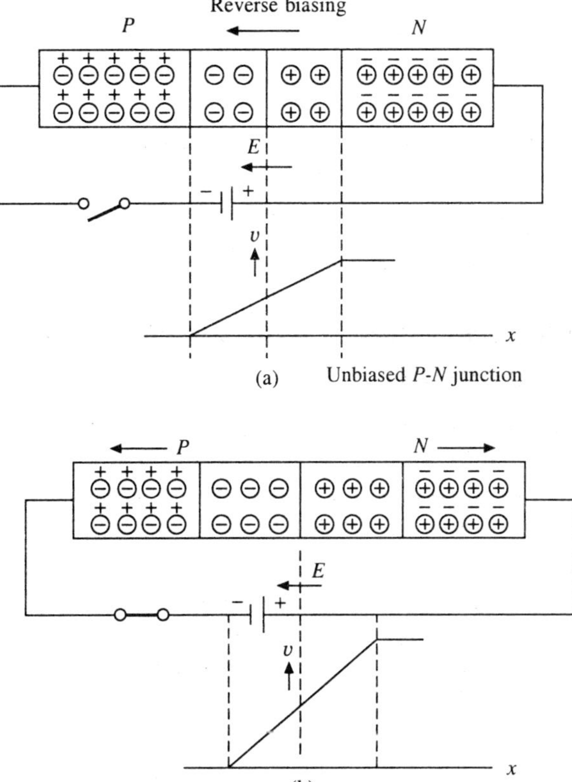

(a)

(b)

Fig. 3.4 Reverse biasing.

due to the breaking of covalent bonds at room temperature. This small current is called *minority carrier current.*

3.4 Diffusion Capacitance

In a forward biased *P–N* Junction (Fig. 3.5) the holes and electrons move towards the junction resulting in fall of electrons into the holes near the junction. The average time, the electrons and holes exist before they recombine, is the life time (about 1 μ sec). The holes and electrons move continuously towards the junction. Because of temporary storage of charges at the junction, the capacitance exists. This capacitance is called *diffusion capacitance* (C_D) which depends upon the life time of the charges.

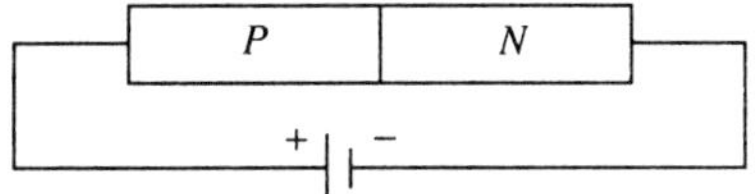

Fig. 3.5 *P-N* junction.

We know that $q = cv$ and $dq = c\ dv$.

$$C = \frac{dq}{dv} = T\frac{dI}{dv}$$

$$C_D = Tg \tag{3.4}$$

where g is the incremental conductance of the diode. T is the average life time.

$$r = \frac{1}{g}$$

where r is the incremental resistance of diode.

3.5 Transition Capacitance C_T (or Junction or Depletion Layer Capacitance)

The depletion layer has +ve and –ve ions separated by a small distance with no free charges available. Thus, it acts like a dielectric. The doped P and N semiconductors act as conductors. Hence the capacitance exists in a reverse biased P-N junction. This capacitance is called depletion layer capacitance.

Let N_A = Acceptor impurity concentration density
N_D = Donor impurity concentration density
W_p = Width of depletion layer in the P-region
W_N = Width of depletion layer in the N-region
q = Charge
p = Charge density

The waveforms for reverse biased P-N junction are shown in Fig. 3.6.

Assumptions

(1) The total width of the depletion layer is approximately equal to W_N.
(2) At $x = 0$, potential $V = V_i$. Since V_i is very small, it is neglected.

From Poisson's equation

$$\frac{d^2V}{dx^2} = \frac{-\rho}{\varepsilon} = \frac{-qN_D}{\varepsilon}$$

On integration

$$\frac{dv}{dx} = \frac{-qN_Dx}{\varepsilon} + C_1 \tag{3.5}$$

From Fig. 3.6, when $x = W_N$, $E = 0$

Therefore, $$\frac{dv}{dx} = 0$$

$$0 = \frac{-qN_D}{\varepsilon}W_N + C_1$$

$$C_1 = \frac{qN_DW_N}{\varepsilon} \tag{3.6}$$

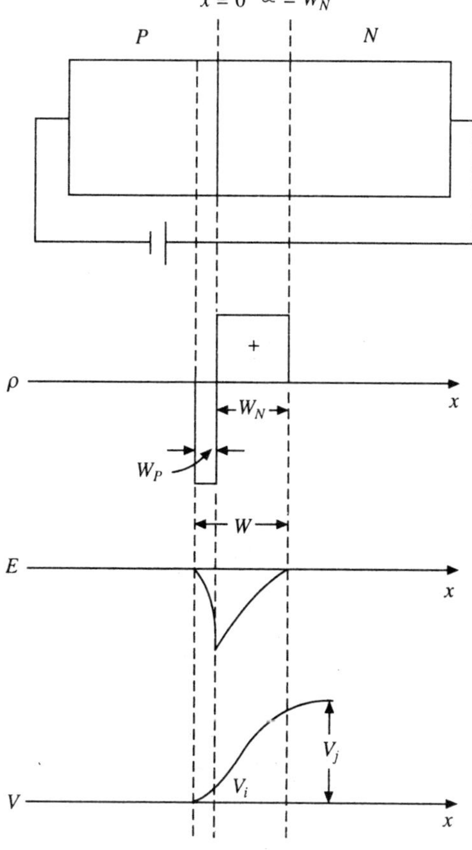

Fig. 3.6 Waveforms.

Substituting equation (3.6) in (3.5), we get

$$\frac{dV}{dx} = \frac{-qN_D}{\varepsilon} x + \frac{qN_D W_N}{\varepsilon}$$

Again integrating,

$$V = \frac{-qN_D}{\varepsilon} \frac{x^2}{2} + \frac{qN_D W_N}{\varepsilon} x + C_2 \tag{3.7}$$

When $x = 0$; $V = 0$.

Therefore,

$$C_2 = 0 \tag{3.8}$$

Substituting equation (3.8) in (3.7)

$$V = \frac{-qN_D x^2}{2\varepsilon} + \frac{qN_D W_N x}{\varepsilon}$$

When $x = W_N$; $V = V_j$ (from Fig. 3.6)

$$V_j = \frac{-qN_D W_N^2}{2\varepsilon} + \frac{qN_D W_N}{\varepsilon} W_N$$

$$V_j = \frac{+qN_D W_N^2}{2\varepsilon}$$

But $W_N \simeq W$. Therefore,

$$V_j = \frac{qN_D W^2}{2\varepsilon}$$

Differentiating on both sides, we get

$$dV_j = \frac{qN_D}{2\varepsilon} 2 \cdot WdW$$

$$\frac{dV_j}{dW} = \frac{qN_D W}{\varepsilon} \tag{3.9}$$

$$\text{Charge } Q = q \cdot N_D (A \cdot W)$$

where q is the charge of electron, N_D = concentration of atoms per m^3 and $A \cdot W$ = volume in mm^3

$$dQ = qN_D AdW \tag{3.10}$$

Find the ratio of equations (3.10) to (3.9)

$$\frac{dQ}{dV_j} = \frac{\varepsilon qN_D A\, dW}{qN_D W\, dW} = \frac{A\varepsilon}{W}$$

But
$$\frac{dQ}{dV_j} = C_T$$

Therefore,
$$C_T = \frac{A\varepsilon}{W} \tag{3.11}$$

Capacitance is directly proportional to area A and inversely proportional to width W. The final result is similar to capacitance of a parallel plate capacitor.

3.6 Characteristics of *P-N* Diode

The connections are made as shown in Fig. 3.7(a). The voltage applied to the diode can be increased by moving the jockey point of the rheostat upwards. No current flows through the diode until the applied voltage is less than knee potential (V_k). If the voltage is increased beyond V_k, the diode starts conducting. The voltage across the diode is 0.7 volts when it is conducting. The current through the diode can be limited by the series resistance R. If the series resistance is not enough to limit the current to safe value, the current increases. The heat produced by the device increases. This results in breaking few more covalent bonds, this process continues, current increases and the device gets damaged. This is called thermal run away. When the diode is reverse biased, small leakage current flows due to minority carriers. This is of the order of micro Amperes.

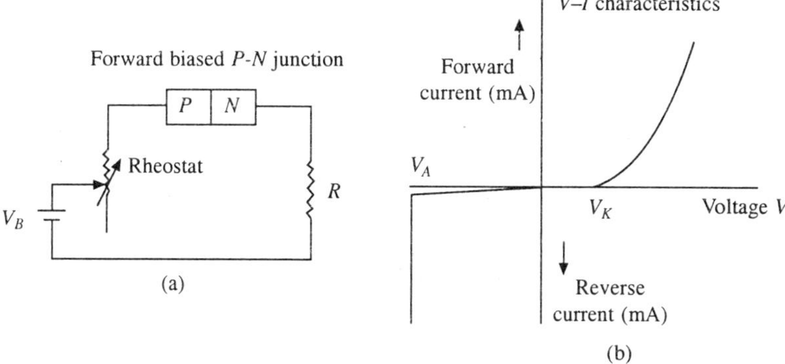

Fig. 3.7 Diode characteristics.

If the voltage is increased beyond V_A the electric field increases. Force experienced by the minority carrier increases. It can bump (knock) one electron by breaking the covalent bond. Two electrons can knock off two more and this process continues. The current increases. If the current is not limited, the diode breaks down. This is called avalanche breakdown. A normal *P-N* diode is not designed to operate in the avalanche breakdown region.

3.7 Diode Current Equation

Diode characteristics is a nonlinear characteristic. It is described by Shockley as follows:

$$I = I_0 [e^{\frac{V}{KV_T}} - 1] \tag{3.12}$$

where V is the voltage applied,

I the current through forward biased diode,

I_0 the reverse saturation current and V_T the temperature volt constant (constant that depends on temp)

$$V_T = \frac{T}{11{,}600} \tag{3.13}$$

T is the temperature in °K and K a constant.

Germanium diode

When $K = 1$, $V_T = 26$ mV

$$I = I_0\,(e^{40V} - 1)$$

For silicon diode

$K = 2$; $V_T = 26$ mV

$$I = I_0\,(e^{\frac{V}{2\times 26\times 10^{-3}}} - 1) = I_0\,(e^{19.23} - 1) = I_0\,(e^{20V} - 1)$$

In the problems if T is given, use the above formula,

$$V_T = \frac{T}{11{,}600}$$

If temperature is not given assume $V_T = 26$ mV

Worked Problem

Ex. 3.1 Determine the Ge *P–N* junction diode current for a forward diode voltage of 0.35 volt at room temperature 25°C with a reverse saturation current of 10 μA.

$$V_T = \frac{T}{11{,}600} = \frac{25 + 273}{11{,}600} = 25.68 \text{ mV}$$

$K = 1$ for Ge

$$I = I_0 (e^{\frac{V}{KV_T}} - 1) = 10 \times 10^{-6} (e^{\frac{0.35}{1\times 25.68\times 10^{-3}}} - 1) = 12 \text{ A}$$

3.8 Zener Diode

Zener is a specially designed diode to operate in the breakdown region and is heavily doped compared to the ordinary *P-N* junction diode. The width of the depletion layer is inversely proportional to impurity concentration. The width is equal to 1000 Å since it is heavily doped. A large electric field is possible with less voltage since the width is lesser.

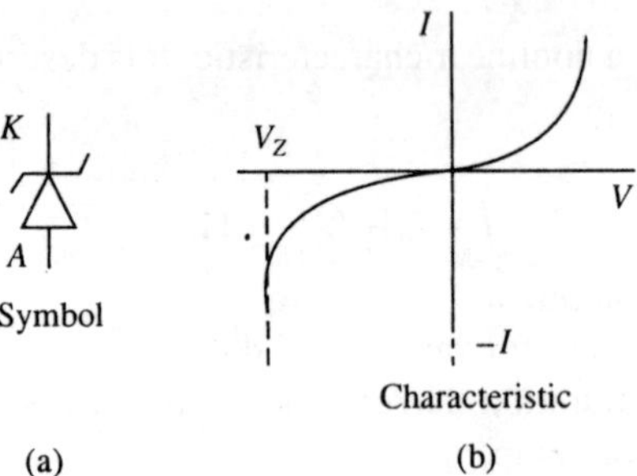

Fig. 3.8 Zener diode.

The electric field is around 30 kV/cm. This field can break the covalent bonds since the electrons acquire very high velocity. One electron bumps another electron, two can bump another two electrons. This process continues and current increases. From the characteristics it can be seen that the voltage remains constant even though current varies. Thus zener diode acts as a voltage source in the breakdown region. This property is used in voltage regulator circuits and meter protection circuits. The forward characteristic of zener diode is similar to forward characteristic of ordinary *P-N* diode. Avalanche breakdown voltage is greater than zener breakdown voltage.

3.9 Tunnel Diode or ESAKI Diode

In an ordinary *P-N* diode the impurity concentration is one atom per 10^8 silicon atoms. The width of the depletion layer is about 10^{-3} m. If the junction is heavily doped, the width of the depletion layer decreases as this is inversely proportional to the impurity concentration. If the impurity atoms are added at the rate of one impurity atom per 10^3 silicon atoms, the width reduces to 10^{-9} m. This doping alters the forward characteristics of the diode. The tunnelling effect can be understood from the EB (energy band) diagrams. The $V - I$ characteristics are shown in Fig. 3.9(b) and (c).

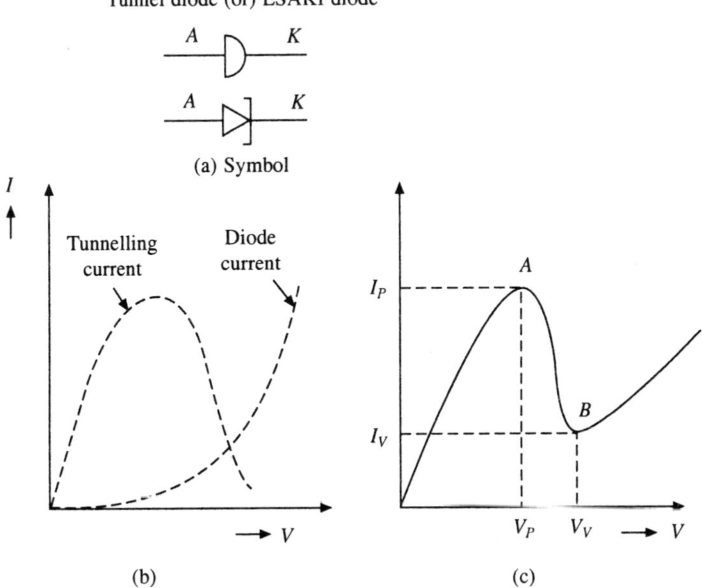

Fig. 3.9 Characteristics.

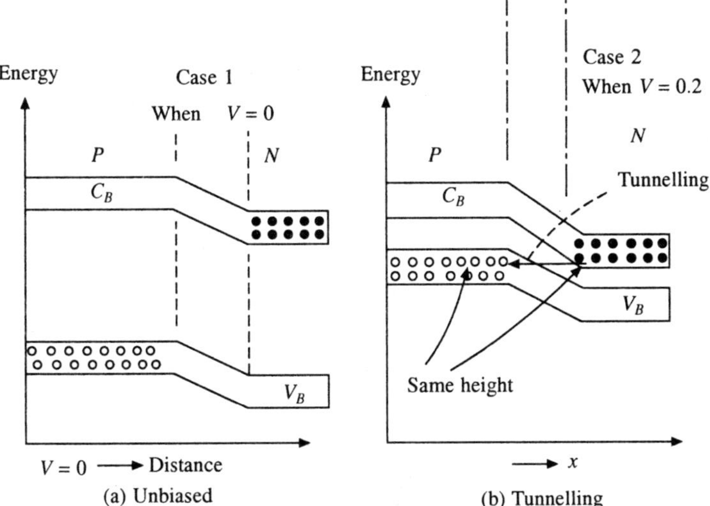

Fig. 3.10 Tunnel diode.

The energy level depends upon the radius of the orbit. The first energy band diagram shown in Fig. 3.10(a) is without any forward applied voltage ($V = 0$). There is a large mismatch between the energy levels of the electrons and the holes. As the forward voltage is increased, the energy level of the electrons shifts. At a particular voltage V_P the energy levels of the electrons and the holes coincide and the electrons travel with the velocity of light and falls into the holes. The effect of electrons directly falling into the holes without climbing the potential hill is known as tunnelling. As the voltage is further increased, the mismatch increases and the current decreases. At a particular voltage V_V the current drops to a minimum value. The characteristics of the tunnel diode can be obtained by superimposition of the tunnelling characteristics and the conventional diode characteristic. In the region AB the voltage increases from V_P to V_V, the current decreases from I_P to I_V (Fig. 3.9(b)). This is the negative resistance region and is used in the oscillators.

3.10 Varactor Diode

A varactor diode is, basically, a reverse biased *P-N* junction, which utilizes the inherent capacitance of the depletion layer. It is also known as varicap, voltcap or Tunnelling diode. It is used as a variable capacitor.

We have already discussed that the depletion layer created by the reverse bias acts as a capacitor dielectric, whereas the *P* and *N* regions act as the capacitor plates.

When the reverse bias voltage increases, the depletion layer widens. This increases the dielectric thickness, which in turn, reduces the capacitance. When the reverse bias voltage decreases, the depletion layer narrows down. This decreases the dielectric thickness, which in turn increases the capacitance. Fig. 3.11(b) shows the variation of capacitance with the reverse voltage. This indicates that the variation capacitance is maximum when the reverse voltage is equal to zero. It reduces in a non-linear manner, as the value of reverse voltage is increased.

In a varactor diode, the capacitance parameters are controlled by the method of doping in the depletion layer, or the size and geometry of the diode construction. For example. Fig. 3.11(a) shows the doping profile for an abrupt junction diode (i.e., the normal *P-N* junction). In this type, the doping is uniform on both sides of the junction. The range of capacitance variation (tuning range) of an abrupt

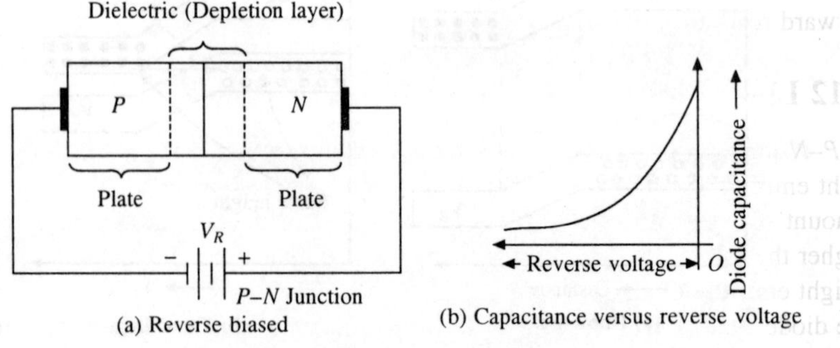

Fig. 3.11 Varactor Diode.

junction diode is 4:1. It means that if its maximum transition capacitance is 100 pF and the minimum is 25 pF, then its tuning range is 4:1.

3.11 *PIN* Diode

A *PIN* diode consists of heavily doped *P* and *N* regions separated by an intrinsic (i.e., undoped) *I*-region as shown in Fig. 3.12. The intrinsic region offers high resistance to the current flowing through it. The *PIN* diode has two advantages over the normal *P–N* junction diode:

1. The capacitance between the *P* and *N* regions decreases because of the increased separation between the *P* and *N* regions. This advantage allows the *PIN* diode to have fast response time. Hence, these diodes are useful at very high-frequencies (i.e., above 300 MHz).
2. There is a greater electron-hole pair generation because of the increased electric field between the *P* and *N* regions. This advantage allows the *PIN* diode to process even weak signals.

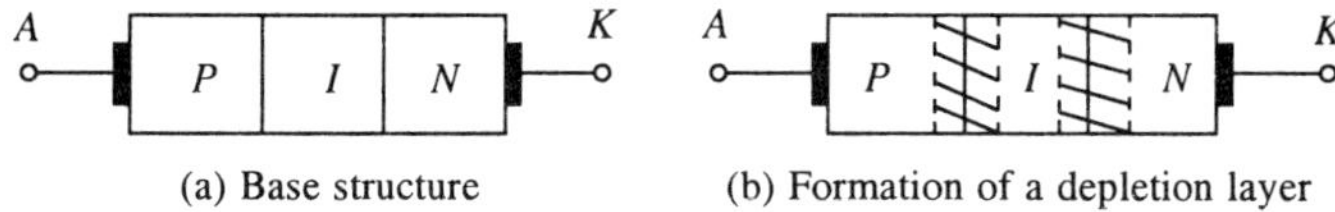

Fig. 3.12 *PIN* diode.

We know that when a *PIN* diode is unbiased (i.e., no voltage is applied across the diode), there is a diffusion of electron and holes across the junction due to the different concentration of atoms in the *P*, *I* and *N* regions. The diffusion of electrons and holes produce a depletion layer across the *PI* and *IN* junctions as shown in Fig. 3.12(b). The depletion layers penetrate to a little distance in the *P*-type and *N*-type semiconductor regions but to a larger distance in the *I*-region. Under such condition, the device has high value of resistance.

When the *PIN* diode is forward biased, the width of depletion layers decrease. As a result of this, more carriers are injected into the I-region. This reduces the resistance of the I-region. If the depletion layers are not thick, then the I-region becomes flooded with the carriers at a suitable bias. Thus, when a *PIN* diode is forward biased, it acts like a variable resistance as shown in Fig. 3.12. The forward resistance of an intrinsic region decreases with the increasing current.

3.12 Light Emitting Diode (LED)

A *P–N* junction diode, which emits light when forward biased, is known as a light emitting diode (LED). The emitted light may be visible or invisible. The amount of light output is directly proportional to the forward current. Thus, higher the forward current, higher is the light output. The schematic symbol of a light emitting diode is shown in Fig. 3.13(a). The arrows pointing away from the diode symbol represent the light, which is being transmitted away from the junction.

Figure 3.13(b) shows the basic structure of a light emitting diode. Here, an *N*-type layer is grown on a *P*-type substrate (not indicated in figure) by a diffusion process. Then a thin *P*-type layer is grown on the *N*-type layer. The metal connecting both the layers make anode and cathode terminals as indicated. The light energy is released at the junction, when the recombination of electrons with holes take place. After passing through the *P*-region, the light is emitted through the window provided at the top of the surface.

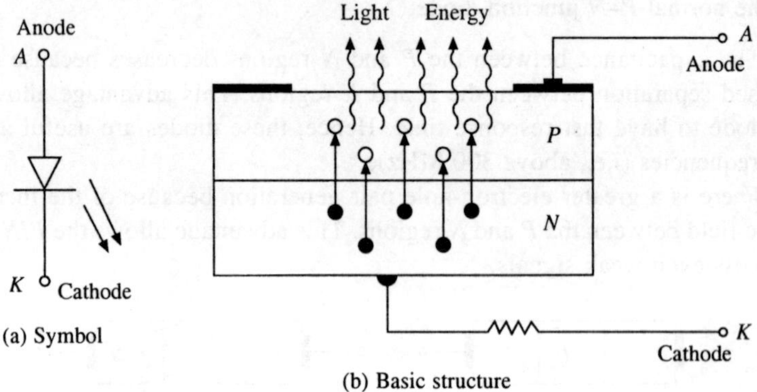

Fig. 3.13 LED.

It will be interesting to know that when the LED is forward biased, the electrons and holes move towards the junction and the recombination takes place. After recombination, the electrons, lying in the conduction bands of *N*-region, fall into the holes lying in the valence band of a *P*-region. The difference of energy between the conduction band and valence band is radiated in the form of light energy. In ordinary diodes, this energy is radiated in the form of heat.

The semiconducting materials used for manufacturing light emitting diodes are gallium arsenide and gallium arsenide phosphide. Silicon and germanium is not used for manufacturing light emitting diodes because these are heat producing materials. Moreover, these materials are very poor in emitting light radiations.

The LED's radiate light in different colours such as red, green, yellow, blue, orange etc. Some of the LED's emit infrared (i.e., invisible) light also. The colour of the emitted light depends upon the type of the semiconductor used. Thus gallium arsenide emits infrared radiations, gallium arsenide phosphide produces either red or yellow light, gallium phosphide emits red or green light and gallium nitride produces blue light.

3.13 Clipping Circuits

Clipper circuits remove part of the waveform and are of four types, viz. positive, negative, biased and combination.

3.13.1 Positive Clipper

Figure 3.14(a) shows the circuit of a positive clipper. It consists of a diode (D) and a resistor (R) with output taken across the resistor. The diode acts as an ideal switch between the source and the load. It acts as a closed switch, when the input voltage is negative (i.e. $v_i < 0$) and as an open switch when the input voltage is positive or zero (i.e., $v_i > 0$). The purpose of resistor (R) is to limit the current through diode, when it acts as closed switch. The operation of the circuit may be explained as follows.

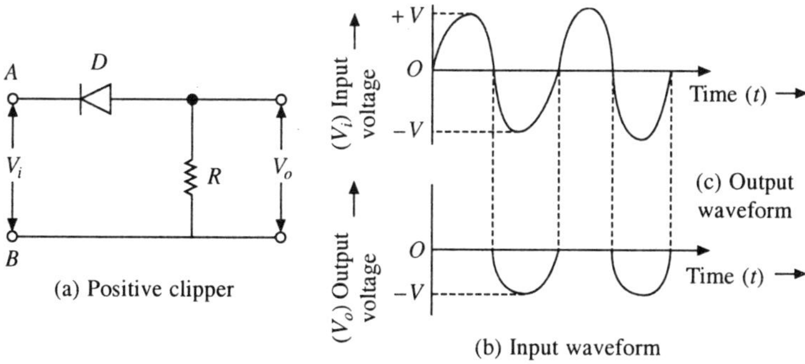

Fig. 3.14

Figure 3.14(b) shows the waveform of an input voltage. During the positive half cycle of the input voltage, the terminal A is positive with respect to B. This reverse biases the diode and it acts as an open switch. Therefore all the applied voltage drops across the diode and none across the resistor. As a result of this, there is no output voltage during the positive half cycle of the input voltage as shown in Fig. 3.14(c). During the negative half cycle of the input voltage, the terminal B is positive with respect to A. Therefore it forward biases the diode and it acts as a closd switch. Thus there is no voltage drop across diode during the negative half cycle of the input voltage. All the input voltage is dropped across the resistor as shown in the output waveform.

It is evident from the above discussion as well as the output waveform that the positive clipper has allowed to pass the negative half cycle of the input voltage and clipped the positive half cycle completely. Hence the circuit of Fig. 3.14(a) is called a positive clipper. It may be noted in Fig. 3.14(a) that the diode acts as a series switch between the source and the load. Due to this reason, the circuit is called series positive clipper. It is also possible to connect a diode as a shunt switch between the source and the load as shown in Fig. 3.15. Such a circuit is known as shunt positive clipper, in this circuit, the diode acts as a closed switch when the input voltage is positive (i.e. $v_i > 0$) and as an open switch when the input voltage is negative (i.e. $v_i < 0$). The output waveform is the same as that of a series positive clipper.

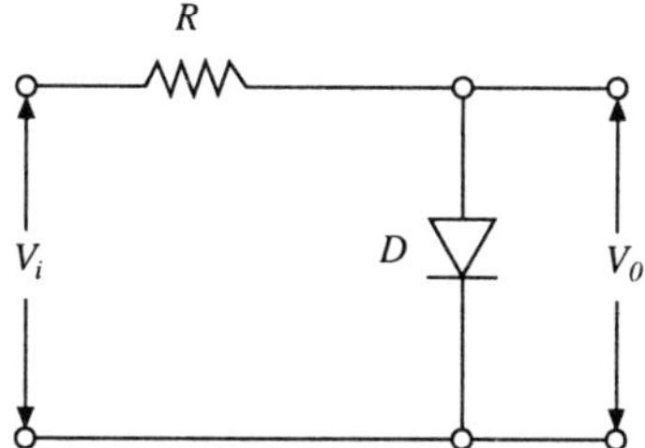

Fig. 3.15 Shunt positive clipper.

3.13.2 Negative Clipper

Figure 3.16(a) shows the circuit of a negative clipper. Here the diode is shown connected in a direction opposite to that of a positive clipper. The function of the resistor (R) is to limit the current when the diode is forward biased (i.e. when it acts as a closed switch). The diode acts as a closed switch for a positive input voltage, i.e. $v_i > 0$ and as an open switch for a negative input voltage, i.e. $v_i < 0$. The operation of the circuit may be explained in the same way as discussed for a positive clipper and is as given below.

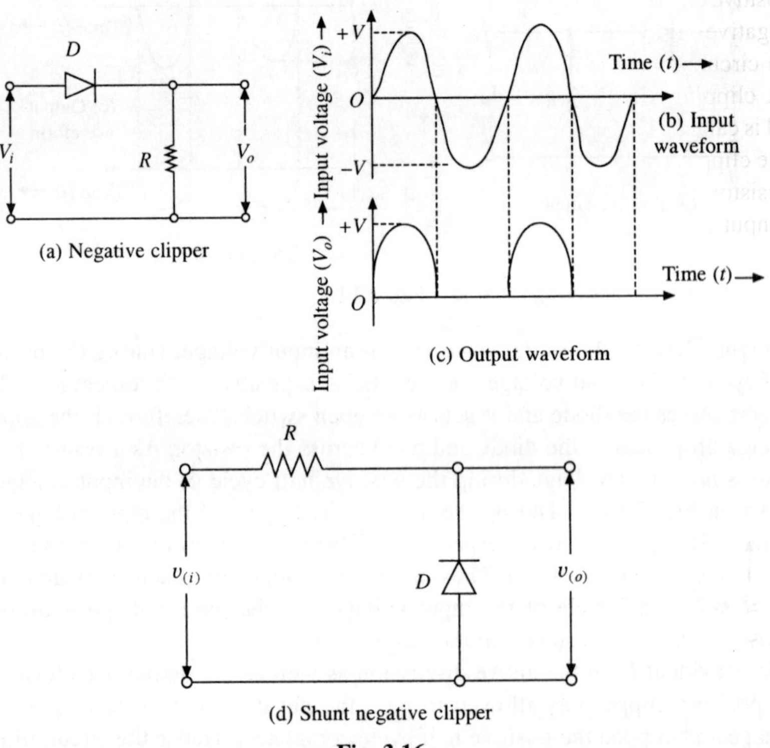

Fig. 3.16

Figure 3.16(b) shows the waveform of the input voltage. During the positive half cycle of the voltage, the terminal A is positive with respect to the terminal B. Therefore the diode is forward biased and it acts as a closed switch. As a result, all the input voltage appears across the resistor as shown in Fig. 3.16(c). During the negative half cycle of the input voltage, the terminal B is positive with respect to the terminal A.

Therefore the diode is reverse biased and it acts as an open switch. Thus there is no voltage drop across the resistor during the negative half cycle as shown in the output waveform.

It is evident from the above discussion that the negative clipper has allowed to pass the positive half cycle of the input voltage and clipped the negative half cycle completely. The negative clipper circuit of Fig. 3.16(a) is called series negative clipper, because the diode acts as a series switch between the source and the load. It is also possible to connect a diode as a shunt switch between the source and the load as shown in Fig. 3.16(d). Such a circuit is known as shunt negative clipper. In this circuit, the diode acts as a closed switch for a negative input voltage (i.e. $v_i < 0$) and as an open switch for a positive input voltage (i.e. $v_i > 0$). The output waveform of the circuit is the same as that of series negative clipper.

3.13.3 Biased Clipper

A positive clipper clips off the entire positive portions of the input signal. Similarly, a negative clipper clips off the entire negative portions of the input signal. In such circuits, there is no provision for adjustment of the clipping level.

A clipping circuit, which has a provision for the adjustment of a clipping level is called a biased clipper. The name bias is designated because the adjustment of the clipping level is achieved by adding a bias voltage in series with the diode or resistor. Fig. 3.17(a) shows the circuit of a biased positive clipper along with the input and output voltage waveforms.

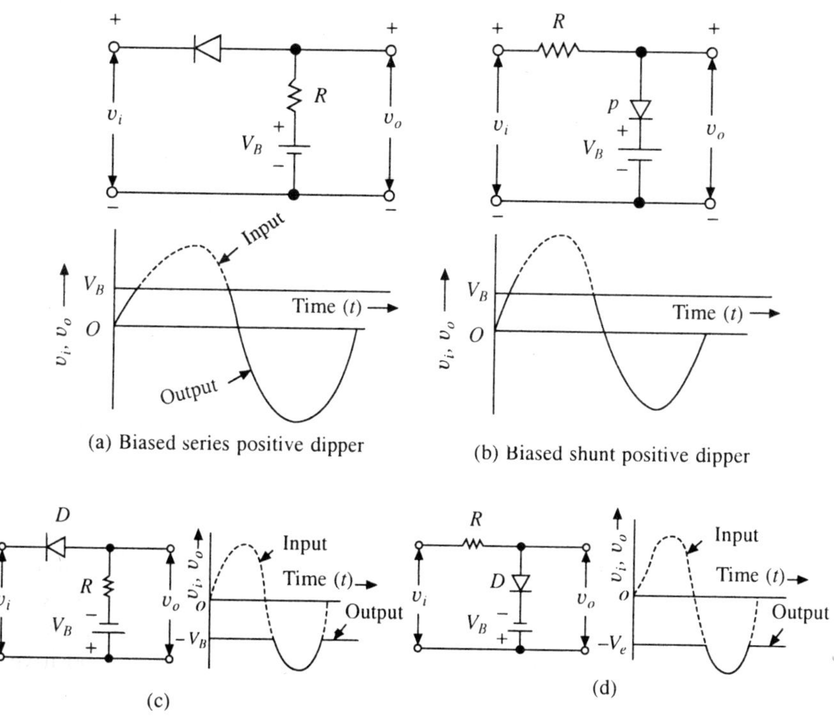

Fig. 3.17 Biased clipper with reversed polarity battery V_B.

The circuit of Fig. 3.17(a) is a biased series positive clipper, while that of Fig. 3.17(b) is a biased shunt positive clipper. It can be noted that the clipping takes place during the positive cycle only when the input voltage is greater than battery voltage (i.e. $v_i > V_B$). The clipping level can be shifted up or down by varying the bias voltage (V_B). It will be interesting to know that if the polarity of the bias voltage of the circuits shown in Fig. 3.17 is reversed, then the resulting circuits will be as shown in Fig. 3.17(c) and (d). Here the input signal lying above the voltage level – V_B is clipped. The resulting waveforms of the output voltage are shown in figures.

3.14 Clamping Circuits

The circuits with which the waveform can be shifted in such a way that a particular part of it (say positive or negative peak) is maintained at a specified voltage level, is called a clamping circuit (or simply a clamper). As a matter of fact, a clamping circuit introduces (or restores) a D.C. level to an A.C. signal. Thus a clamping circuit is also known as D.C. restorer. Such circuits are used in television receivers to restore the original D.C. reference signal to the video signal.

Now consider an A.C. signal waveform as shown in Fig. 3.18(a). It is a sine wave with equal positive and negative swings of 5 V above and below the zero level. It has no D.C. level. On the other hand, the signal waveform shown in Fig. 3.18(b) is lifted up just to touch the horizontal axis. It is said to have a D.C. level of +5 V and is known as a waveform positively clamped at 0 V. Similarly, the waveform shown in Fig. 3.18(c) has a D.C. level of –5 V and is known as a waveform negatively clamped at 0 V.

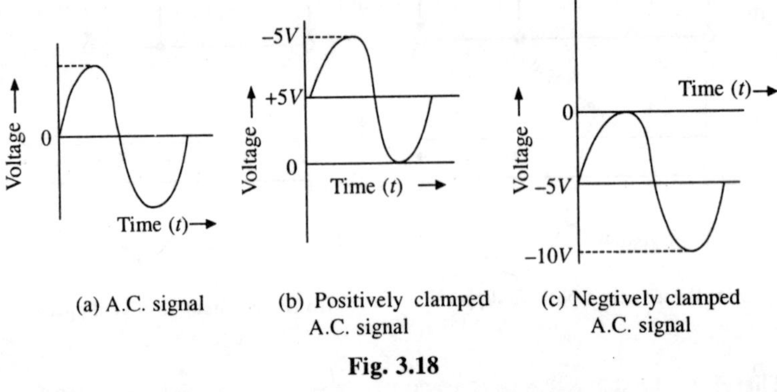

Fig. 3.18

It is possible to clamp a waveform positively at 0 V by connecting a D.C. battery in series with the source, whose e.m.f is equal to the peak value of the sine wave ($v_i = V_m \sin \omega t$) as shown in Fig. 3.19. However, it is a very crude way of achieving the required clamping action and therefore is not used in actual practice.

A correct way of clamping the input signal, positively at 0 V, is to use a circuit as shown in Fig. 3.19(b). This circuit consists of a diode and capacitor and is popularly known as a diode clamper. The clamping action of the circuit may be understood from the fact that during the negative half cycle of the input voltage (v_i), the diode is forward biased and the current flows through the circuit. As a result of this, the capacitor (C) is charged to a voltage equal to the negative peak value (i.e., $-V_m$). Once the capacitor is fully charged to $-V_m$, it cannot discharge

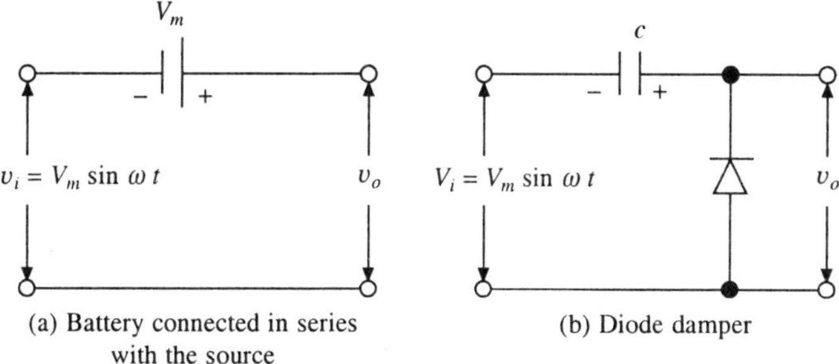

(a) Battery connected in series with the source

(b) Diode damper

Fig. 3.19 Circuits for clamping input signal positively at 0 V.

because the diode cannot conduct in the reverse direction. It means that this capacitor acts as a battery with an e.m.f. equal to the $-V_m$. The polarity of this voltage is such that it adds to the input signal. Therefore the output voltage is equal to the sum of the A.C. input signal and the capacitor voltage V_m, i.e.

$$v_0 = v_i + V_m = V_m \sin \omega t + V_m$$

It is evident from the above relation, that a D.C. voltage equal to V_m is added to input signal. This causes the waveform to clamp positively at 0 V.

Now suppose it is required to clamp the input signal waveform ($v_i = V_m \sin \omega t$) negatively at 0 V. This can be done by reversing the diode connections as shown in Fig. 3.20. Here the diode is forward biased during the positive half-cycle of the input signal.

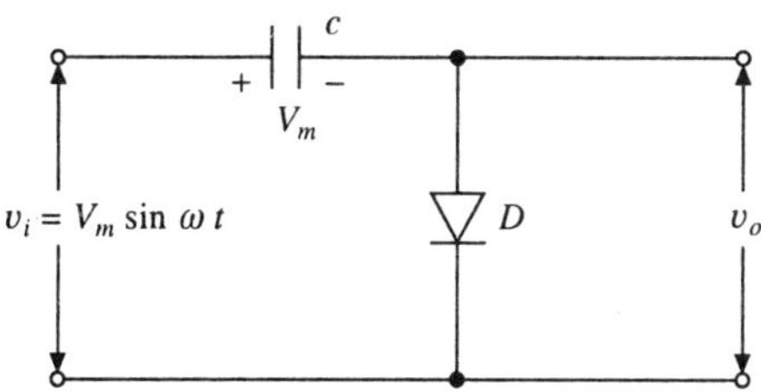

Fig. 3.20 Clamper Circuit

Therefore the capacitor is charged to a voltage equal to the positive peak of the input signal (i.e., eV_m), the polarity of this voltage is such that it is subtracted from the input signal. Hence, the output voltage

$$v_0 = v_i - V_m = V_m \sin \omega t - V_m$$

It is evident from the above relation that a D.C. voltage equal to $-V_m$ is added to the input signal. This causes the waveform to clamp negatively to 0 V.

The diode clamper circuits shown in Fig. 3.19(b) are also known as positive clamper and negative clamper, respectively.

3.15 Biased Clampers

Like a biased clipper, it is also possible to have a biased clamper. A biased clamper means that the clamping can be done at any voltage level other than zero. Now consider a circuit shown in Fig. 3.21(a). Here a battery of 5 V is added in such a way that the clamping takes palce positively at –5 V as shown in Fig. 3.21(b).

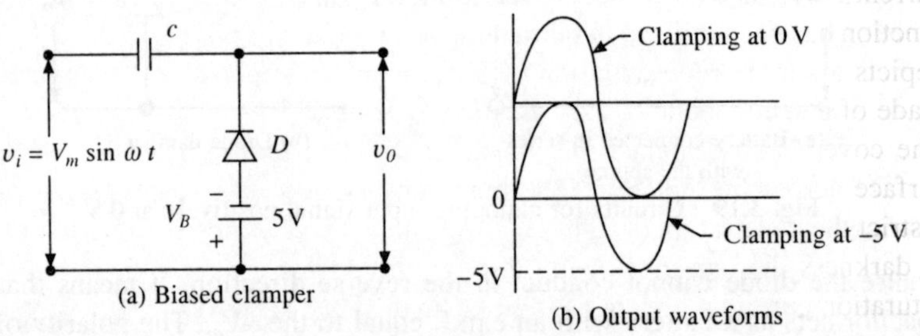

(a) Biased clamper

(b) Output waveforms

Fig. 3.21 Biased clamper.

Similarly, it is also possible to clamp the input waveform positively at +5 V by reversing the battery connections.

3.16 Voltage Multipliers

Voltage doubler circuit is shown in Fig. 3.22. A voltage multiplier is a rectifier circuit that gives a D.C. output voltage approximately equal to a multiple of the peak input A.C. voltage.

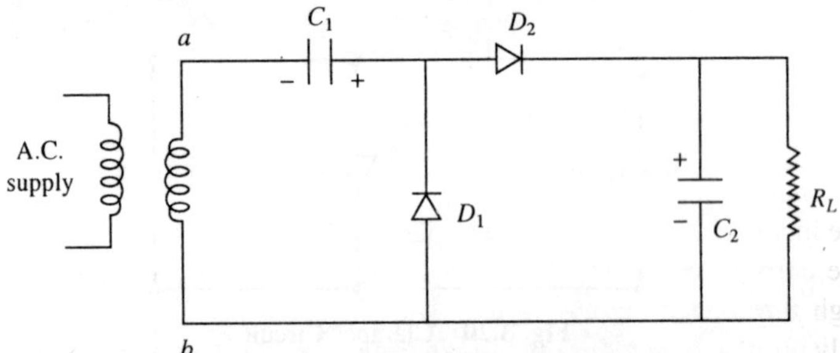

Fig. 3.22 Voltage doubler circuit.

Figure 3.22 shows the circuit diagram of a half-wave voltage doubler. During the negative half cycle of the input A.C. voltage when the point b is at a positive potential with respect to the point a the diode D_1 conducts and charges the capacitor C_1 to the peak voltage V_P of the transformer secondary. The diode D_2 does not conduct during this portion of the cycle. When the point a attains a positive potential with respect to the point b, the diode D_2 conducts and D_1 does not. Since the voltage across C_1 is V_P, clearly the capacitor C_2 is charged to $2V_P$. To maintain the output D.C. voltage at $2V_P$, the time constant of the load circuit has to be large. Thus the load resistance R_L and the capacitor C_2 should be large. Since the output D.C. voltage is twice the peak value V_P of the input A.C. voltage, the circuit is called a voltage doubler. The advantage of the circuit is that it can produce a high output voltage avoiding the use of bulkier transformers.

3.17 Photo Diodes

It has been observed that, the output current of a reverse biased *P-N* junction changes if the device is exposed to illumination. The variation in the output current is linear with respect to the luminous flux. This characteristic of a *P-N* junction has been utilised in fabricating semiconductor photodiodes. Figure 3.23 depicts a simple representation of a semiconductor photodiode. The device is made of a semiconductor *P-N* junction kept in a sealed plastic or glass casing. The cover is so designed that the light rays are allowed to fall only on one surface across the junction. The remaining sides of the casing are painted to restrict the penetration of light rays. When the reverse biased photodiode is kept in darkness the current flowing through the device corresponds to the reverse saturation current which is negligibly small. This current is due to the minority charge carriers crossing over the junction. On the other hand, when the device is illuminated by allowing light rays falling on its surface, additional electron hole pairs are generated which cross over the junction to reach the conduction band.

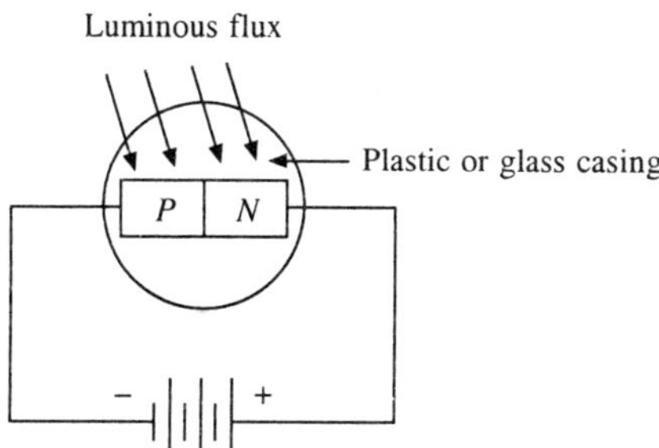

Fig. 3.23 Construction of a semiconductor photodiode.

These charge carriers boost the conduction and hence the current through the device increases. The conduction of the device is due to the diffusion of minority charge carriers. Thus, by changing the illumination level, the current flowing through the device can be varied. The current flowing through the device is directly proportional to the quantity of light flux. The volt-ampere characteristics of a semiconductor photodiode have been shown in Fig. 3.24. A set of *V-I* characteristics for a semiconductor photodiode, as shown in Fig. 3.24 depicts that as the luminous flux φ is increased the diode current increases.

The first curve corresponds to the dark condition, i.e., when φ is zero. When the luminous flux is zero, the diode current is negligibly small. Curves II, III, IV, V correspond to the various levels of light intensities, say luminous flux being φ_1, φ_2, φ_3 and φ_4, respectively. It is seen from the set of *V-I* characteristics that as the value of φ increases, the diode current also increases. It is also seen that except for the characteristic corresponding to the zero light flux, all other characteristics do not pass through the origin. This indicates that for some illumination level even if the reverse bias voltage is zero, there will be some very small amount of current flowing through the device. Only in case of dark condition when the light flux is zero, the diode current will be zero if the reverse

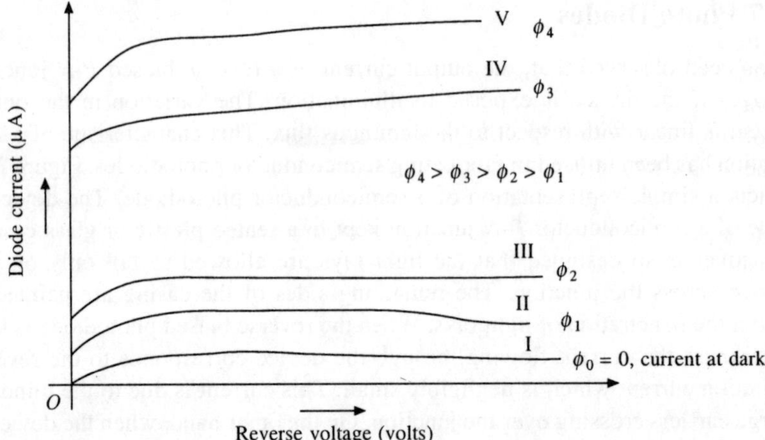

Fig. 3.24 Set of volt-ampere characteristics for a semiconductor photodiode.

bias voltage is also zero. It is also clear from the *V-I* characteristics that the diode current is directly proportional to the amount of reverse bias voltage.

Another factor which governs the amount of current flowing through the device is the distance of the source of illumination from the junction. We know that the amount of current flowing in a reverse biased semiconductor photodiode is due to the diffusion of minority charge carriers to the junction. Obviously, if the distance is increased, the conduction will be less and the amount of current flowing will be decreased bacause there will be a chance of recombination for the minority charge carriers. To increase the amount of current flowing through the device one has to bring the illuminating source nearer to the *P-N* junction. The current flowing through the photodiode is also dependent on the temperature. The working temperature limit specified by the manufacturer must be strictly maintained for efficient performance of the device.

A semiconductor photodiode finds its application in computer card punching and tapes, light operated switches, sound track films, systems for detecting light, electronic control circuits, etc.

3.18 Photo Transistors

A photo transistor is a type of semiconductor *N-P-N* transistor which is generally used in common emitter configuration i.e., the bias voltage is applied between the emitter-collector circuit with the base left open. The transistor action of this device is controlled by light flux. Figure 3.25 shows a simple representation of a photo transistor. The construction of a photo transistor is just like a conventional semiconductor *N-P-N* transistor with a little hole made on the surface near the collector-base junction. A small lens is fixed on this hole for allowing a focused light beam to concentrate on the collector base junction. In the modern methods of fabrication, highly light effective materials are used instead of making a hole and fixing a lens on it.

From Fig. 3.25 it is clear that the emitter base junction J_E is forward biased, whereas the collector base junction J_C is reverse biased. When the transistor is kept in darkness there will be very few minority charge carriers (thermally generated) which will cause the flow of reverse saturation collector current. This current, for obvious reasons, will be negligibly small. On light being focused at the collector base junction, additional photo generated minority charge carriers will be available which will add to the reverse saturation current. Thus, as soon as the light source is applied, the transistor starts conducting and amplified current starts flowing through the reverse biased junction. Thus, owing to the transistor amplifier action the current caused by the luminous flux will increase a lot.

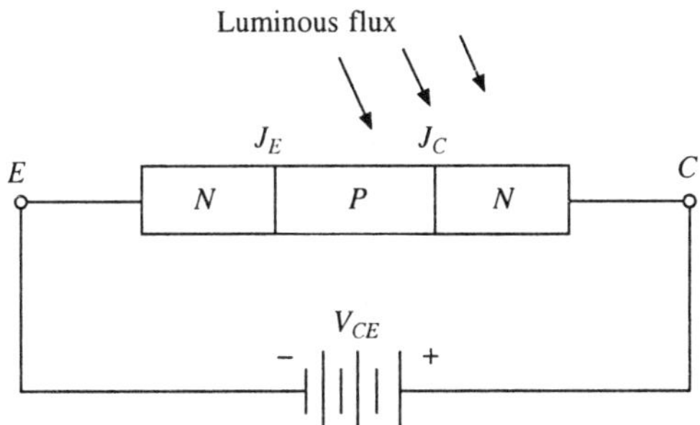

Fig. 3.25 Photo transistor.

3.19 Photovoltaic Cells

A photovoltaic cell is a photodiode with no reverse bias voltage applied. If a photodiode is illuminated without any reverse bias voltage being applied, the device starts generating a potential across its terminals. This generated voltage is called the photovoltaic e.m.f. and this phenomenon of generating voltage is known as the photovoltaic effect. When the photodiode generates photovoltaic e.m.f. it is called a photovoltaic cell. We have seen in case of a photodiode that the reverse saturation current becomes constant at large reverse bias voltages. This current is constituted by the injection of only minority charge carriers. Hence, when the current becomes saturated, its value remains constant even if the reverse voltage is changed. Reducing the reverse voltage only results in lowering of the potential barrier which does not effect the value of flow of reverse current. However, if the reverse voltage is reduced substantially, it allows some majority charge carriers also to cross over the barrier, hence causing flow of forward current. The forward current has a tendency to reduce the value of reverse current already flowing. Now, as the reverse voltage is further decreased, there will be a tendency for the reverse current to decrease rapidly. If forward voltage is applied the potential barrier is further lowered and gradually a position arises when the forward current due to the majority charge carriers become equal to the reverse current due to the minority charge carriers.

This results in neutralization of the two opposite currents and the net current becomes zero. The voltage at which the resultant current in a semiconductor photodiode becomes zero, is called the photovoltaic e.m.f. Since no current flows through the device, the total voltage (photovoltaic e.m.f.) appears across the terminals of the device. The *V-I* characteristics of a photovoltaic cell are represented as shown in Fig. 3.26.

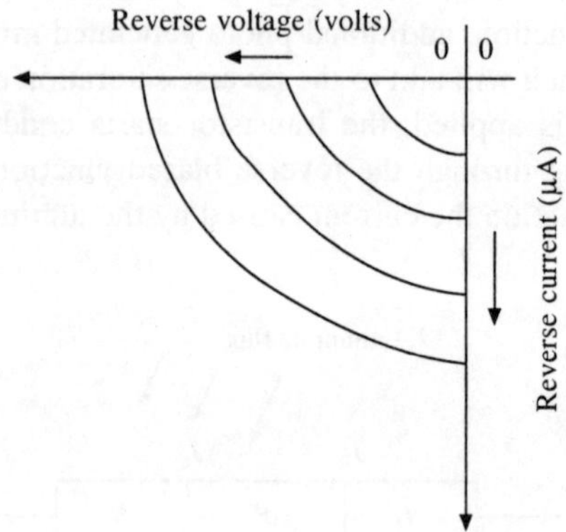

Fig. 3.26 Set of *V-I* characteristics of a photovoltaic cell.

It is clear from the *V-I* characteristics that the reverse current decreases as the reverse voltage is decreased. The no load voltage of a photovoltaic cell varies non-linearly with the amount of luminous flux falling on the surface of the device. This characteristic between the no load voltage and luminous flux has been shown in Fig. 3.26 In case the terminals of a photovoltaic cell are short circuited in the presence of the luminous flux, the photovoltaic e.m.f. present across it will cause the flow of a large amount of current, which is known as the short circuit current of the device. The short circuit current varies linearly with the luminous flux as shown in Fig. 3.26 Seeing the variations of (i) no load voltage and (ii) the short circuit current, with the luminous flux, it can be concluded that the internal resistance of a photovoltaic cell decreases with increase in luminous flux.

Photovoltaic cells are mainly used in laboratory and research applications. BPY 10 is a commonly used photovoltaic cell.

3.20 Photoconductive Cells

A photoconductive cell is basically a light-dependent resistor (LDR) made of some semiconductor material. The device when kept in darkness poses very high resistance of the order of several hundred mega ohms. When the device is kept in darkness and supplied with some external potential, owing to the very high resistance, almost a negligible amount of current flows in the circuit. This is called the dark current. On illuminating the device, its resistance decreases to a low value and hence, immediately a large amount of current starts flowing.

The IR drop of the device appears as the voltage drops across it. Sensitivity of this device is found to be of the value of a few milliamperes per unit lux. This is greater than the sensitivity of gas filled vacuum photo cells. The heat dissipation of the device is of the order of 100 mW to 1 W.

Figure 3.27 shows voltage and short circuit current characteristics with respect to luminous flux.

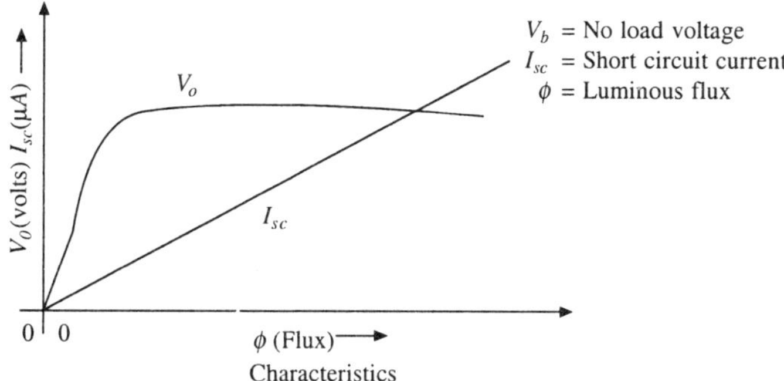

Fig. 3.27 Characteristics.

Photoconductive cells are used for the following purposes:

1. Recording situations where fast changes take place
2. Measuring light quantity (light metre)
3. As a light operated ON-OFF relay
4. Recording modulating light intensity
5. Light sensitive trips and alarms

Cadmium sulphide cell is the photoconductive cell most widely used. The surface of this cell has a coating of cadmium sulphide. This coating also contains some impurities like antimony, indium nitrate etc, which increases the efficiency of the device. The cadmium sulphide cells have a very high dissipation capability and sensitivity. A lead-sulphide cell is another photoconductive cell which is used for infrared spectrums, viz. infrared deduction or infrared absorption measurements.

Worked Problems

Ex. 3.1 The reverse saturation current at 300° K of P-N germanium diode is 5 μA. Find the voltage to be applied to obtain forward current of 50 mA.

$$I = I_0\left(e^{\frac{V}{\eta V_T}} - 1\right); \quad \eta = 1; \quad V_T = 26 \text{ mV}$$

$$I = I_0(e^{40V} - 1)$$

$$50 \times 10^{-3} = 5 \times 10^{-6}(e^{40V} - 1)$$

$$\frac{50 \times 10^{-3}}{5 \times 10^{-6}} = e^{40\,V} - 1$$

$$V = 0.23$$

Ex. 3.2 Calculate ratio of current for a forward bias of 0.06 V to the current for the same value of reverse bias applied to a germanium P-N diode at 27°C.

$$V_T = \frac{T}{11{,}600} = \frac{27 + 273}{11{,}600} = \frac{300}{11{,}600}$$

forward current $$i_1 = I_0(e^{40V} - 1) \qquad (1)$$

reverse current $$i_2 = I_0(e^{-40V} - 1) \quad (2)$$

$$\frac{(1)}{(2)} = \frac{i_1}{i_2} = \frac{I_0\,(e^{40V} - 1)}{I_0\,(e^{-40V} - 1)}$$

$$= \frac{i_1}{i_2} = \frac{(e^{40\times 0.06} - 1)}{(e^{-40\times 0.06} - 1)}$$

$$\frac{i_1}{i_2} = -\ 11.0229$$

Ex. 3.3 The current flowing through a silicon diode is 60 mA. With the forward bias of 0.9 V determine static and dynamic resistance of the diode.

Static resistance $$r_{DC} = \frac{V}{I} = \frac{0.9}{60 \times 10^{-3}}$$

Dynamic resistance $$r_{AC} = \frac{26\,n}{I}$$

$$r_{AC} = \frac{0.026\,n}{I} \quad (n = 2 \text{ for silicon})$$

$$= \frac{26 \times 2}{60} = \frac{52}{60}$$

$$r_{AC} = 0.87\ \Omega$$

Short Questions and Answers

Q. 1. What is the breakdown mechanism in a lightly doped *P-N* diode?
Ans. Avalanch break down

Q. 2. What is the thickness of the unbiased depletion region?
Ans. 0.5 μm

Q. 3. What are the parametes on which Hall effect depends?
Ans. Hall voltage depends on B, I, l, d and Q.

Q. 4. What is the transient current in a reverse biased *P-N* junction?
Ans. The depletion layer keeps growing until its voltage is equal to the applied voltage. A small current flows in the transient period. Such current is called "transient current".

Q. 5. What is reverse saturation current of *P-N* diode?
Ans. The minority carriers in reverse biased *P-N* junction produce a small current. This current is called reverse saturation current.

Q. 6. What is the charge storage in a forward biased diode?
Ans. In a forward biased *P-N* diode, continuously the holes come to the left side and electrons come to the right side of the junction. They are stored temporarily before they recombine. This is called charge storage.

Q. 7. Explain whether the dynamic resistance of zener diode varies with current.
Ans. Slope is constant. Dynamic resistance is constant.

Q. 8. Define voltage-temperature constant of *P-N* junction.

Ans. $$V_T = \frac{T}{11{,}600}$$

Temperature T is in degrees Kelvin.

Q. 9. A transistor has I_C as 100 mA and I_B of 0.5 mA. What is its α?

Ans.

$$\alpha = \frac{I_C}{I_E} = \frac{I_E}{I_B + I_C}$$

$$= \frac{100}{100.5} = 0.995$$

Q. 10. What is the speciality of a silicon controlled switch?

Ans. The device can be turned ON and turned OFF using anode gate or cathode gate

Q. 11. Give the applictions of SCR.

Ans. (i) Heating, welding

(ii) battery charge

(iii) speed control of motors

Q. 12. Define (i) Intrinsic concentration (ii) Fermi energy (iii) Mean life time

Ans. (i) Number of electrons/m^3 in the intrinsic semiconductor is intrinsic concentration.

(ii) Maximum energy of an electron inside the crystal at 0°K.

(iii) τ, average time for which the electrons and holes exist before the recombination is called mean life time.

Q. 13. What is step graded function and abrupt function?

Ans. In step graded function the impurity concentration is uniform throughout.
In linear graded function the impurity concentration increases linearly.

Q. 14. What are band gap semiconductors?

Ans. A small gap is present in the C.B. of gallium arcinide. Such semiconductors are called band gap semiconductors.

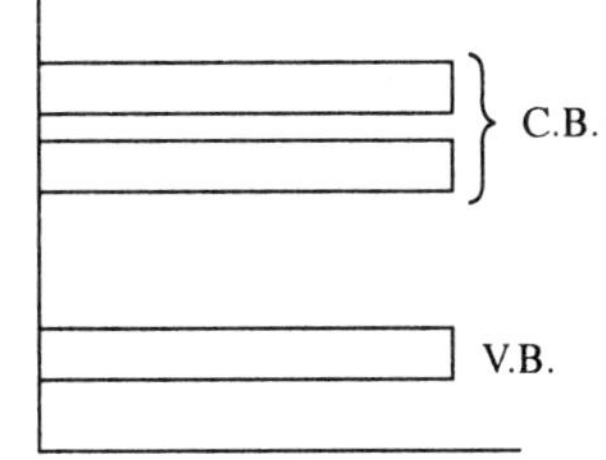

Q. 15. Why is diffusion capacitance negligible for a reverse biased diode?

Ans. There is no charge storage in reverse biased diode.

Q. 16. Draw *V-I* characteristic of D/AC.

Ans.

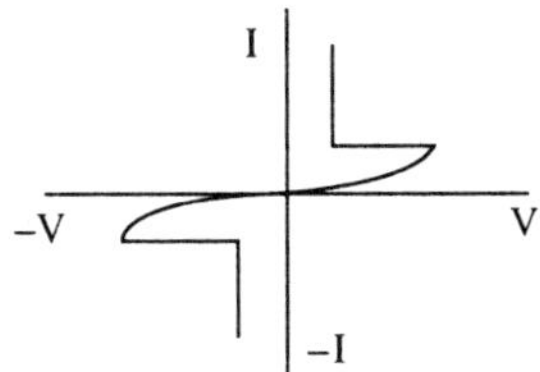

Q. 17. Define latching current of SCR.

Ans. Minimum current required to "turn ON" the SCR is called latching current.

4

Bipolar Junction Transistor

This chapter deals with low frequency and high frequency models of BJT.

4.1 Manufacturing Process of BJT

Two technologies used in the manufacture of BJT are alloy junction and silicon planar.

4.1.1 Alloy Junction Transistor

It is made by alloying two indium pellets on the opposite sides of *n*-type germanium wafer. The assembly is kept in the furnace having a temperature of 500°C. The indium melts and combines with germanium. Acceptor impurity (indium) is dissolved into the germanium giving P-regions at the top and bottom. Thus we obtain P.N.P. structure as shown in Fig. 4.1(a). Leads for collector and emitter are soldered to the dots making non-rectifying contacts. Large area collector junction helps in collecting most of the holes emitted from the emitter ensuring that the collector current almost equals the emitter current. The spacing between two junctions inside germanium wafer is very small. This determines the electrical properties of the transistor.

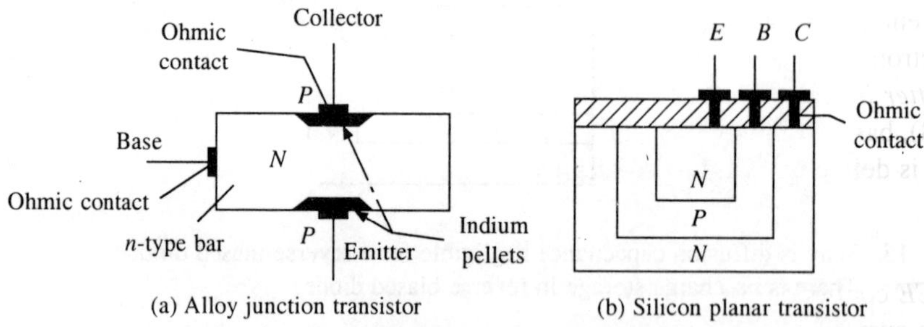

Fig. 4.1 Manufacturing methods.

4.1.2 Silicon Planar Transistor

A *P*-type cylinder is sliced into many thin wafers. The wafer can be divided into number of sections. One of these sections can be used to form *P*-layer. A gas mixture of pentavalent atoms are passed over silicon crystal. This forms a layer of *N*-type semiconductor. To prevent contamination, oxygen is blown over the surface. Holes are provided to insert the metal contacts. Emitter, base and collector terminals are brought outside as shown in the Fig. 4.1(b).

4.2 Bipolar Junction Transistor (BJT)

The regions of the transistor are emitter, base and collector. The emitter is heavily doped, the base is lightly doped. The doping of collector is in between the doping of emitter and base. The emitter junction is forward biased by the battery V_{EE}. The collector junction is reverse biased by the battery V_{cc}. Consider a stream of electrons starting from the negative terminal of the battery V_{EE}. They travel through the emitter region and they diffuse into the base region. Since the base is very thin and is lightly doped, only a few electrons recombine by falling into the holes. The remaining electrons enter into the depletion layer of the collector junction. The depletion layer field pushes these electrons into the collector region and these electrons are attracted by the positive terminal of V_{CC}. Thus, surprisingly a large current flows in the collector during forward reverse biased condition. This action is known as transistor action. The condition is that the emitter junction must be forward biased and the collector junction must be reversed biased as shown in Fig. 4.2.

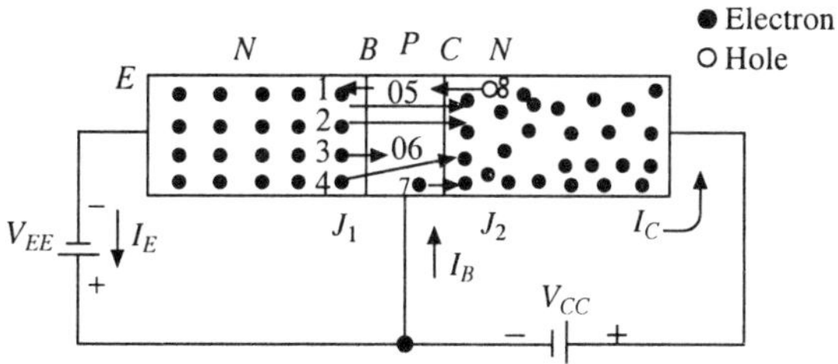

Fig. 4.2 Bipolar junction transistor.

The junction J_1 is forward biased. The electrons and the holes diffused are numbered as 1, 2, 3 and 4. The hole diffused is numbered as 5. The total emitter current is the sum of electron current and the hole current. The ratio of the electron current to the total emitter current is defined as the *emitter efficiency* or *emitter injection ratio*. Typical value is 0.995. In common base configuration, (CB), base is directly connected to ground without any resistance. Current gain (α) is defined as ratio of collector current to the emitter current.

$$\text{Current gain in } CB = \alpha = \frac{I_C}{I_E}$$

In *CE* configuration, current gain (B) is defined as ratio of collector current to base current.

$$\beta = \frac{I_C}{I_B}$$

If $I_C = 0.99\text{A}$ and $I_B = 0.01$ A, then

$$\beta = \frac{0.99}{0.01} = 99$$

4.2.1 Relation between α and β

We know $I_E = I_C + I_B$. Dividing throughout by I_C

$$\frac{I_E}{I_C} = \frac{I_C}{I_C} + \frac{I_B}{I_C}$$

$$\frac{1}{\alpha} = 1 + \frac{1}{\beta} = \frac{\beta + 1}{\beta} \tag{4.1}$$

$$\alpha = \frac{\beta}{1 + \beta} \tag{4.2}$$

From (4.1)

$$\frac{1}{\beta} = \frac{1}{\alpha} - 1 = \frac{1 - \alpha}{\alpha}$$

$$\beta = \frac{\alpha}{1 - \alpha} \tag{4.3}$$

4.3 Characteristics of CE Circuit

The input characteristics are drawn with input voltage on X-axis and input current on Y-axis. The input characteristics are drawn with constant output voltage. The output voltage is kept constant by keeping the position of potential divider R_2 shown in Fig. 4.3(a) at a particular value. The input voltage can be increased by moving the jockey point of the potential divider R_1 upwards. The readings of input voltage V_{EE} and input current I_B are tabulated. The characteristics are similar to the characteristics of forward biased P-N diode. They are shown in Fig. 4.3(b).

The output characteristics are drawn with output voltage on X-axis and output current on Y-axis. Initially the input circuit is kept open. The current flowing

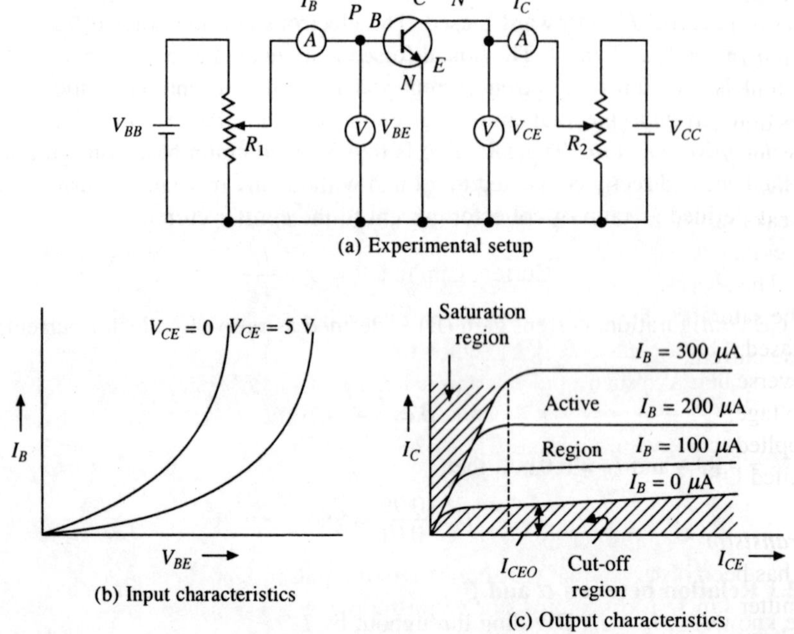

Fig. 4.3 Characteristics of CE configuration.

from collector to emitter with base terminal open is denoted as I_{CEO} (leakage current from collector to emitter when base is open). The input current is set to a fixed value of 100 μA using the potential divider R_1. The output voltage is gradually increased and the corresponding current values are noted down. The voltage V_{CE} is brought to the minimum value. Now the base current is set to 200 μA and the above procedure is repeated. Thus a family of output characteristics are drawn for different values of base current. The characteristics are divided into three regions. In the cutoff region, the current through the transistor is negligible. In the saturation region, the voltage across the transistor is very less. In the active region the collector current is almost constant. The output characteristics are shown in Fig. 4.3(c).

4.4 Characteristics of Common Base Configuration

The experimental set-up used for obtaining the characteristics is shown in Fig. 4.4(a).

Input characteristics: For obtaining input characteristics, the output voltage is kept constant. Zero voltage can be maintained by short circuiting collector and base terminals. The input voltage is varied by varying the jockey point of the potential divider R_1. The readings of input voltage and current are tabulated. Input characteristic is drawn with V_{EB} on X-axis on I_E on Y-axis. The input characteristic is similar to forward characteristic of the P-N diode. They are shown in Fig. 4.4(b).

Output characteristics: For obtaining output characteristic, the input current has to be maintained constant. Output voltage is varied by varying the jockey point of R_2. For a particular value of emitter current (1 mA), the reading of voltmeter and ammeter in the output circuit are noted down. Output characteristics are drawn with V_{CB} on X-axis, I_C on Y-axis. Similar procedure has to be repeated for an emitter current value of 2 mA. Thus a family of characteristics can be drawn. Beyond a particular value of V_{CB} the reverse biased collector junction breaks down and the current increases rapidly. A transistor is not designed to operate in the breakdown region. Output characteristics are shown in Fig. 4.4(c).

The switch operates between two states namely saturation and cut-off state. The saturation state occurs when both the junctions of a transistor are forward biased. On the other hand, the cut-off state occurs when both the junctions are reverse biased. When a pulse is applied at the base, the transistor is saturated, the voltage V_{CE} is approximately equal to zero, it is the on state. When no pulse is applied, current through the BJT is zero, voltage across the device is V_{cc}. It is called OFF state. The corresponding circuit is shown in Fig. 4.5(a).

Transistor Switching Times

It has been observed that the application of a pulse input at the input (i.e., base-emitter junction) of a transistor is not followed directly by a change in output voltage. In other words, there is always some delay before the output changes as compared to the input.

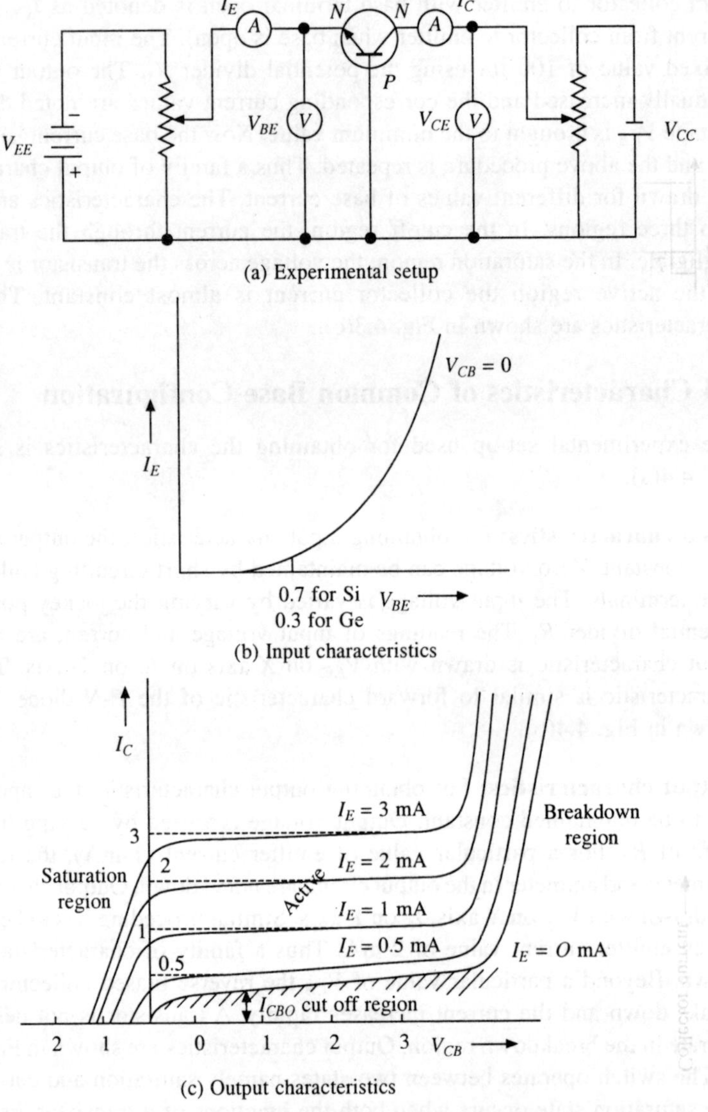

(a) Experimental setup

(b) Input characteristics

(c) Output characteristics

Fig. 4.4 Common base characteristics.

Figure 4.5b (i) shows the waveform of the input pulse applied to the transistor and Fig. 4.5b (ii) indicates the resulting waveform of collector current I_c along with various time delays involved. Following are the important time delays in the collector current waveform, which may be noted from the figure.

1. *Delay time* (t_d). It may be noted that the collector current does not quickly respond to the input signal at time $t = T_1$. Instead, there is a delay indicated by T_1. The time that elapses during this delay, together with the time required for the collector current to rise to 10% of its maximum (i.e. saturation) value $I_{c(sat)}$ equal to V_{cc}/R_c is called the delay time t_d.

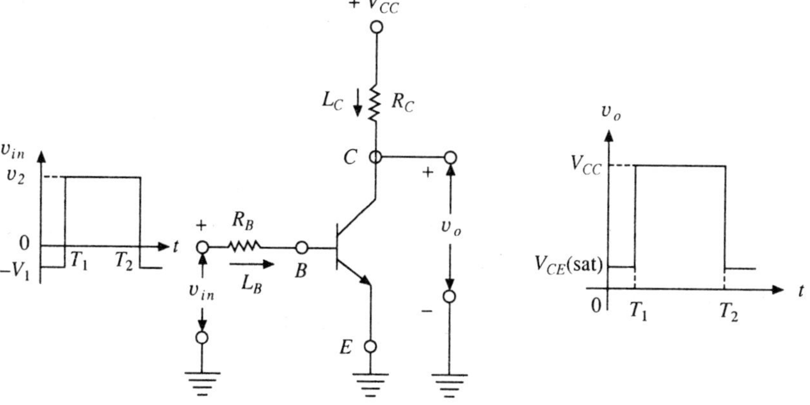

A transistor switch with input and output waveforms.
(a) Circuit

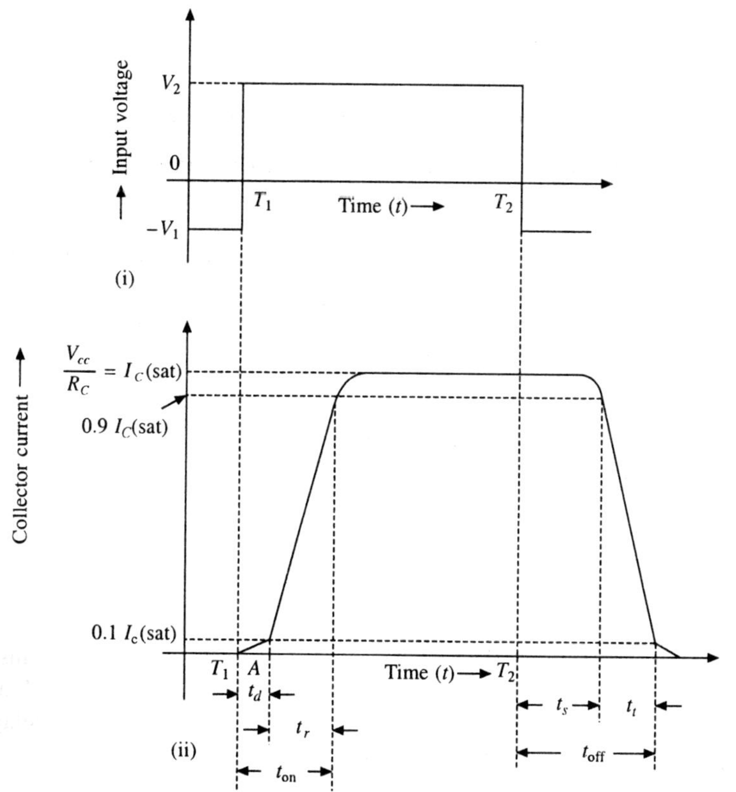

(b) (i) Input pulse waveform, (ii) Collector current waveform with various time delays.

Fig. 4.5 Transistor as a switch

2. *Rise time* (t_r). The time required for the collector current to rise from 10% to 90% of its maximum value is called rise time t_r.

3. *Turn-on time* (T_{ON}). The sum of the delay time t_d and the rise time t_r is called the turn-ON time. Mathematically, the turn-ON time,

$$t_{ON} = t_d + t_r$$

4. *Storage time* (t_s). It may be noted from the figure that when the input signal returns back to its initial state, i.e. $-V_1$ at time $t = T_2$, the collector current gain fails to respond immediately. The interval, which elapses between the transition of the input voltage waveform and the time when collector current has dropped to 90% of its maximum value is called the storage time. Storage time is about 225 nanosecond.

5. *Fall-time* (t_f). The storage interval is followed by fall time t_f. It is the time required for the collector current to fall from 90% to 10% of its maximum value. Its value is about 60 nanosecond.

6. *Turn-Off time* (t_{OFF}). It is the sum of storage time t_s and the fall time t_f. Mathematically, the turn-off time

$$t_{OFF} = t_s + t_f$$

4.6 Linear Model of BJT (Hybrid Parameters)

BJT has three terminals. If one terminal is taken as a common terminal to input and output, then there are four terminals. Therefore BJT can be treated as a 2-port network.

In defining h-parameters, V_1 and I_2 are expressed as a function of I_1 and V_2.

$$V_1 = f(I_1, V_2)$$

$$I_2 = f(I_1, V_2)$$

I_1 I_2

2-Port net work

V_1 V_2

$$\begin{pmatrix} V_1 \\ I_2 \end{pmatrix} = \begin{pmatrix} h_{11} & h_{12} \\ h_{21} & h_{22} \end{pmatrix} \begin{pmatrix} I_1 \\ V_2 \end{pmatrix}$$

$$V_1 = h_{11}I_1 + h_{12}V_2 \quad (4.4)$$

$$I_2 = h_{21}I_1 + h_{22}V_2 \quad (4.5a)$$

Short circuit port 2

$$\text{Input impedance } h_{11} = \frac{V_1}{I_2} \text{ when } V_2 = 0$$

$$\text{Forward current gain } h_{21} = \frac{I_2}{I_1} \text{ when } V_2 = 0$$

Open circuit port 1

$$\text{Reverse voltage gain } h_{12} = \frac{V_1}{V_2} \text{ when } I_1 = 0$$

$$\text{Output Admittance } h_{22} = \frac{I_2}{V_2} \text{ when } I_1 = 0$$

$h_{11} = h_i$ (i stands for input)
$h_{21} = h_f$ (f stands for forward)
$h_{22} = h_o$ (o stands for output)
$h_{12} = h_r$ (r stands for reverse)

i.e
$$\begin{pmatrix} h_{11} & h_{12} \\ h_{21} & h_{22} \end{pmatrix} = \begin{pmatrix} h_i & h_r \\ h_f & h_o \end{pmatrix}$$

$$\begin{vmatrix} i & r \\ f & o \end{vmatrix}$$

When the suffixes are read in anticlockwise direction, we get *if or.* This can be used as a thumb rule to remember the matrix.

Replacing $V_1 = V_b$, $V_2 = V_c$, $I_1 = I_b$, $I_2 = i_c$

$$V_b = f_1(i_b, v_c)$$

$$i_c = f_2(i_b, v_c)$$

$$\begin{pmatrix} V_b \\ I_c \end{pmatrix} = \begin{pmatrix} h_i & h_r \\ h_f & h_o \end{pmatrix} \begin{pmatrix} I_b \\ V_c \end{pmatrix}$$

$$V_b = h_i I_b + h_r V_c \tag{4.5b}$$

$$I_c = h_f I_b + h_o V_c \tag{4.5c}$$

These equations are represented by the circuit shown in Fig. 4.6.

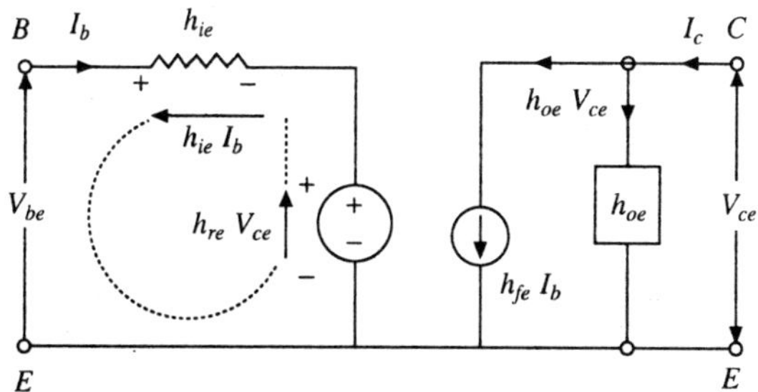

Fig. 4.6 Low frequency model.

If we apply KVL on the input loop for Fig. 4.6 we get (4.5b). If we apply KCL at collector we get (4.5c).

$$h_i = \frac{V_b}{I_b} \text{ when } V_c = 0$$

$$h_f = \frac{I_c}{I_b} \text{ when } V_c = 0$$

$$h_r = \frac{v_b}{V_c} \text{ when } I_b = 0$$

$$h_o = \frac{I_c}{V_c} \text{ when } I_b = 0.$$

h_i is expressed in ohms. h_f has no dimensions h_r is dimension less. h_o is expressed in mhos. They are called hybraid parameters since the dimensions are mixed. Where h_i is the input impedance; h_f the forward current gain; h_o the output admittance and h_r the reverse voltage ratio

The circuit shown in Fig. 4.6 is called low frequency model.

4.7 Graphical Determination of *h*-Parameters

Input and output characteristics of BJT in common emitter configuration are shown in Fig. 4.7.

$$h_{ie} = \frac{\Delta v_{be}}{I_b} \tag{4.6}$$

$$h_{re} = \frac{V_{b2} - V_{b1}}{V_{ce2} - I_{ce1}} \tag{4.7}$$

$$h_{oe} = \frac{\Delta I_c}{\Delta v_{ce}} \tag{4.8}$$

$$h_{fe} = \frac{I_{c2} - I_{c1}}{I_{b2} - I_{b1}} \tag{4.9}$$

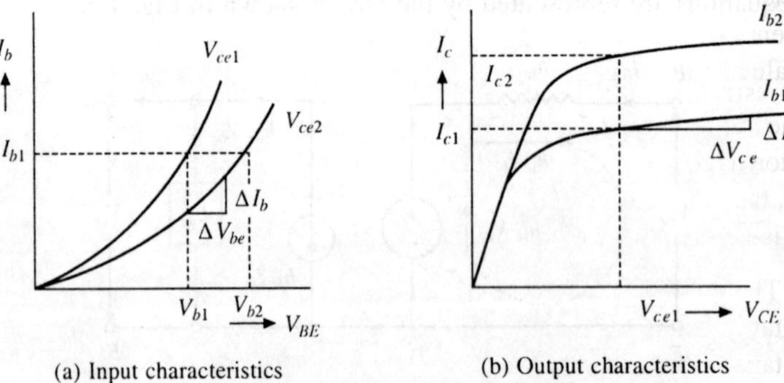

(a) Input characteristics (b) Output characteristics

Fig. 4.7 Characteristics of BJT

Input characteristics are drawn by keeping output voltage constant. Consider a small variation in the input voltage ΔV_{be} as shown in Fig. 4.7a. The corresponding variation in the input current is (ΔI_b) and h_{ie} is the ratio of change in input voltage to the change in input current for a constant value of output voltage. This definition is given in equation (4.6).

Consider a constant value of base current I_{b1}, the corresponding input voltages from the graph are v_{b1} and v_{b2} and h_{re} is defined as the ratio of change in input voltage to the change in output voltage for a constant base current as given by equation (4.7).

Output characteristics are drawn for constant values of base current. Consider a small variation in the output voltage ΔV_{ce}. The corresponding change in the output current is ΔI_c. h_{oe} is defined as the ratio of change in output current to the change in output voltage for a constant value of base current as given in equation (4.8).

Consider a constant output voltage V_{ce1}, the corresponding output currents from the graph are I_{c1} and I_{c2}. Forward current gain h_{fe} is defined as the ratio of change in output current to the change in input current for a constant value of V_{ce} (equation (4.9)).

4.8 High Frequency Model of BJT

High frequency model is shown in Fig. 4.8. Low frequency model (Fig. 4.6) cannot be used for high frequency analysis. Various elements of high frequency model are:

$r_{b'b}$: The width of the *p*-channel reduces as the depletion layers occupy some space in the base region. The *P* channel is holding the positive charges. The resistance of this *P*-channel is defined as base spread resistance. The value is about 100 Ω.

$r_{b'c}$: There is a leakage current in the transistor from the collector to the base. To represent the leakage current, a high resistance is shown between the collector and the virtual base. The value is about 60 $k\Omega$.

r_{ce}: There is a leakage current from the collector to emitter namely I_{CEO}. To represent this current, a high resistance is shown between collector and emitter. The value is about 80 kΩ.

g_m: The transconductance or the mutual conductance is defined as the ratio of the collector current to the input voltage at a constant output voltage. Due to transistor action, the collector current increases when the input voltage is increased. This effect is shown by a voltage dependent current source $g_m\ v_{b'e}$.

$C_{b'e}$: The emitter junction of the transistor is forward biased and a diffusion capacitance exists at this junction due to the temporary storage of charges. This capacitance is shown as $C_{b'e}$.

$C_{b'c}$: The collector junction of the transistor is reverse biased. The base and the collector act as conductors and the depletion layers act as the dielectric. Hence a depletion layer capacitance exists at the output junction.

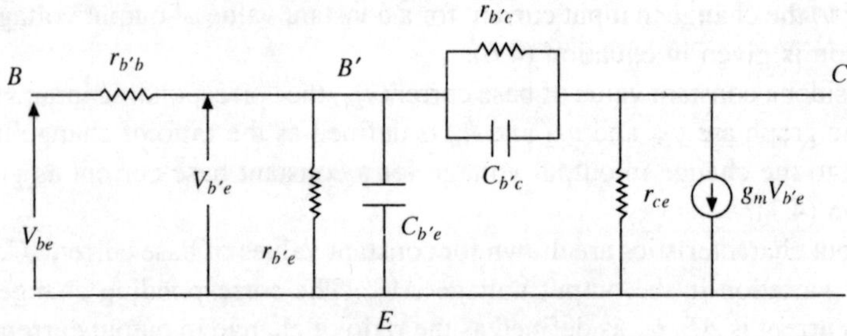

Fig. 4.8 High frequency model of BJT.

4.9 Relation between *h*- and High Frequency-Parameters

We know that

$$I_c = g_m \, V_{b'e}$$

Hence

$$\boxed{g_m = \frac{I_c}{V_{b'e}}} \qquad (4.10)$$

mutual conductance is ratio of output current to input voltage

$$\boxed{\therefore gm = \frac{I_c}{26}}$$

In the above equation, I_c is expressed in mill Amperes.

4.9.1 $r_{b'e}$ in terms of *h*-parameters

In D.C. analysis the capacitors have to be open circuited. The circuit shown in Fig. 4.8 gets reduced to Fig. 4.9(a).

$$I_c = h_{fe} \, I_b + h_{oe} \, V_{ce}.$$

$$h_{fe} = \left. \frac{I_c}{I_b} \right|_{V_{ce} = 0}$$

$V_{ce} = 0$ indicates that the output side has to be short circuited. It is shown in Fig. 4.9(b)

$$I_c = g_m \, V_{b'e}$$

$$I_c = h_{fe} \, I_b$$

Hence

$$h_{fe} = \left. \frac{I_c}{I_b} \right|_{V_{ce} = 0}$$

Short circuit in parallel with r_{ce} reduces to a short circuit. It is shown in Fig. 4.9(c). Parallel combination of $r_{b'c}$ and $r_{b'e}$ is approximately equal to $r_{b'e}$ since $r_{b'e}$ is the smaller of the two (Fig. 4.9(d)).

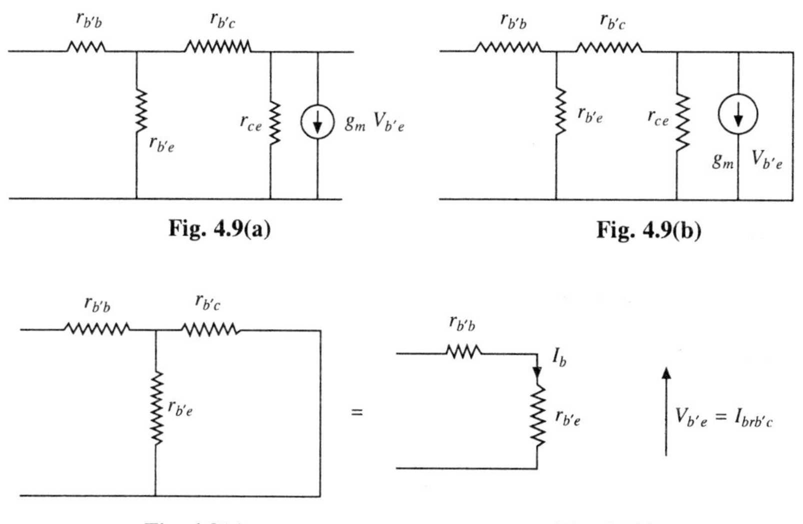

Fig. 4.9(a) Fig. 4.9(b)

Fig. 4.9(c) Fig. 4.9(d)

$I_c = g_m V_{b'e}$. Substituting $V_{b'e} = I_b r_{b'e}$

$$h_{fe} I_b = g_m I_b r_{b'e}$$

$$\boxed{r_{b'e} = \frac{h_{fe}}{g_m}} \tag{4.11}$$

4.9.2 *Expression for $r_{b'b}$* $V'_{be} = h_{ie} I_b + h_{re} V_{ce}$

$$h_{ie} = \left.\frac{V_{b'e}}{I_b}\right|_{V_{ce}=0}$$

From Fig. 4.9(d)

$$V'_{be} = I_b (r_{b'b} + r_{b'e})$$

$$\frac{V'_{be}}{I_b} = h_{ie} = r_{b'b} + r_{b'e}$$

$$\boxed{r_{b'b} = h_{ie} - r_{b'e}} \tag{4.12}$$

4.9.3 $r_{b'c}$ in terms of *h*-parameters

$$I_c = h_{fe} I_b + h_{oe} V_{ce}$$

$$h_{oe} = \left.\frac{I_c}{V_{ce}}\right|_{I_b=0}$$

$$V_{ce} = I(r_{b'e} + r_{b'c}) \tag{4.13}$$

$$V_{b'e} = I_b r_{b'e}$$

$$I_c = g_m V_{b'e}$$

$$h_{oe} = \frac{g_m \; V_{b'e}}{V_{ce}}$$

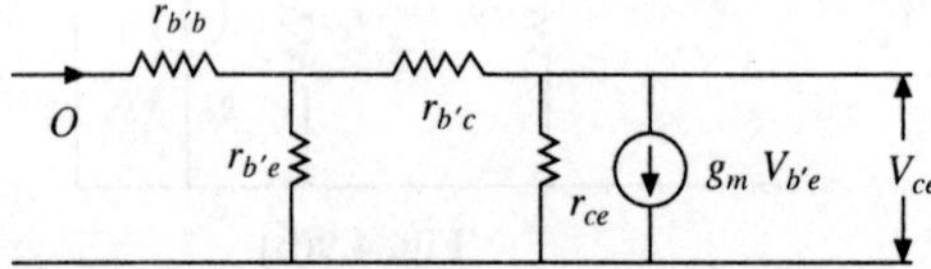

Fig. 4.10(a)

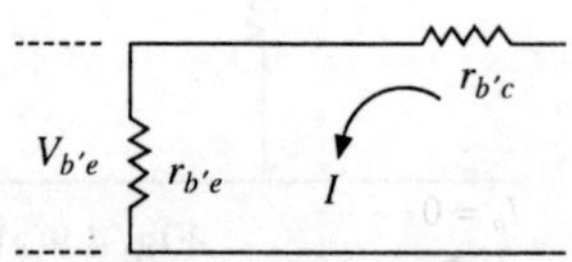

Fig. 4.10(b)

Substitute (4.13) here

$$h_{oe} = \frac{g_m \; I r_{b'e}}{I(r_{b'e} + r_{b'e})} \tag{4.14}$$

$$r_{b'e} + r_{b'c} = \frac{g_m \; r_{b'e}}{h_{oe}}$$

$$r_{b'e} = \frac{h_{fe}}{g_m} \tag{4.15}$$

Substituting (4.15) in (4.14), we get

$$h_{oe} = \frac{g_m \times \dfrac{h_{fe}}{g_m}}{r_{b'e} + r_{b'c}}$$

$$h_{oe} = \frac{h_{fe}}{r_{b'e} + r_{b'c}}$$

$$r_{b'e} + r_{b'c} = \frac{h_{fe}}{h_{oe}}$$

$$\boxed{r_{b'c} = \frac{h_{fe}}{h_{oe}} - r_{b'e}} \tag{4.16}$$

4.9.4 Expression for r_{ce}

$$V_{be} = h_{ie} \, I_b + h_{re} \, V_{ce}$$

$$h_{re} = \left. \frac{V_{be}}{V_{ce}} \right| I_b = 0$$

$$I_c = \frac{V_{ce}}{r_{b'e} + r_{b'c}} + \frac{V_{ce}}{r_{ce}} + g_m \; V_{b'e}$$

Dividing each term by V_{ce} and neglecting $r_{b'e}$

$$\Rightarrow \quad \frac{I_c}{V_{ce}} = \frac{1}{r_{b'c}} + \frac{1}{r_{ce}} + g_m \frac{V_{b'e}}{V_{ce}}$$

$$h_{oe} = g_{b'c} + g_{ce} + g_m h_{re}$$

$$\boxed{g_{ce} = h_{oe} - g_{b'c} - g_m h_{re}} \tag{4.17}$$

$$r_{ce} = \frac{1}{g_{ce}}$$

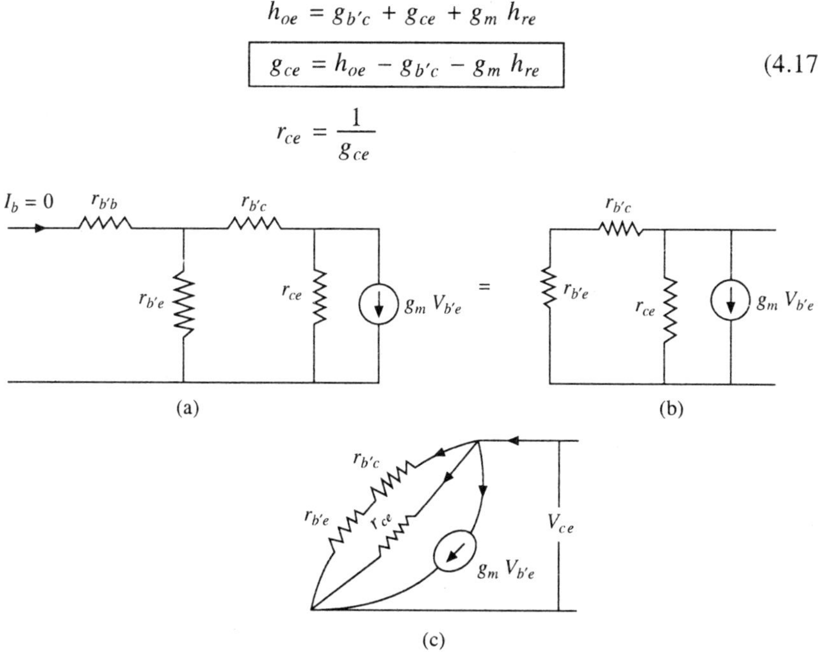

Fig. 4.11 High Frequency Model with $I_b = 0$.

4.10 Short Circuit Current Gain

Short circuit current gain is defined as the ratio of out put current to input current when the output port is short circuited (Fig. 4.12(a)).

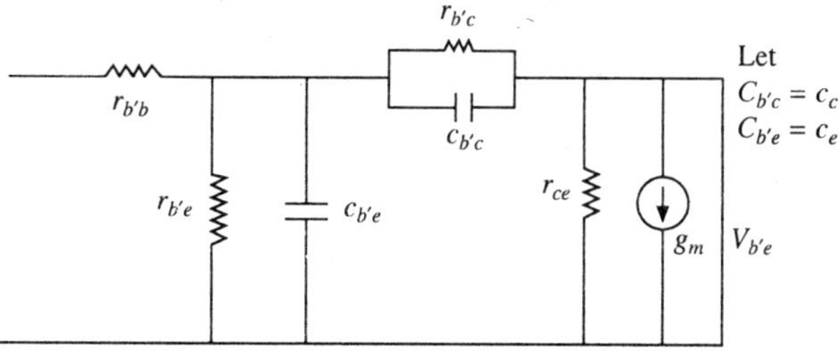

Fig. 4.12(a) High frequency model of BJT

Assumptions

(a) $r_{b'c}$ can be treated as open circuit since it is a high resistance.
(b) $r_{b'b}$ can be neglected.

In Fig. 4.12(b), $r_{b'c}$ is shown as an open circuit. r_{ce} in parallel with short circuit is a short circuit. In Fig. 4.12(c), one short circuit is shown as two short circuits in parallel. In Fig. 4.12(d), two capacitors are reduced to single capacitor. Let $C_{b'c} = C_e$ and $C_{b'c} = C_c$

From Fig. 4.12(d), $I_o = -g_m V_{b'e}$

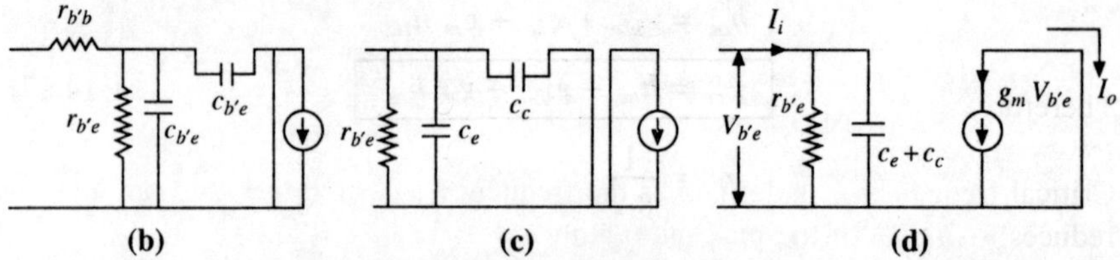

Fig. 4.12 Simplified circuits

$$A_I = \frac{I_o}{I_i} \qquad I_i = V_{b'e}\, g_{b'e} + V_{b'e}\, j\omega c$$

$$A_I = \frac{-g_m V_{b'e}}{V_{b'e}\, g_{b'e} + j\omega c\, V_{b'e}}$$

$$A_I = \frac{-g_m V_{b'e}}{V_{b'e}(g_{b'e} + j\omega c)}$$

$$A_I = \frac{-g_m}{g_{b'e} + j\omega c} \tag{4.18}$$

$$r_{b'e} = \frac{h_{fe}}{g_m}$$

$$g_m = \frac{h_{fe}}{r_{b'e}}$$

$$g_m = h_{fe}\, g_{b'e} \tag{4.19}$$

Substituting (4.19) in (4.18)

$$A_I = \frac{-h_{fe}\, g_{b'e}}{g_{b'e} + j\omega c} = \frac{-h_{fe}\, g_{b'e}}{g_{b'e}\left(1 + \dfrac{j\omega c}{g_{b'e}}\right)}$$

$$A_I = \frac{-h_{fe}}{1 + \dfrac{j\omega c}{g_{b'e}}} = \frac{-h_{fe}}{1 + \dfrac{j2\pi f c}{g_{b'e}}}$$

$$A_I = \frac{-h_{fe}}{1 + j2\pi f c \times r_{b'e}}$$

Critical frequency $\boxed{f_b = \dfrac{1}{2\pi r_{b'e} c}}$

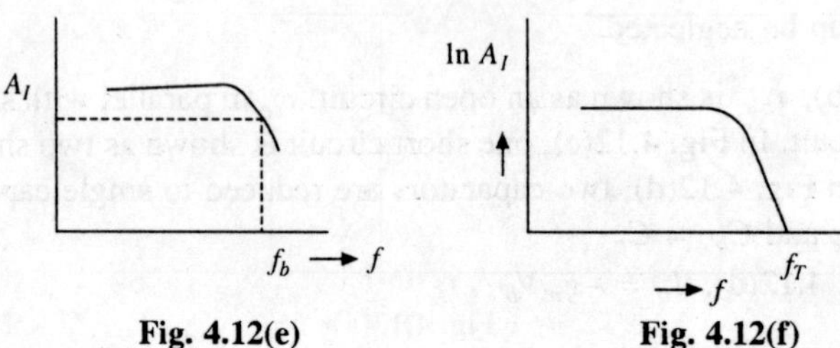

Fig. 4.12(e) **Fig. 4.12(f)**

Therefore,
$$A_I = \frac{-h_{fe}}{1 + j\dfrac{f}{f_b}}$$

Critical frequency f_b is defined as the frequency at which the short circuit gain reduces to 70.7% of the maximum gain.

$$|A_I| = \frac{h_{fe}}{\sqrt{1 + \left(\dfrac{f}{f_b}\right)^2}}$$

f_T is the frequency corresponding to unity current gain.
when $f = f_T$; $A_I = 1$.

$$1 = |A_I| = \frac{h_{fe}}{\sqrt{1 + \left(\dfrac{f_t}{f_b}\right)^2}}$$

$$1 \times \sqrt{1 + \left(\frac{ft}{f_b}\right)^2} = h_{fe}$$

$$1 + \left(\frac{f_t}{f_b}\right)^2 = h_{fe}^2$$

$$\left(\frac{f_t}{f_b}\right)^2 = h_{fe}^2 - 1 \simeq h_{fe}^2$$

Hence
$$\frac{f_t}{f_b} = h_{fe}$$

$$\boxed{f_T = h_{fe} f_b} \tag{4.20}$$

Worked Problems

Ex. 4.1 Prove that transfer function $\dfrac{V_{ce}}{V_{b'e}}$ is $\dfrac{g_{b'c} - g_m}{g_{b'c} + g_L + g_{ce}}$ for the circuit shown in Fig. 4.13(a) given $g_L = \dfrac{1}{R_L}$.

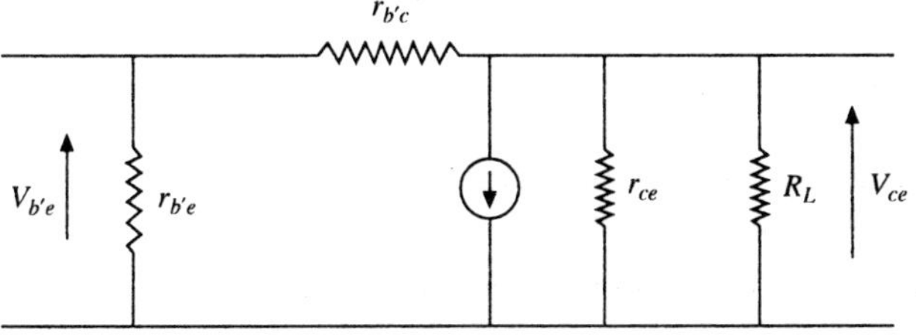

Fig. 4.13(a)

Reduce r_{ce} and R_L into single resistance

Apply *KCL* at *P*

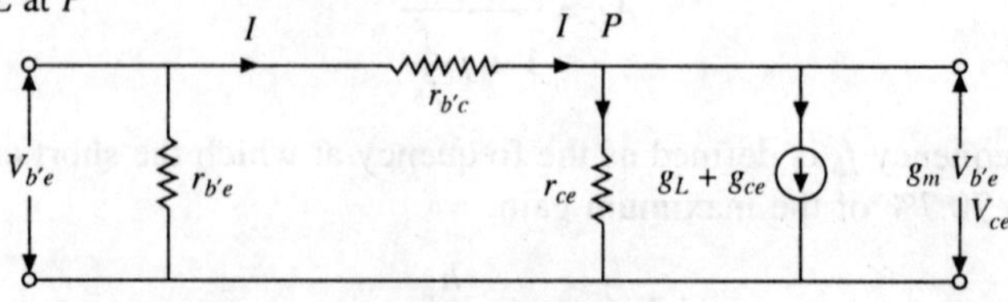

Fig. 4.13(b)

$$I = \frac{V_{ce}}{r_{ce}} + g_m V_{b'e} + \frac{V_{ce}}{R_L}$$

$$I = V_{ce}(g_L + g_{ce}) + g_m V_{b'e} \tag{4.21}$$

$$V_{b'e} = V_{ce} + I r_{b'c} \tag{4.22}$$

Substituting (4.21) in (4.22), we get

$$V_{b'e} = V_{ce} + V_{ce}(g_L + g_{ce}) r_{b'c} + g_m V_{b'e} r_{b'c}$$

$$V_{b'e}(1 - g_m r_{b'c}) = V_{ce}\left(1 + \frac{g_L + g_{ce}}{g_{b'c}}\right)$$

$$V_{b'e}(1 - g_m r_{b'c}) = V_{ce}\left(\frac{g_{b'c} + g_L + g_{ce}}{g_{b'c}}\right)$$

$$\frac{(1 - g_m r_{b'c})}{(g_{b'c} + g_L + g_{ce})} = \frac{V_{ce}}{V_{b'e}} \cdot \frac{1}{g_{b'c}}$$

$$\frac{\left(\frac{g_{b'c} - g_m}{g_{b'c}}\right) g_{b'c}}{g_{b'c} + g_L + g_{ce}} = \frac{V_{ce}}{V_{b'e}}$$

$$\frac{V_{ce}}{V_{b'e}} = \frac{g_{b'c} - g_m}{g_{b'c} + g_L + g_{ce}}$$

Ex. 4.2 Show that low frequency hybrid II-model with $r_{b'c}$ and r_{ce} taken as infinity reduces to approximate common emitter hybrid model.

In Fig. 4.13b, substitute $r_{b'c} = \infty$

$$r_{ce} = \infty$$

We know that $r_{b'b} + r_{b'e} = h_{ie}$.

The circuit is nothing but approximate model of BJT.

Short Questions and Answers

Q. 1. What are the regions used when BJT is used as a switch?

Ans. Saturation and cut-off regions

Q. 2. Gate current of FET is of the order of *nano amperes.*

Q. 3. Define latching current of SCR.

Ans. Minimum current required to take the SCR from OFF-state to ON-state is called latching current.

Q. 4. What is thermal resistance of power BJT?

Ans. Thermal resistance is the resistance to the flow of heat. Heat flows from the junction to the surrounding air. Larger the transistor case smaller the thermal resistance and vice versa. Thermal resistance is reduced by providing *heat sink* with the transistor.

Q. 5. What is the punch through (or) reach through?

Ans. Width of the depletion layer depends on the applied voltage. If the voltage applied to the output junction is increased, width of the depletion layer increases. At a particular output voltage, *both the depletion layers touch.* A large current flows through transistor. Such effect is called punch through or reach through. The process of variation of base width is called base width modulation.

5

Junction Field Effect Transistor (JFET) and Thyristors

5.1 Introduction

The devices discussed in this chapter are: (i) JFET, (ii) MOSFET, (iii) UJT, (iv) SCR, (v) Diac and (vi) Triac.

JFET is shown in Fig. 5.1. The gates of the devices are permanently connected inside. The battery V_{GG} reverse biases the gate source junctions. The battery V_{DD} drives a current through the *N*-channel. The electrons in the N-bar are repelled by the negative polarity of the battery V_{DD}. These electrons travel from source to drain and towards the positive terminal of V_{DD}. In other words, a conventional drain current flows from drain to source. Thus the FET is a unipolar device unlike BJT. In *N*-channel FET the conduction is due to only electrons. The width of the depletion layer depends upon the amount of reverse bias. As the gate is made more negative, the width of the depletion layer increases and the width of the conducting channel decreases. Thus FET is a voltage controlled device i.e. the output current is controlled by controlling the input voltage.

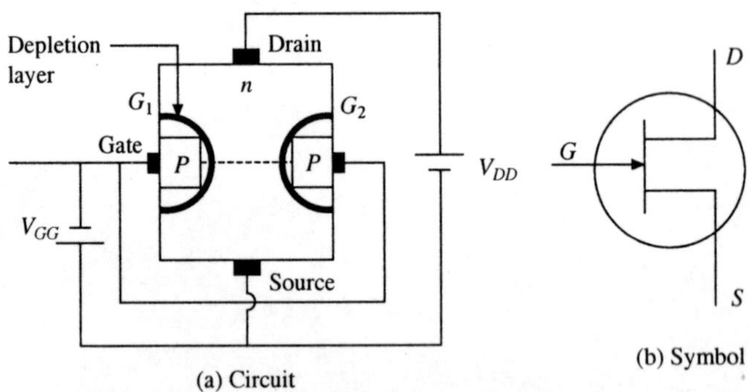

Fig. 5.1 Junction field effect transistor.

The action of FET is similar to the vacuum triode. When the gate voltage of FET is made more and more negative, the drain current becomes less and less. For a particular negative gate voltage, the drain current reduces to zero since the depletion layers are touching. The width of the conducting channel reduces to zero. This value of gate voltage is noted as $V_{GS\,(Off)}$.

The terminology of FET is based on the analogy of over head tank system. In over head tank system, the flow of water between tank (source) and tube (drain) is controlled by controlling the position of gate valve. In FET, electron flow from source to drain is controlled by varying the potential of the gate.

5.1.1 Characteristics of JFET

The experiemental set up used to obtain the characteristics of FET is shown in Fig. 5.2. The drain characteristics are plotted with drain to source voltage V_{DS} on X-axis and I_d on Y-axis. The input voltage V_{GS} is kept constant and the output voltage is varied by varying the potential divider. A family of drain curves are

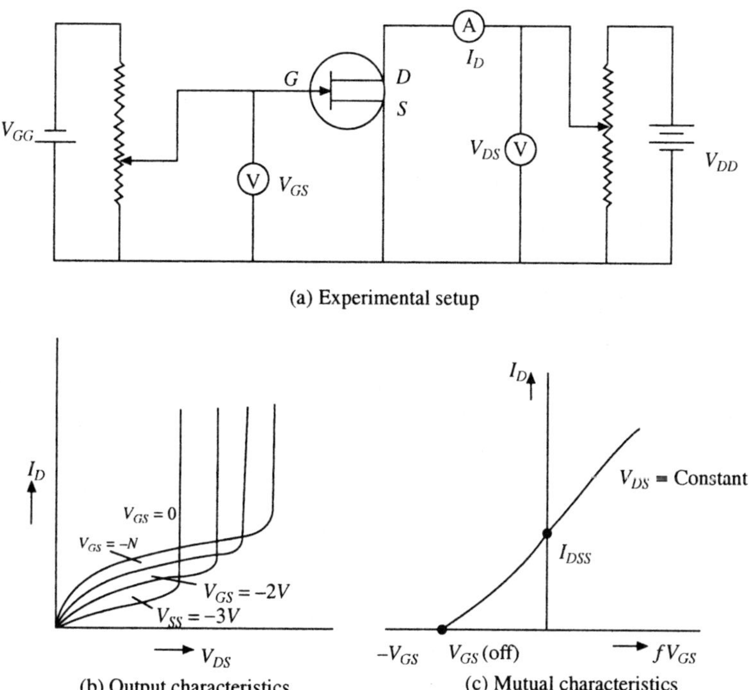

Fig. 5.2 Characteristics of JFET.

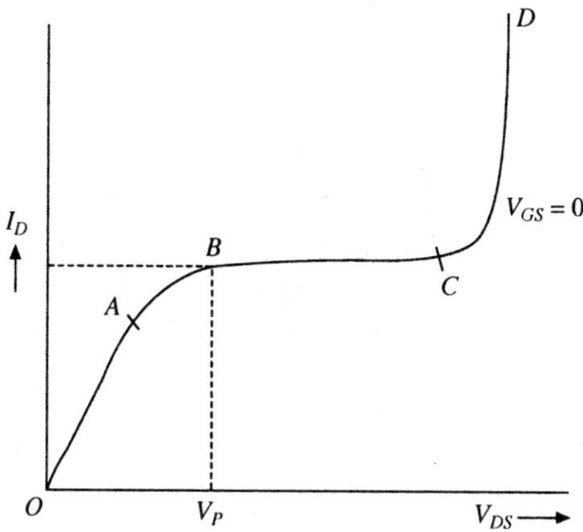

Fig. 5.3 Output characteristics.

drawn for constant values of input voltage like 0, –1, –2, –3 V etc. From the drain characteristics we observe that the current initially increases with the increase in voltage and then the current levels off. After some voltage the current shoots up. The regions of the characteristics are explained below:

Region OA: In this region the current linearly increases with the increase in the voltage. The FET behaves like a resistance.

Region AB: In this region, the current varies as square of the applied voltage.

Region BC: In this region the current is almost constant. This region is called as pinch off region or amplifier region. The FET behaves like current source.

Region CD: As the voltage V_{DS} is increased, at a particular voltage, the device breaks down by avalanche breakdown mechanism.

The current abruptly increases whereas the voltage remains constant. The device behaves like a voltage source. In the normal operation, the FET is not used in this region as the current value reaches dangerously high.

Pinch Off Voltage: This is the drain to source voltage beyond which the current remains almost constant. In other words the rise of drain current is pinched off.

5.1.2 Drain Current with Short Circuited Gate Source Connection

This is the maximum current that can be obtained under normal operation of the FET with the gate source voltage being maintained at zero by short circuiting gate and source. The value is shown in the drain characteristics and in the mutual characteristics.

Let ΔV_{DS} be the small change in the output voltage and ΔI_D be the corresponding change in the drain current. The drain resistance r_D in the pinchoff region is defined as the ratio of ΔV_{DS} to ΔI_D.

5.1.3 Mutual Characteristics

The characteristics drawn with Input voltage V_{gs} on X-axis and output current I_D on Y-axis is known as transistory characteristics or mutual characteristics. From the characteristics we observe that the drain current decreases as the gate voltage is made more and more negative.

5.2 MOSFET

Metal oxide semiconductor field effect transistor (MOSFET) is used as a switch at very high frequencies (1 to 10 MHz) since it has low turn-on and turn-off time values. The structure is shown in Fig. 5.4.

Depletion Mode: In depletion mode, the gate is made negative with respect to source. Conductive coating of gate acts as one conductor. N-layer acts as another conductor. Oxide coating acts as dielectric. By electrostatic induction, a positive

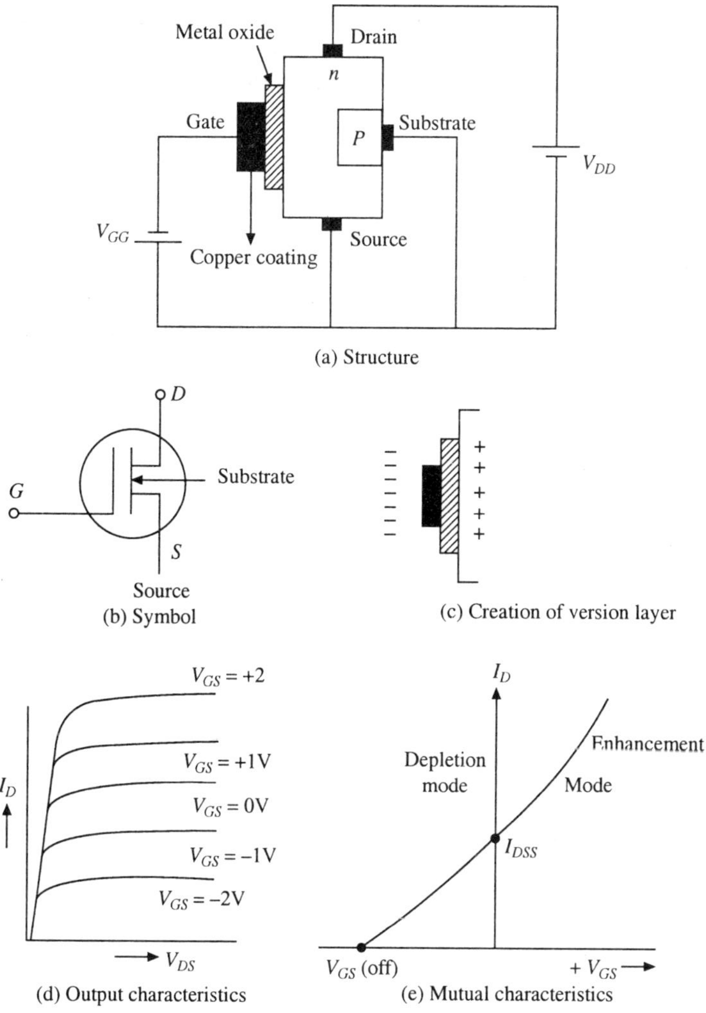

Fig. 5.4 MOSFET.

charge is induced in the N-layer when gate is made negative. The N-layer near the oxide coating losses its identity as N-layer. It behaves as a virtual P-layer (Fig. 5.4(c)). The junction J_1 is reverse biased by the supply voltage V_{DD}.

The region near oxide coating is depleted (removed) with electrons. Hence this mode is called depletion mode. If the gate is made more negative, the junction J_1 blocks through current since it is reverse biased. The current through the device (I_D) can be controlled by controlling the gate voltage. As the gate voltage is made less and less negative, the drain current becomes larger and larger.

Enhancement Mode: When gate is made positive, negative charge is induced in the region near oxide coating. The electrons are increased (or) enhanced. Larger the value of gate voltage, larger the number of conduction electrons in *N*-layer and larger the current. Thus the drain current can be increased by increasing the gate voltage. This can be seen from the characteristics shown in Fig. 5.4 (d) and 5.4(e). Hence this mode is called enhancement mode.

5.3 EMOSFET

In enhancement only MOSFET, the *P*-layer extends fully to the left as shown in Fig. 5.5a. A large positive voltage is applied at the gate. This will induce a large negative charge in the *P*-layer. In other words, large conduction electrons are injected. The *P*-layer loses its identity as *P*-region. It behaves as virtual *N*-layer. This layer is called *N*-type inversion layer. The junction J_1 is reverse biased by the supply. The electrons around J_1 acquire high velcoity and break down the junction J_1. The current flows from drain to source.

Initially the drain current is zero when the gate voltage is zero. The device starts conducting when $V_{GS} = V_{GS\,(on)}$. As the gate voltage is increased, the number of electrons in the *P*-layer increases and hence the drain current increases. The layer formed by the electrons in the *P*-region is called *N*-type inversion layer. The MOSFET is also called an insulated gate FET since the gate is insulated from the device by the coating. The output and mutual characteristics are shown in Fig. 5.5.

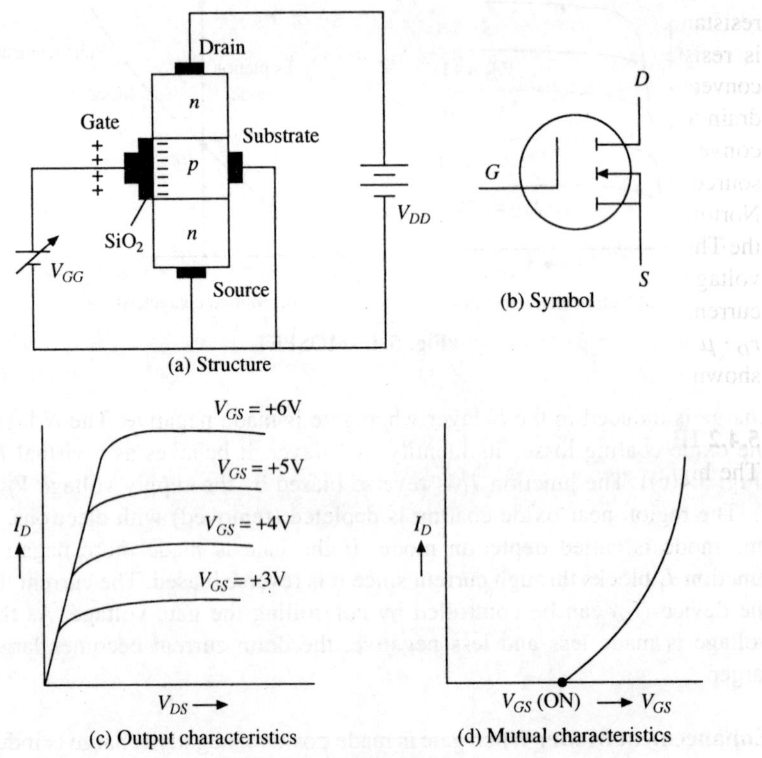

(a) Structure

(b) Symbol

(c) Output characteristics

(d) Mutual characteristics

Fig. 5.5 EMOSFET.

5.3.1 Differenc between BJT and FET

BJT	FET
1. Current controlled device	Voltage controlled device
2. Low input impedance	High input impedance
3. More noise	Less noise
4. Thermal run away problem is present	It is not present
5. Total current is due to holes and electrons (bi-polar device)	Total current is either due to electrons or holes (unipolar device)
6. Change in voltage by 0.1 V produce a change in current of about 20 milli amps, i.e. more sensitive to voltage variations	Change in voltage by 0.1 V produce a change in current of < 10 mA, i.e. less sensitive to voltage variations

5.4 Model of JFET

5.4.1 Small Signal Model of JFET

The input junction of FET is reverse biased. The resistance between gate and source is about infinity. Therefore gate source junction is represented as open circuit. From the output characteristic of FET, it can be seen that current is almost constant during pinch off region. Therefore, we model the output side as a Norton equivalent circuit having a current source of '$g_m V_{GS}$' in parallel with resistance r_D. The source $g_m V_{GS}$ is called voltage controlled current source. 'r_D' is resistance of the channel. The current source direction looks down since conventional current in JFET flows from drain to source. Current source can be converted into voltage source using source transformation technique. (i.e) Norton's circuit can be converted into the Thevinin's circuit. The direction of voltage source is same as direction of current source. The value of voltage source is product of current and resistance $r_D \cdot \mu V_{GS}$ is called voltage dependant voltage source. Low frequency model is shown in Fig. 5.6.

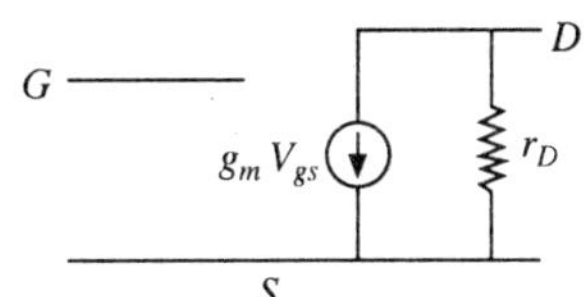

Fig. 5.6 Low frequency model of JFET.

5.4.2 High Frequency FET Model

The high frequency model is shown in Fig. 5.7. It is similar to low frequency

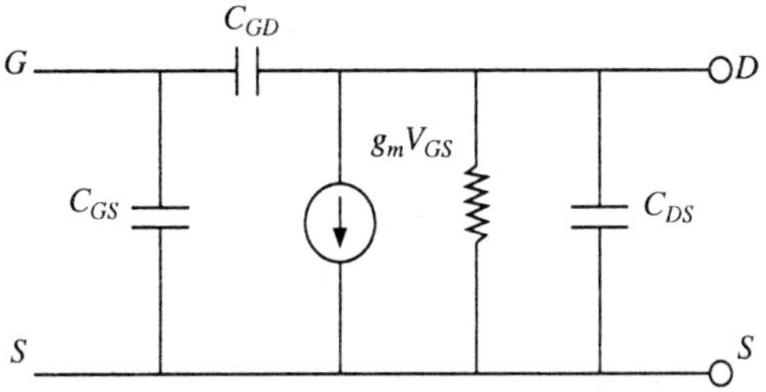

Fig. 5.7 High frequency model of JFET.

model except for the addition of 3 capacitances between each pair of terminals. The capacitance C_{GS} represent the barrier capacitance between gate and source. Its value lies between 1 PF to 10 PF. The capacitance C_{GD} represents barrier capacitance between gate and drain. Its value also lies between 1 and to 10 PF. The capacitance C_{DS} represents the barrier capacitance between drain and source. Its value lies between 0.1 and 1 PF. The effect of these capacitances can be ignored at low frequencies. Their effect cannot be neglected at high frequencies since they offer considerable reactance at high frequencies.

5.5 Biasing of JFET

5.5.1 Requirements of Biasing

Following are the requirements of biasing:

1. To apply proper polarity to drain and gate, i.e. drain should be made positive with respect to source and gate should be made negative with respect to source.

2. To select an operating point so that a signal can be applied with respect to that point.

Various types of biasing are: (i) Gate bias (ii) Self bias and (iii) Voltage Divider bias.

5.5.2 Gate Bias

Gate bias circuit is shown in Fig. 5.8. We know that JFET has to be operated with gate source junction reverse biased. Hence a negative gate to source voltage is required for *N*-channel JFET. In gate bias, the battery V_{GG} applies the required negative voltage at the gate. The battery V_{DD} makes drain positive with respect to source. By applying KVL in input loop, we get $V_{GS} = - V_{GG} \cdot I_G$ is very small because of reverse bias condition. This method is not used in practice since it requires two batteries.

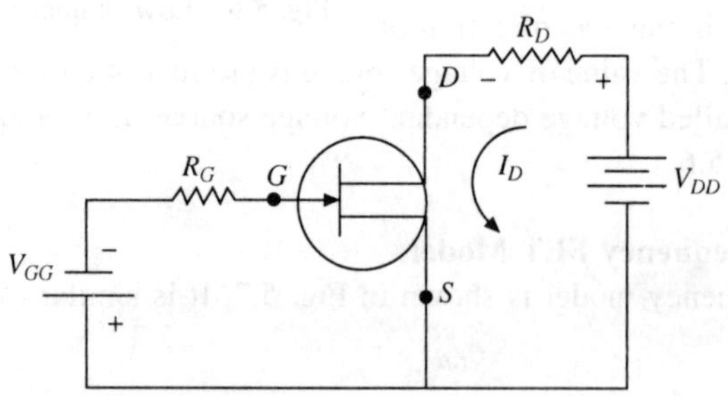

Fig. 5.8 Gate bias.

5.5.3 Self Bias

Self bias circuit is shown in Fig. 5.9. In this biasing a resistance R_s is introduced between source and ground. Voltage across R_s is applied between gate and source through the resistance R_G. The drop across R_G is negligible since I_G is very less. The resistor R_G acts as a short as far as voltage is concerned. Therefore voltage between gate and source is approximately equal to drop across R_s. From the figure, it can be seen that negative polarity of the drop across R_s is given to gate terminal. This circuit is superior to previous circuit since it uses one battery. If the drain current increases due to increase in temperature, the drop across R_s increases. This will increase the magnitude of negative gate voltage. Drain current cannot increase further, since gate is made more negative. Thus self-bias circuit prevents shifting of operating point due to variation in temperature etc.

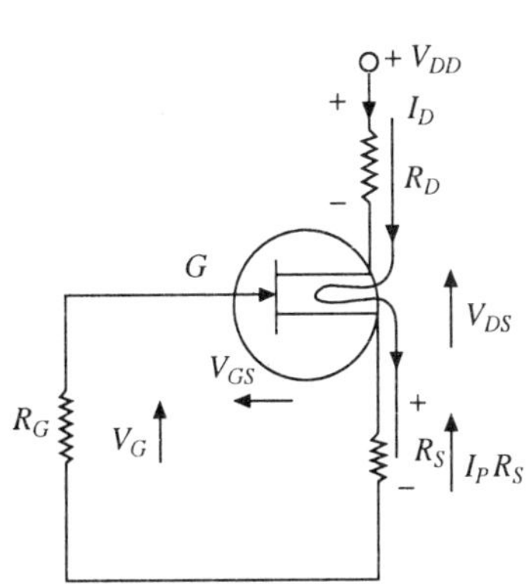

Fig. 5.9 Self bias.

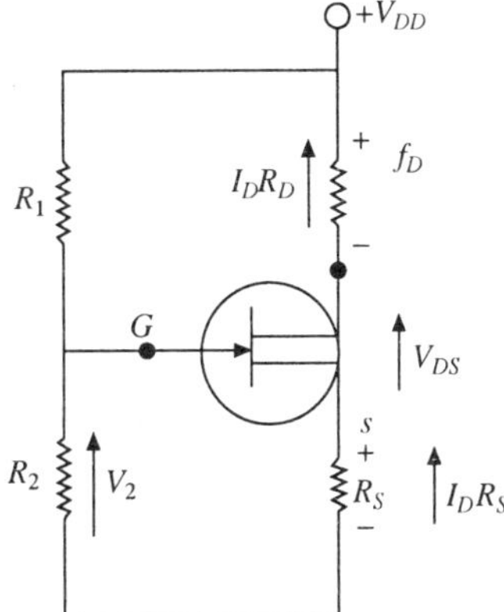

Fig. 5.10 Voltage divides bias.

5.5.4 Voltage Divider Bias

Voltage divider bias circuit is shown in Fig. 5.10. This biasing circuit is similar to voltage divider bias of BJT, except that the values of R_1 and R_2 are very high. The name voltage divider comes from the fact that R_1 and R_2 form voltage divider circuit. This type of bias is best suitable with BJTs than JFETs. Voltage divider biasing circuit can provide 0.7 V required between base and emitter of BJT.

5.6 Unijunction Transistor

The structure of UJT (Fig. 5.11a) consists of three terminals, viz. base 1, base 2 and emitter. Base 1 and base 2 are taken from the ends of *N*-bar. A heavily doped *P*-region of the device forms *P–N* junction. The emitter terminal is taken from the *P*-layer. Resistance between base 1 and base 2 varies from 4 to 10 kΩ. During the ON-state, the resistance is 4- kΩ and it is 10 k Ω during OFF state. Intrinsic stand off ratio is defined as ratio of off-state voltage across R_{B1} and supply voltage, where R_{B1} is internal resistance of *P-N* junction formed by the lower part of *N*-bar. When the voltage V_E is less than V_P, the diode is off. The current is negligible. This region is called off - region of UJT. The emitter characteristics are shown in Fig. 5.11(d). If the voltage increases beyond V_P, diode conducts. When the diode starts conducting the voltage reduces while the current increases. This region is called negative resistance region. When the resistance value R_{B1} settles to low valve (40 Ω) the voltage settles to valley

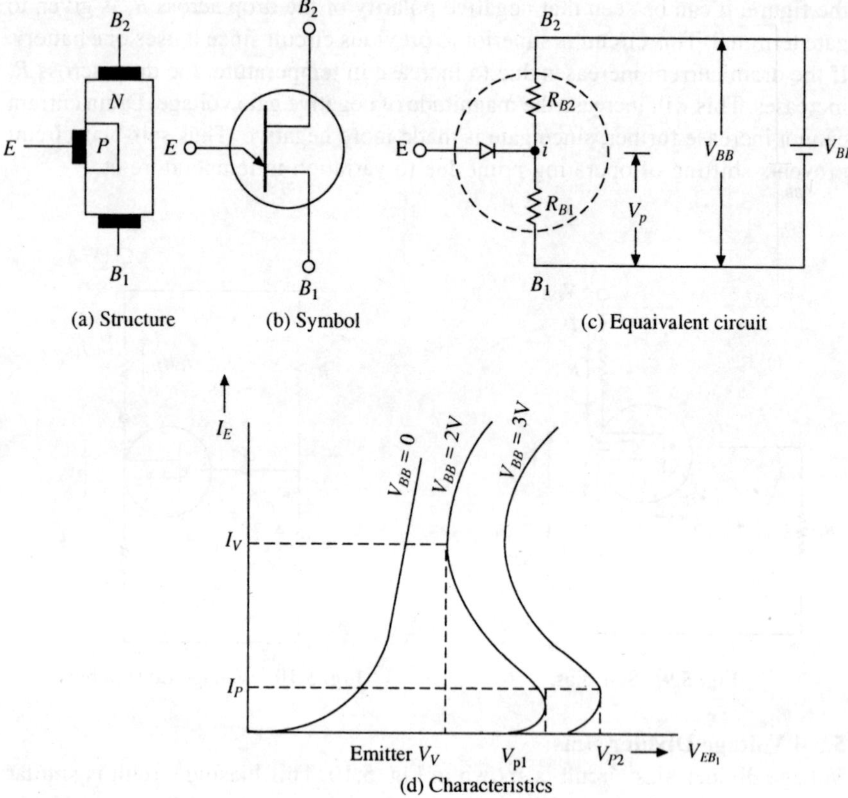

Fig. 5.11 Unijunction transistor.

voltage. This is called on-region since the voltage across the device settles to a low value with increase in the current. UJT is said to be latched (closed) in this condition.

5.6.1 UJT Relaxation Oscillator

The UJT relaxation oscillator circuit is shown in Fig. 5.12(a) where the UJT is replaced by its equivalent circuit in Fig. 5.12(b). Initially UJT is off. The capacitor gets charged through the high resistance R. Voltage across capacitor varies exponentially. Anode of the internal diode of the UJT is at capacitor potential. Cathode is at intrinsic potential. Diode turns on if V_c is greater than intrinsic potential. Diode remains off if V_c is less than intrinsic potential. Beyond peak point, capacitor voltage is more than intrinsic potential. Diode turns on. In this condition UJT is said to be latched (closed). The capacitor discharges very fast since the values of R_1 and R_{B1} (on state) are very less. Beyond valley point capacitor voltage is less than intrinsic potential, diode turns off. Again capacitor starts charging slowly through the resistance R and the above process gets repeated. Voltage across capacitor is approximately a saw-tooth voltage. The voltage across R_1 has pulse of smaller width whenever UJT conducts as shown in Fig. 5.12(c).

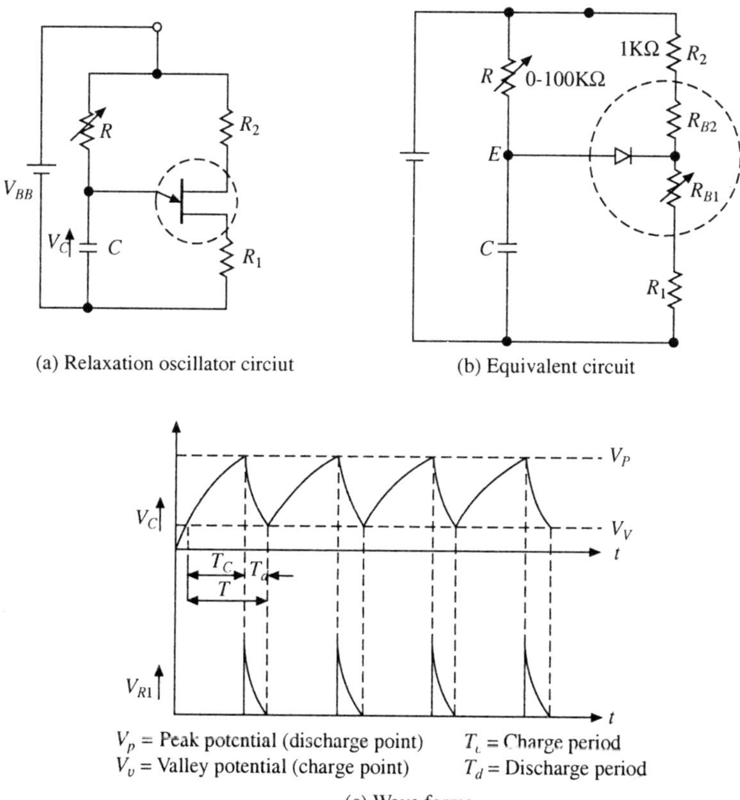

(c) Wave forms

Fig. 5.12 Unijunction transistor.

5.7 Silicon Controlled Rectifier (SCR)

Different symbols used for SCR are shown in Fig. 5.13(a) The structure of SCR is shown in Fig. 5.13(b). SCR is a four layer device having three junctions namely J_1, J_2, and J_3. When the SCR is forward biased, the junctions J_1 and J_3 act as closed switches since they are forward biased. J_2 acts as a open switch since it is reverse biased. There cannot be a current from anode to cathode. The SCR shown in Fig. 5.13c is said to be in forward blocking state. The characteristics are shown in Fig. 5.13e.

If the voltage is increased beyond V_{FB}, the junction J_2 breaks down and the SCR goes to conduction state. The voltage across the device reduces to 1 V since most of the applied voltage drops across the resistance in the anode circuit.

When the anode of the SCR is connected to the negative terminal and the cathode to the positive terminal of the battery (Fig. 5.13(d), the junctions J_1 and J_3 are reverse biased and J_2 is forward biased. There cannot be a current from cathode to anode. The SCR is said to be in reverse blocking state. The equivalent circuit of the SCR in the reverse bias condition has two diodes in series with reverse voltage being applied. Therefore the characteristic of the thyristor in the reverse biased condition is similar to that of diode.

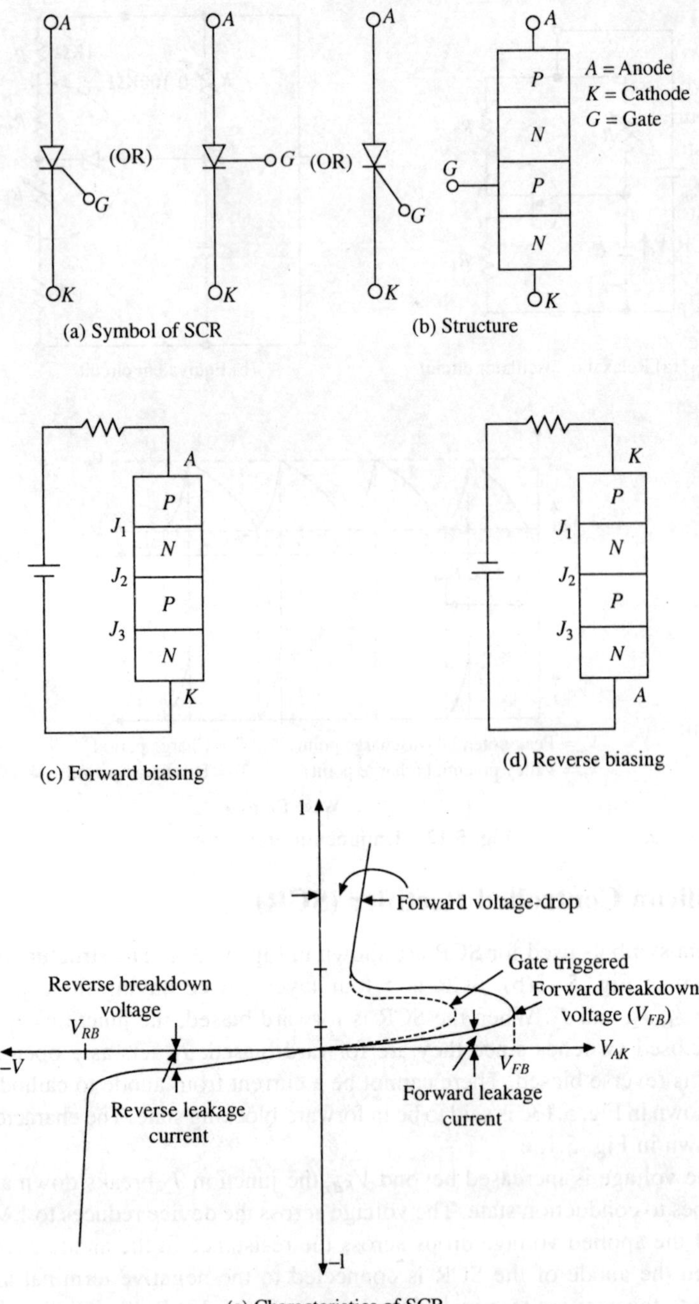

Fig. 5.13 Silicon controlled rectifier.

The process of changing the SCR from OFF state to ON state is called turn on. The process of bringing the SCR from ON state to OFF state is called turn-off.

5.7.1 Two Transistor Analogy of SCR

The two transistor analogy of SCR is shown in Fig. 5.14. When gate is made positive with respect to cathode, the NPN transistor conducts and this provides base current for the PNP transistor. The output of PNP is the input to the NPN transistor Even though the gate voltage is removed the transistors continue to conduct since they are connected back to back as shown in Fig. 5.15. The two transistors finally get saturated and thus the SCR acts as a closed switch when it is forward biased and a positive pulse is applied at the gate. Without gate voltage (pulse), a large voltage has to be applied between anode and cathode. With gate pulse, a small voltage is sufficient to make the SCR conduct.

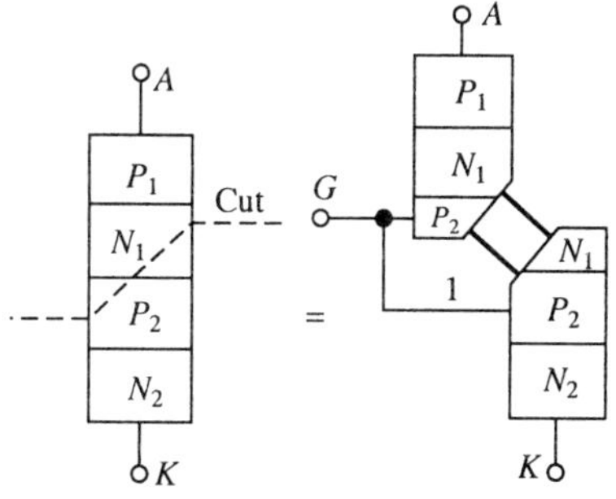

Fig. 5.14 Two transistor analogy.

The anode voltage required for conduction is a function of the gate current. From the characteristic it can be seen that larger the value of gate current, smaller the anode voltage required for conduction.

For transistor (1),

$$I_{E1} = I_{c1} + I_{B1}$$

Dividing throughout by I_{E1}, we get

$$1 = \frac{I_{C1}}{I_{E1}} + \frac{I_{B1}}{I_{E1}}$$

Substituting $\frac{I_{C1}}{I_{E1}} = \alpha_1$, we get $1 = \alpha_1 + \frac{I_{B1}}{I_{E1}}$

$$1 - \alpha_1 = \frac{I_{B1}}{I_{E1}}$$

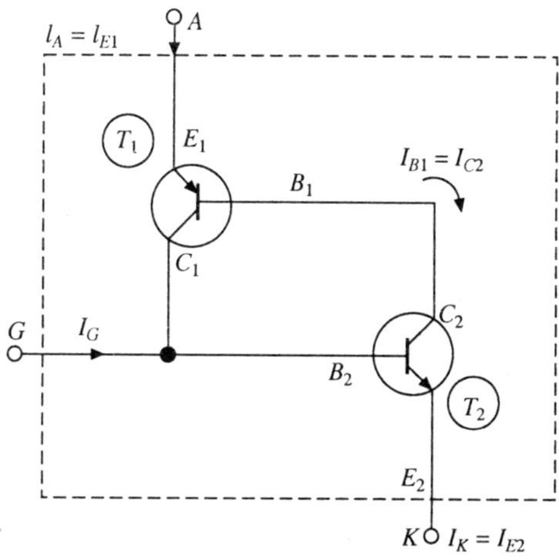

Fig. 5.15 Equivalent circuit.

$$I_{B1} = (1 - \alpha_1)\, I_{E1} \tag{5.1}$$

From Fig. 5.15, $I_{C2} = (1 - \alpha_1)I_A$

For transistor (2)

$$\alpha_2 = I_{C2}/I_{E2};\ I_{C2} = \alpha_2 I_{E2}$$

From Fig. 1.8(b), $I_{E2} = I_K$

$$I_{C2} = \alpha_2 I_K \tag{5.2}$$

Equations (5.1) and (5.2) are equal.

Therefore, $$(1 - \alpha_1)\, I_A = \alpha_2 I_K \tag{5.3}$$

From the figure according to KCL

$$I_A + I_G = I_K \tag{5.4}$$

Substituting (5.4) into (5.3)

$$(1 - \alpha_1)\, I_A = \alpha_2(I_A + I_G);\quad I_A(1 - \alpha_1 - \alpha_2) = \alpha_2 IG$$

Therefore $I_A = \alpha_2 I_G/[1 - (\alpha_1 + \alpha_2)]$

$$\frac{I_A}{I_G} = \frac{\alpha_2}{1 - (\alpha_1 - \alpha_2)} \tag{5.5}$$

If $\alpha_1 + \alpha_2 = 1$, then I_A/I_G tends to ∞. This is not possible as $\alpha_1 + \alpha_2$ is around 0.99. From equation (5.5), it can be seen that using SCR, very large anode current can be controlled by using small gate current.

Let us consider that the SCR is in on state. The anode current can be reduced by increasing the resistance in the anode circuit. As the resistance in the anode circuit increases, the anode current decreases. Below a particular current value I_h, the SCR fails to remain in the ON sate. The minimum current below which SCR goes to OFF state is called holding current. Typical value is 20 mA. Let us consider the SCR to be in the OFF state. The minimum current required to keep the SCR latched after it is turned on and the gate signal is removed is called latching current.

5.7.2 Reverse Bias Condition of SCR

When an SCR is reverse biased as shown in Figs 15.5(b) positive voltage should not be applied at the gate. The junctions J_1 and J_3 are reverse biased and J_2 is forward biased. When gate is made positive with respect to cathode the junction J_3 is also forward biased. Hence the entire voltage appears across J_1. The junction J_1 might break down by avalanche breakdown principle. There will be a current from the cathode to anode. But the SCR is supposed to conduct only when it is forward biased and positive voltage is applied at the

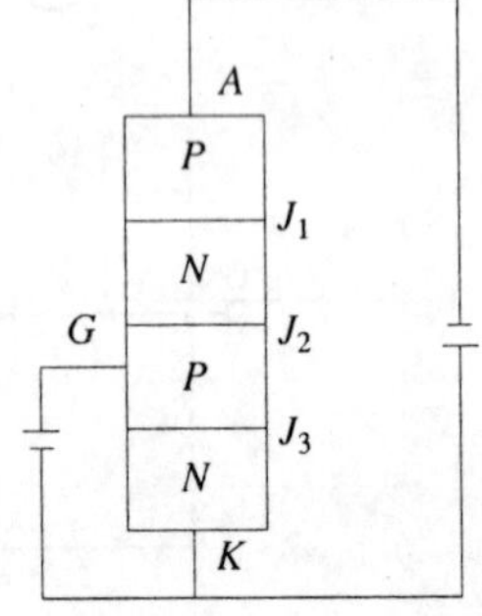

Fig. 5.15(b)

gate. Hence the SCR gate should not be given positive voltage when it is reverse biased.

5.7.3 Difference between Transistor and SCR

Transistor acts as a controlled switch. It acts as a closed switch as long as base drive is available at the input. It acts as an open switch when there is no base drive. To keep the transistor in ON state, a continuous base drive is needed at the input. In the case of SCR, a continuous D.C. voltage is not required at the gate to keep the SCR ON. Only a pulse is needed at the gate since the SCR is self latching device.

5.8 Triggering Circuits for SCR

The basic triggering circuits used for turning the SCR on are:

(1) *R*-Triggering
(2) *RC*-Triggering
(3) UJT-Triggering

5.8.1 *R*-Triggering

The circuit of half controlled rectifier with *R*-triggering is shown in Fig. 5.16(a). During the positive half cycle, the SCR is triggered at α. It conducts from α to π. During the negative half cycle, it does not conduct since it is reverse biased. In the next positive half cycle, the SCR is triggered at $(2\pi + \alpha)$. It conducts from $(2\pi + \alpha)$ to 3π. The voltage across the load is positive segments of supply voltages shown in Fig. 5.16(b).

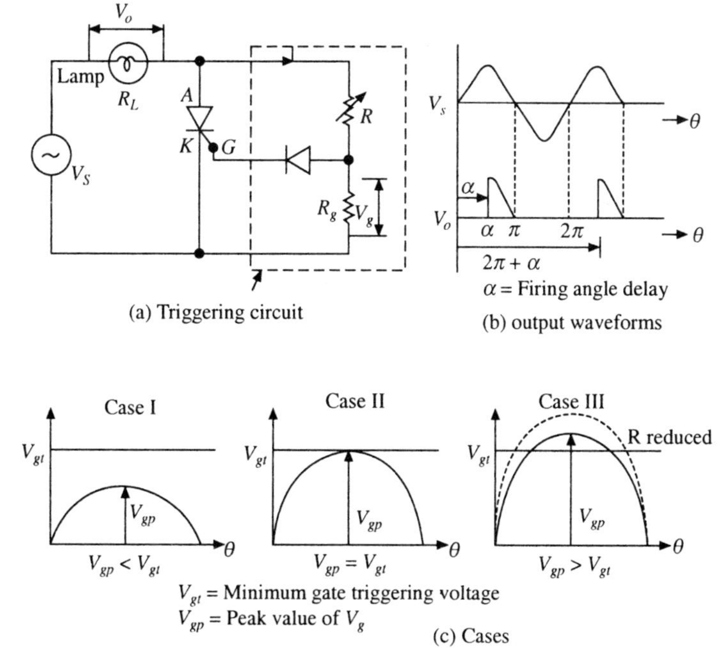

(a) Triggering circuit

(b) output waveforms

(c) Cases

Fig. 5.16 *R*-triggering.

$$I = \frac{V_s}{R_L + R + R_g} \tag{5.6}$$

$$V_g = I \cdot R_g$$

Let V_{gt} be the minimum gate triggering voltage and V_{gp} the peak value of V_g.

The current through R_g can be varied by varying the resistance R. If the resistance is large, the current is less and V_g is less. In case (1), $V_{gp} < V_{gt}$. Therefore the SCR cannot be turned "ON". If the resistance R is moderate, V_g is such that $V_{gp} = V_{gt}$. Therefore the SCR turns ON at $\alpha = 90°$. If the resistance R is further reduced, V_g increases. Therefore the SCR gets triggered at an angle α. The value of α can be varied by varying the resistance R. The firing angle variation is from 0 to 90°. The diode will ensure that only positive voltage is applied at the gate. During the negative half cycle, the diode blocks the negative voltage.

5.8.2 *R-C* Triggering

R-C triggering circuit is shown in Fig. 5.17(a). To understand this, we need the fundamentals of series *R-C* circuit. Consider the vector diagram of *R-C* circuit shown in Fig. 5.17(b). From this figure it can be seen that capacitor voltage lags the supply voltage by an angle that can be varied by varying the resistance R. During the positive half cycle, current flows through R and C. At $\theta = \alpha$, v_c reaches V_{gt}, SCR turns ON. It conducts from α to π. At $\theta = \pi$, $i = 0$, SCR turns OFF. The waveforms are shown in Fig. 5.17(c).

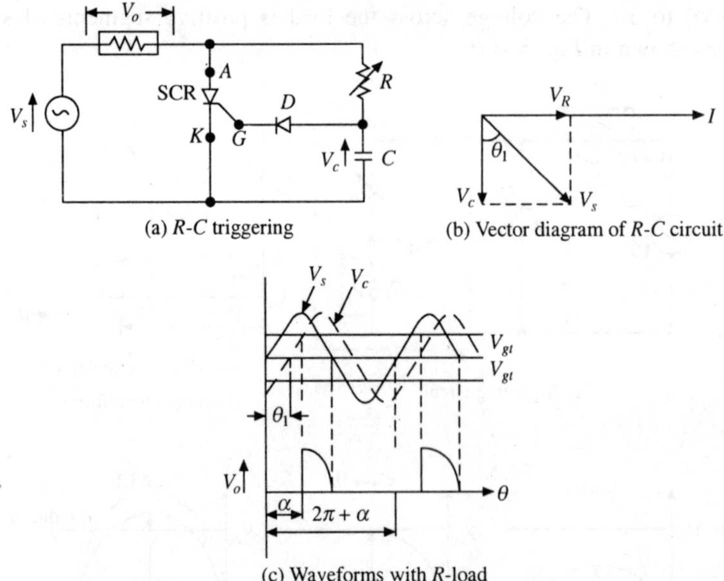

(a) *R-C* triggering

(b) Vector diagram of *R-C* circuit

(c) Waveforms with *R*-load

Fig. 5.17 R–C triggering circuit

During the negative half cycle, the SCR is reverse biased. The capacitor voltage is such that the diode is reverse biased. Therefore SCR cannot be turned ON during negative half cycle.

The firing angle delay is varied by varying R. In an R–C circuit, the time delay is RC. Therefore the delay can be varied by varying R or C. With resistance alone a variation from 0° to 90° can be obtained. With capacitance another 90° variation can be achieved. Therefore the total firing angle delay can be in the range of 0 to 180°. In R–C triggering, there is no isolation between power circuit and control circuit.

5.8.3 UJT Triggering

UJT relaxation oscillator circuit is shown in Fig. 5.18(a). The resistance R_1 is replaced by a pulse transformer in Fig. 5.18(b). The pulses in the secondary of this transformer are used for triggering the SCR.

When the battery in Fig. 5.18(b) is connected, the capacitor charges exponentially through the resistance R. The emitter of UJT is at capacitor voltage. The intrinsic

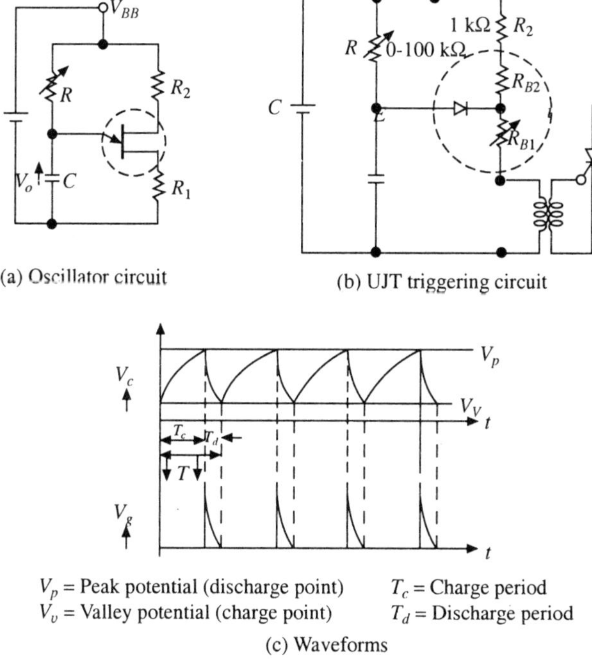

Fig. 5.18 UJT triggering circuit.

point is at a potential V_p (peak voltage). As long as the anode voltage is less than the peak potential, the diode is OFF. When the capacitor voltage goes more than V_p the diode is ON. The capacitor quickly discharges through (R_1) 40 Ω resistor. The discharging is very fast since the resistance is very less. When the capacitor voltage goes below V_V (valley voltage), the diode is OFF since its anode potential is less than the cathode potential. The capacitor again gets charged and the above process repeats. When the capacitor is charging there is no current through R_1. It has a discharging current only during the ON period. Thus the voltage across R_1 is a train of pulses which can be used for triggering the SCR. The firing angle delay is varied by varying the value of the resistance R. This circuit is called relaxation oscillator circuit.

$$\eta = \frac{V_{RB1}}{V_{BB}} = \frac{I R_{B1}}{I(R_{B1} + R_{B2})}; \quad \eta = \frac{R_{B1}}{R_{B1} + R_{B2}} \tag{5.7}$$

Value of η lies between 0.6 to 0.7.

In *R-C* circuit, capacitor voltage follows the equation

$$v_c = V_{BB}(1 - e^{-t/RC})$$

When $t = T_c$, $v_c = V_P$

$$V_P = V_{BB}(1 - e^{-Tc/RC}); \quad \frac{V_P}{V_{BB}} = (1 - e^{-Tc/RC})$$

$$\frac{V_P}{V_{BB}} = \eta; \quad T \simeq T_c$$

Neglect discharging period T_d. since it is very small, therefore $\eta = 1 - e^{-T/RC}$

$$e^{-T/RC} = 1 - \eta$$

Take log on both sides

$$\frac{T}{RC} \ln e = \ln(1 - \eta); \quad \ln_e e = 1$$

$$-T = RC \ln(1 - \eta'); \; T = -RC \ln(1 - \eta)$$

$$T = RC \ln 1/(1 - \eta); \; T = \text{Time period; Frequency } (f) = 1/T$$

Worked Problems

Ex. 5.1 Estimate the minimum and maximum charging resistor of UJT circuit for control of α between 20° to 160° of 50 Hz supply. Assume $C = 0.1\ \mu F$ and $\eta' = 0.7$.

$$T = \frac{1}{f} = \frac{1}{50} = 20 \text{ msec}$$

The time period for 50 Hz supply is 20 ms. The time period for half a cycle is 10 ms. Let time corresponding to 20° be T_1.

$$T_1 \rightarrow 20° \qquad \frac{T_1}{10} = 20/180$$

$$10 \text{ ms} \rightarrow 180°$$

Therefore, $T_1 = 1.11\ \mu$sec

$$\frac{T_2}{10} = \frac{160}{180}; \; T_2 = 8.88 \text{ msec}; \; T_1 = R_1 C \ln \frac{1}{1 - \eta'}$$

$$1.11 \times 10^{-3} = R_1 \times 0.1 \times 10^{-6} \frac{1}{10(1 - 0.7)}; \; R_1 = 9.25 \text{ k}\Omega$$

(ii) $T_2 = R_2 C \ln \frac{1}{1 - \eta'}$; $8.88 \times 10^{-3} = R_2 \times 0.1 \times 10^{-6} \ln \frac{1}{1 - 0.7}$; $R_2 = 74 \text{ k}\Omega$

5.9 Diac

The symbol and structure of diac are shown in Fig. 5.19(a) and (b), respectively. Consider the structure shown in Fig. 5.19(b). Diac is a five layer four junction

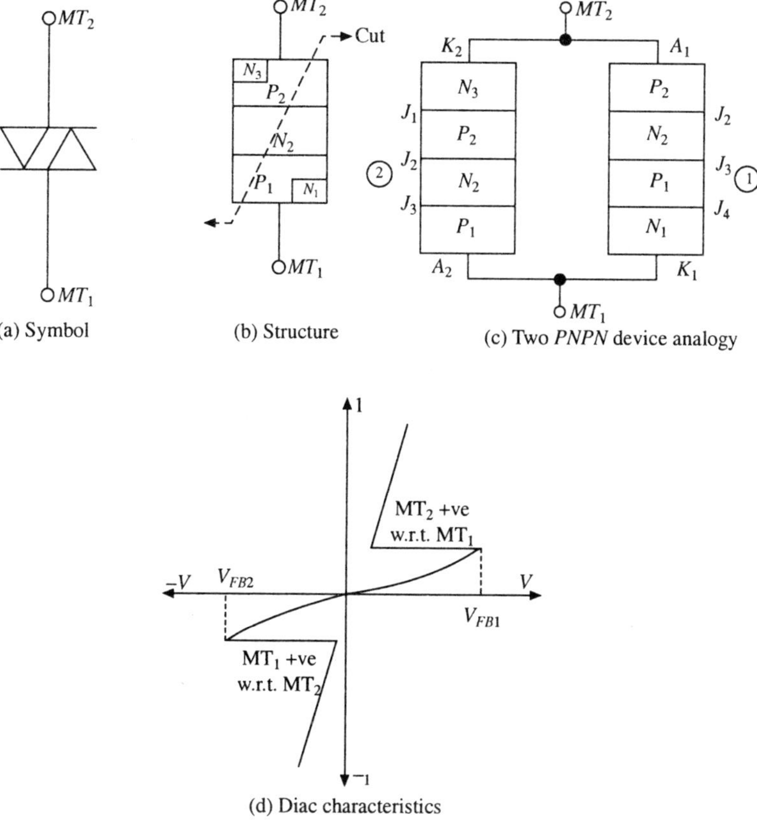

(a) Symbol (b) Structure (c) Two *PNPN* device analogy

(d) Diac characteristics

Fig. 5.19 DIAC.

device. The word diac can be split into DI and AC. DI stands for two electrodes namely M_{T1} and M_{T2}. AC indicates its ability to conduct in both the directions. From the structure the equivalent circuit can be drawn with two PNPN devices connected back to back. When M_{T2} is made positive with respect to M_{T1}, the device 1 is forward biased. Junctions J_2 and J_4 are forward biased and J_3 is reverse biased. When the voltage is increased beyond the break over voltage the junction J_3 breaks down and the device 1 goes from high impedance state to low impedance state. The characteristic is similar to that of SCR forward characteristic without gate facility.

When M_{T1} is made positive with respect to M_{T2}, the device 2 is forward biased. The junctions J_1 and J_3 are forward biased and J_2 is reverse biased. J_2 breaks down when the voltage is increased beyond break over voltage and the current flows from M_{T1} to M_{T2}. The characteristic in the third quadrant is similar to that of first quadrant. The diac is used in the triggering circuit of TRIAC.

5.10 TRIAC

The equivalent circuit and characteristic of TRIAC are shown in Fig. 5.20 (a) and (b) respectively. TRIAC has three terminals namely M_{T1}. M_{T2} and gate (G).

The word TRIAC can be split into *TRI* and A.C. *TRI* stands for three and A.C. for its ability to conduct in both directions. The TRIAC is equivalent to two thyristors connected back to back with their gate terminals tied up. When M_{T2} is positive with respect to M_{T1}, the *SCR* 1 is forward biased. If the gate is made positive with respect to M_{T1}, the *SCR* 1 conducts and the device goes from high impedance state to low impedance state. When M_{T1} is made positive with respect to M_{T2}, *SCR* 2 is forward biased and it conducts when the gate is made positive. Thus the TRIAC can conduct in both the directions. The SCR demands a positive voltage between gate and cathode. But the TRIAC can conduct with either positive or negative voltage at the gate. There are four modes of operation:

(i) MT_1 positive with respect to MT_2, gate + Ve
(ii) MT_1 positive with respect to MT_2, gate – Ve
(iii) MT_2 positive with respect to MT_1, gate + Ve
(iv) MT_2 positive with respect to MT_1, gate – Ve

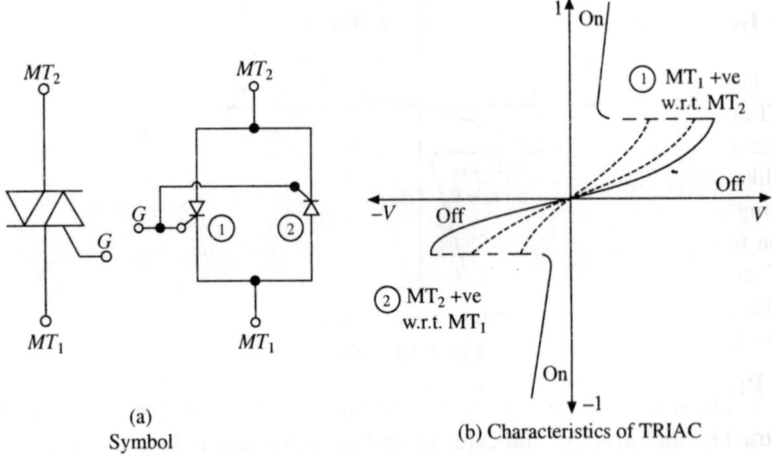

Fig. 5.20 TRIAC.

5.11 Gate Turn Off Thyristor (GTO Thyristor)

The *SCR* cannot be turned OFF using the gate. GTO can be turned OFF by using the gate. The bidirectional arrow in the symbol indicates that the gate current can flow in both the directions. To turn OFF of the GTO, a gate current of about 1/5th of the anode current is required. The turn ON and turn OFF times of GTO are less than those of SCR.

When a negative voltage is applied between gate and cathode, the excess charges in the NPN transistor are attracted by the battery. The collector current of T_2 reduces. In other words the base current of T_1 reduces. Collector current of T_1 reduces. This will reduce base current of T_2. This continues until the two transistors are turned OFF.

The turn ON and turn OFF circuit is shown in Fig. 5.21. When the transistor T_1 is turned ON, the source V_1 is connected between gate and cathode. Gate of GTO is made positive with respect to cathode. GTO turns ON.

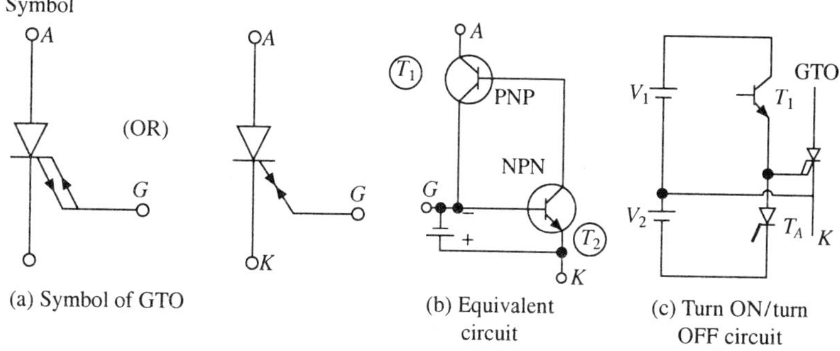

(a) Symbol of GTO (b) Equivalent circuit (c) Turn ON/turn OFF circuit

Fig. 5.21 Gate turn OFF thyristor.

To turn OFF GTO, auxiliary SCR T_A is turned ON. The source V_2 is connected between gate and cathode. Gate is made negative with respect to cathode and GTO gets turned OFF. The characteristic of GTO is similar to that of SCR.

5.12 Insulated Gate Bipolar Transistor (IGBT)

IGBT has the combined characteristics of BJT and MOSFET. It has high input impedance and fast turn ON and turn OFF like MOSFET. It has high power handling capability like BJT.

The function and characteristics of IGBT are same as that of BJT and MOSFET. IGBT acts as a switch. It is used in high power applications.

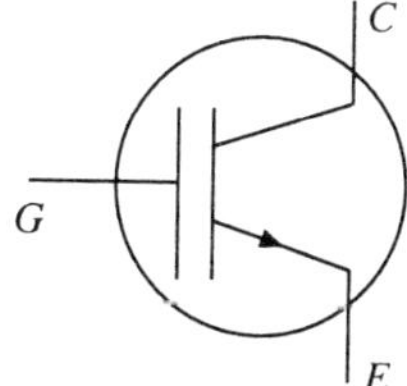

Fig. 5.22 Symbol of IGBT.

5.13 Programmable Unijunction Transistor (PUT)

The structure, symbol and characteristics of PUT are shown in Fig. 5.23. PUT has a structure similar to that of a SCR. The gate terminal is taken from N_1 instead of P_2. This gate is called "anode gate". The characteristics of PUT are similar to that of SCR with largest rating of 200 V and 1 A.

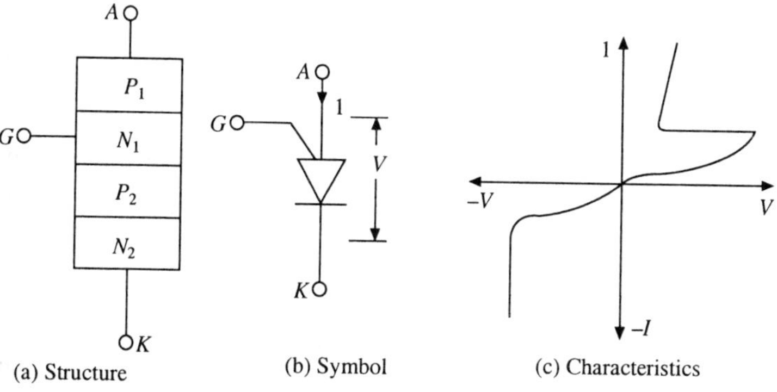

(a) Structure (b) Symbol (c) Characteristics

Fig. 5.23 Programmable unijunction transistor.

5.14 Silicon Controlled Switch (SCS)

The structure, symbol and characteristics of SCS are shown in Fig. 5.24. SCS is a four layer device with two gates namely cathode gate and anode gate. This device is similar to a vacuum device tetrode. The characteristic of SCS are similar to the characteristic of SCR.

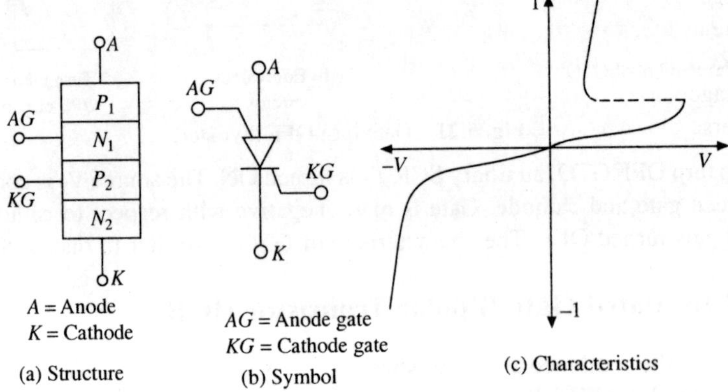

Fig. 5.24 Silicon controlled switch.

SCS can be turned on by either gate. A large reverse current through anode gate can be used to turn off it. Its rating is around 100 V and 200 mA. This is used for timing, logic and triggering circuits.

5.15 Silicon Unilateral Switch (SUS)

A SUS is similar to a "PUT" but with an in built low voltage avalanche diode between gate and cathode as shown in Fig. 5.25. Because of the presence of diode, SUS turns ON for a fixed anode to cathode voltage unlike an SCR whose trigger voltage and current vary widely with changes in ambient temperature. SUS is used mainly in timing, logic and trigger circuits. Its rating is about 20 V, 0.5 A. The structure, symbol and characteristics are shown in Fig. 5.25.

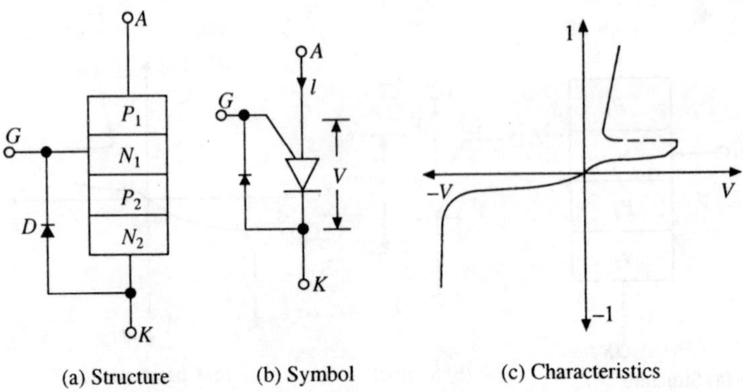

Fig. 5.25 Silicon unilateral switch.

5.16 Reverse Conducting Thyristor (RCT)

The RCT symbol, equivalent circuit and characteristics are shown in Fig. 5.26. In chopper and inverter circuits, diode is connected in antiparallel (i.e back to back) with the SCR. This diode is called feedback diode. RCT is a single packing in which both SCR and diode are present. The forward characteristic of RCT is same as that of SCR. If the RCT is reverse biased the internal diode gets forward biased and it starts conducting. Therefore the reverse characteristic of RCT is same as the forward characteristic of a diode.

RCT is also called "asymmetrical thyristor" (ASCR). The forward blocking voltage varies from 400 to 2000 V and the current rating goes up to 500 A. The reverse blocking voltage is typically 30 to 40 V.

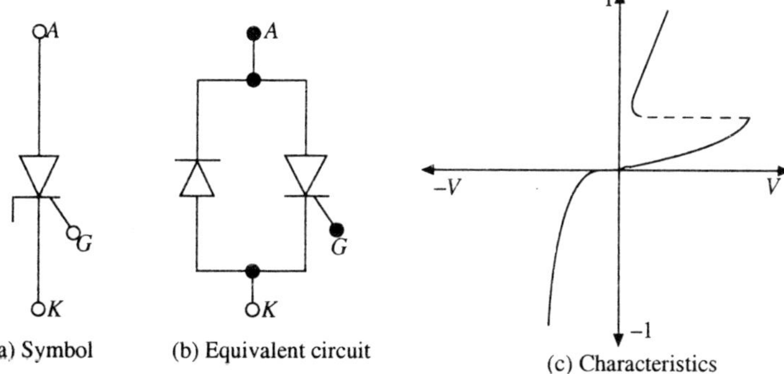

(a) Symbol (b) Equivalent circuit (c) Characteristics

Fig. 5.26 Reverse conducting thyristor.

5.17 Light Activated SCR (LASCR)

The symbol and firing circuit of LASCR are shown in Fig. 5.27. In this SCR the gate is turned ON by using light energy. It can be seen that isolation exists between the power circuit and control circuit. Whenever the SCR has to be turned ON, the control circuit generates a pulse. This pulse makes the LED to conduct. When the LED conducts the light falls on the gate surface of the SCR. The SCR gets turned on. LASCR is used in high voltage direct current transmission system.

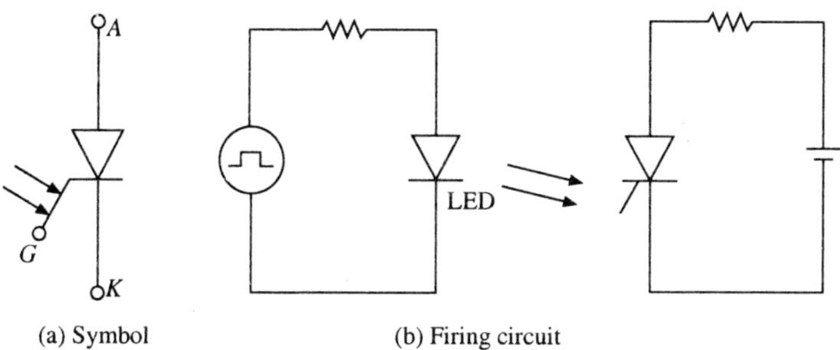

(a) Symbol (b) Firing circuit

Fig. 5.27 LASCR.

Short Questions and Fill in the Blanks

1. Explain that with the increase in gate current the breakdown voltage of an SCR *decreases*.
2. In a UJT what is the relation betwen V_P and V_{BB}?

$$\eta = \frac{V_P}{V_{BB}} = \frac{V_{EB2}}{V_{BB}}$$

3. Number of thyristors connected in parallel provide total rated current less than the sum of *individual ratings*. Explain.
4. Inverter grade SCRs have t_q *less than 25 µs*. Explain.
5. Excess *dv/dt* to a thyristor may cause *mal operation*. Explain.
6. Explain that in case of thyristor in series, both forward and reverse voltages have to be shared.
7. The collector current is 2.9 mA in a certain transistor. If the base current is 100 μA, what is the value of α.

Ans. $\alpha = \frac{I_c}{I_E} = \frac{I_c}{I_c + I_B} = \frac{2.9}{3} = 0.97$

8. An example of solid state device is FET. Explain.
9. GTO can be turned-OFF by applying large negative current. Explain.
10. Turn-OFF time of SCR depends on *temperature and forward current*. Explain.
11. What is a *P–N* junction?

    Ans. If a pure semiconductor is partly doped with 3rd group and partly with fifth group impurity, a *P–N* junction is formed. If offers low resistance in forward direction and high resistance in reverse direction.
12. What are the members of thyristor family?

    Ans. SCR, DIAC, TRIAC, SCS, SUS, GTO, LASCR.
13. What is the effect of negative gate current on a normal SCR?
    Gate has no control over the turn-OFF of the device. Negative gate current produces additional heat at the gate junction.
14. How is the forced turn-OFF of an SCR OFF different from the natural turn-OFF?
    In AC circuits the current naturally reduces to zero. In D.C. circuits the current is forced to zero by using *L* and *C*.
15. The reverse recovery time of a diode is the time required by carriers to recombine and get neutralised. Explain.
16. Compared to that of the thyristor the switching loss of GTO is much lower since turn-ON and turn OFF times of GTO are less. Explain.
17. A transformer is employed in the triggering circuit for isolation purpose. Explain.
18. A diode is usually connected across the primary of a pulse transformer to protect the transistor Explain.
19. UJT is most suitable for being used in an oscillator circuit due to existence of negative resistance region. Explain.
20. Ferrite core is used in pulse transformer to ensure no A.C. saturation. Explain.
21. The most suitable gate signal for SCR is high frequency pulse train. Explain.
22. Is the latching current more than holding current?

Ans. Yes.

23. What is the difference between holding and latching currents?

Ans. Latching current is the minimum current which keeps the device latched during turn ON. Holding current is the minimum current required to keep the device ON.

24. Which device is suitable for high frequency applications?

Ans. MOSFET (100 kHz to 1 mHz)

25. What is the range of firing angle in the case of: (i) *R–T* triggering, (ii) *R–C* triggering and (iii) UJT triggering.

Ans. (i) 0–90°, (ii) 0–180° (iii) 0–180°)

26. What is the function of diode in *R–C* triggering circuit?

Ans. Diode ensures that the SCR is turned ON only in positive half cycle.

27. What is the advantage of UJT triggering circuit over *R-C* triggering circuit?

Ans. UJT triggering circuit provides isolation betwen control circuit and power cirucit using a pulse transformer. Isolation is not present in the case of *R-C* triggering circuit.

6

Rectifiers

6.1 Introduction

Rectifiers convert alternating voltage into uni-directional voltage. There are some applications like speed control of D.C. motor and battery charging which require D.C. supply. Rectifiers can be used to supply D.C. required by these applications. Rectifiers are classified as (i) Half wave rectifier (ii) Full wave rectifier.

6.2 Half Wave Rectifier

Single phase half wave rectifier circuit and its waveforms are shown in Fig. 6.1. A step down transformer is used to produce low voltage A.C. which is converted into low voltage D.C. by using a rectifier. During the positive half cycle, diode is forward biased. It acts as a closed switch. Sinusoidal current flows through the load resistance during the positive half cycle. At $\theta = \pi$, current reduces to zero and diode turns OFF. During the negative half cycle, diode is reverse biased. It acts as an open switch. No current flows through the resistance R_L. Therefore output voltage is zero. During next positive half cycle, diode is forward biased and the above process repeats. It can be seen that current through the load resistance always flows in downward direction. The source current in primary flows in both the directions. Thus rectifier converts alternating current into unidirectional current.

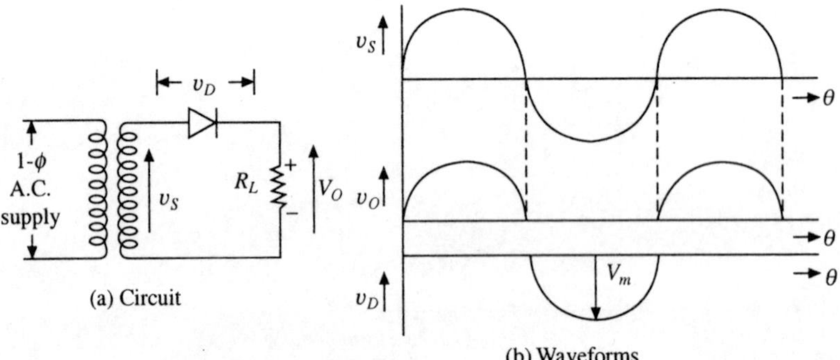

Fig. 6.1 Half wave rectifier.

The waveforms are shown in Fig. 6.1(b). By applying KVL in the circuit, we get $v_s - v_D = 0$; $v_D = v_s$ during negative half cycle. Voltage across diode is replica of supply voltage. Peak inverse voltage is V_m.

6.3 Full Wave Rectifier

Full wave rectifiers are classified as (i) Centre tap rectifier and (ii) Bridge rectifier.

6.3.1 Centre Tap Rectifier

Full wave rectifier using centre tapped transformer and its waveforms are shown in Fig. 6.2. The circuit of centre tap rectifier uses two diodes (D_1 and D_2). Load is connected between centre tap of transformer and common cathode point. During the positive half cycles of secondary voltage, the diode D_1 is forward biased and D_2 is reverse biased. The current flows through the diode D_1, load resistor and the upper half of the winding as shown in Fig. 6.2(a). During negative half-cycles, diode D_2 becomes forward biased and D_1 reverse biased. Now D_2 conducts and D_1 becomes open. The current flows through diode D_2, load resistor R_L and the lower half winding as shown in the Fig. 6.2(b). Note that the load current in both the figures is in the same direction. The waveform of the current i_o and the load voltage V_o are in phase since the load is pure resistive load.

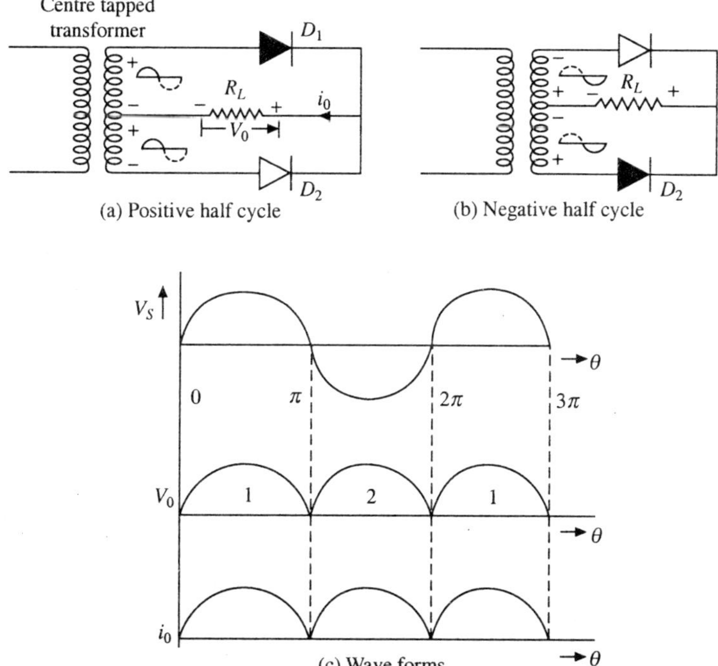

Fig. 6.2 Centre tap full wave rectifier.

Consider centre tap rectifier circuit at the instant of secondary voltage reaching its positive maximum value. The voltage V_m is the maximum voltage across each half of the secondary winding. At this instant, the diode D_1 is conducting and it offers almost zero resistance. The whole voltage V_m appears across the load resistor R_L. Therefore the reverse voltage that appears across the non-conductive diode is the sum of the voltage across the lower half winding and the voltage across the load resistor R_L. From the figure, this voltage is $V_m + V_m = 2V_m$. Thus PIV = $2V_m$. Peak inverse voltage (PIV) is the largest negative voltage across a non-conducting diode.

6.3.2 Bridge Rectifier

Bridge rectifier circuit and its waveforms are shown in Fig. 6.3. It has two legs. Each leg consists of two diodes. Diodes are numbered such that odd numbers are at the top and even numbers are at the bottom. Sum of the suffixs in each leg is equal to 5.

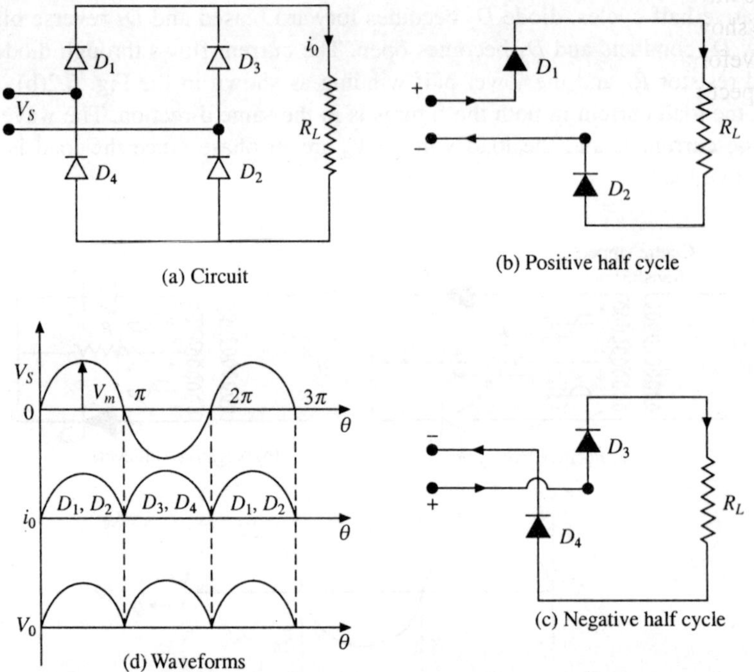

Fig. 6.3 Bridge rectifier.

A more widely used full wave rectifier circuit is the bridge rectifier. It requires four diodes instead of two, but avoids the need for a centre-tapped transformer. During positive half cycle of the secondary voltage, diodes D_1 and D_2 are conducting and diodes D_3 and D_4 are not conducting. Therefore, current flows through the secondary winding, diode D_1, Load resistor R_L, and diode D_2 as shown in Fig. 6.3(b). During negaive half cycles of the secondary voltage, diodes D_3 and D_4 conducts and the diodes D_1 and D_2 does not conduct. The current therefore flows through the secondary winding, diode D_3, load resistor R_L and diode D_4 as shown in Fig. 6.3(c). In both the cases, the current passes through the load resistor in the same direction. Therefore, a fluctuating unidirectional voltage is developed across the load.

Consider the bridge rectifier circuit at the instant of the secondary voltage reaching positive peak value, V_m. The diodes D_1 and D_2 are conducting, whereas diodes D_3 and D_4 are reverse biased and are non-conducting. The conducting diodes D_1 and D_2 have almost zero resistance. The entire voltage V_m across the secondary winding appears across load resistance. The reverse voltage across the non-conducting diode is V_m. Thus PIV is V_m. Now a days, diode bridge is available as a simple black.

6.4 Expressions for Half Wave Recifier

6.4.1 Average Value of Output Current

The waveforms of half wave rectifier are shown in Fig. 6.4(b). Reference directions are shown in Fig. 6.4a. Average value is the ratio of area to the base. In a periodic waveform, base is the width of repetitive segment. The waveform repeats with respect to 2π. Therefore base is 2π.

$$I_{av} = \frac{\text{Area}}{\text{Base}} = \frac{\int dA}{2\pi} = \frac{1}{2\pi}\int_0^{\pi} i\,d\theta = \frac{1}{2\pi}\int_0^{\pi} I_m \sin\theta\, d\theta$$

$$\boxed{I_{av} = \frac{I_m}{\pi}}\text{. Similarly } \boxed{V_{av} = \frac{V_m}{\pi}} \qquad (6.1)$$

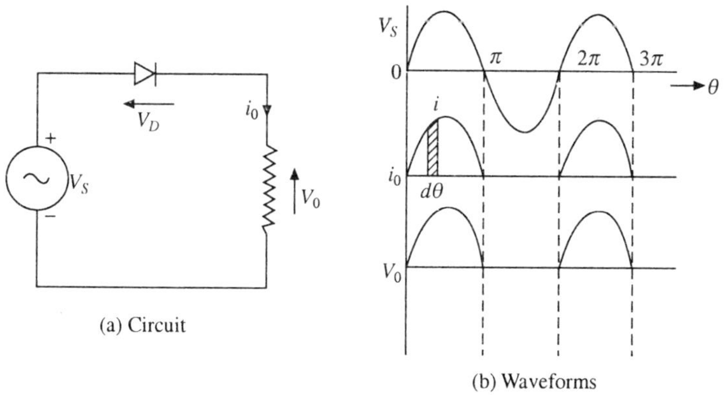

(a) Circuit

(b) Waveforms

Fig. 6.4 Waveforms of half wave rectifier.

6.4.2 RMS Value of Output Current

Root of mean of squares can be written as follows. Limits are taken from 0 to π since there is no current flowing from π to 2π.

$$I_{rms} = \sqrt{\frac{1}{2\pi}\int_0^{\pi} i^2 d\theta}$$

Square on both sides

$$I_{rms}^2 = \frac{1}{2\pi}\int_0^{\pi} I_m^2 \sin^2\theta\, d\theta = \frac{1}{2\pi}\int_0^{\pi} I_m^2 \left(\frac{1-\cos 2\theta}{2}\right) d\theta$$

$$I_{rms} = \frac{2 I_m}{\pi} \tag{6.2}$$

6.4.3 Ripple Factor (*r*)

Output of rectifier is not pure D.C. It contains both D.C. and A.C. Let the rms value of ripple be V_r. Ripple factor is the ratio of A.C. in the output to the D.C. voltage.

$$r = \frac{V_r}{V_d} = \frac{\sqrt{V_{rms}^2 - V_d^2}}{V_d}$$

Similarly,

$$r = \frac{\sqrt{I_{rms}^2 - I_d^2}}{I_d}$$

Divide term by term

$$r = \sqrt{\left(\frac{I_{rms}}{I_d}\right)^2 - 1} \tag{6.3}$$

For H.W.R.

$$r = \sqrt{\left(\frac{0.5 I_m}{(1/\pi) I_m}\right)^2 - 1}; \quad r = 1.21$$

Hence

$$r = \frac{I_r}{I_d} = 1.21; \quad I_r = 1.21\, I_d$$

ripple current in the output of half wave rectifier is 121% of the D.C. current. Half wave rectifier is not used in practice due to high ripple content.

6.4.4 Efficiency (η)

It is the ratio of D.C. power output to A.C. power input.

$$\eta = \frac{\text{D.C. power output}}{\text{A.C. power input}}$$

The conducting diode is represented by its resistance R_f in Fig. 6.5.

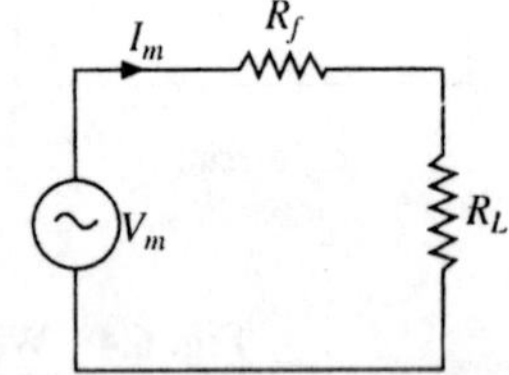

Fig. 6.5 Equivalent circuit.

$$\eta = \frac{I_d^2 R_L}{I_{rms}^2 (R_L + R_f)}$$

Divide numerator and denominator by R_L

$$= \frac{I_d^2}{I_{rms}^2 \left(\frac{R_L + R_f}{R_L}\right)}$$

$$\eta = \frac{I_d^2}{I_{rms}^2 \left(1 + \frac{R_f}{R_L}\right)} = \frac{\frac{I_m^2}{\pi^2}}{\left(\frac{I_m}{2}\right)^2 \left(1 + \frac{R_f}{R_L}\right)}$$

$$\boxed{\eta = \frac{4}{\pi^2\left(1 + \frac{R_f}{R_L}\right)}} \quad \text{this is the exact expression for efficiency.} \quad (6.4)$$

To find approximate value of η, neglect R_f.

So, $$\eta \cong \frac{4}{\pi^2} \qquad \eta \cong 0.405$$

$$\boxed{\eta \cong 40.5\%}$$

6.4.5 Regulation of the Half Wave Rectifier

Regulation is defined as the ratio of drop to the full load voltage

$$\text{Regulation} = \frac{\text{Drop}}{V_{FL}} = \frac{\text{Drop}}{\text{Reference voltage}}$$

In a regulated power suppy, full load voltage is kept constant. Hence full load voltage is taken as reference voltage. Therefore,

$$\text{Regulation} = \frac{V_{NL} - V_{FL}}{V_{FL}} = \frac{V_{NL}}{V_{FL}} - 1$$

$$V_{FL} = V_d = I_d R_L = \frac{I_m}{\pi} R_L$$

Substitute $$I_m = V_m/(R_f + R_L)$$

$$V_{FL} = \frac{V_m}{\pi\,(R_f + R_L)} R_L$$

Divide numerator and denominator by R_L

$$V_{FL} = \frac{V_m}{\pi\left(1 + \frac{R_f}{R_L}\right)}$$

If $$R_L = \infty;\ V_0 = V_{NL}$$

$$V_{NL} = \frac{V_m}{\pi}$$

Hence $$\boxed{\text{Reg} = \frac{V_{NL}}{V_{FL}} - 1}$$

$$\text{Reg} = \frac{(V_m/\pi)}{V_m/\pi\left(1 + \frac{R_f}{R_L}\right)} - 1 = 1 + \frac{R_f}{R_L} - 1 = \frac{R_f}{R_L}$$

$$\boxed{\% \text{ Regulation} = \frac{R_f}{R_L} \times 100} \tag{6.5}$$

Regulation depends on forward resistance of diode and load resistance.

6.4.6 Transformer Utilization Factor (TUF)

"TUF" is defined as the ratio of D.C. power output to the rms A.C. rating of the transformer.

$$\text{TUF} = \frac{\text{D.C. power output}}{\text{rms Rating}}$$

Transformer used in HWR is designed with rms current rating of $I_m/2$ since the rms value of the current flowing through it is $I_m/2$. The voltage available at the secondary of a transformer is a sine wave. The rms value of sine wave is $\frac{V_m}{\sqrt{2}}$

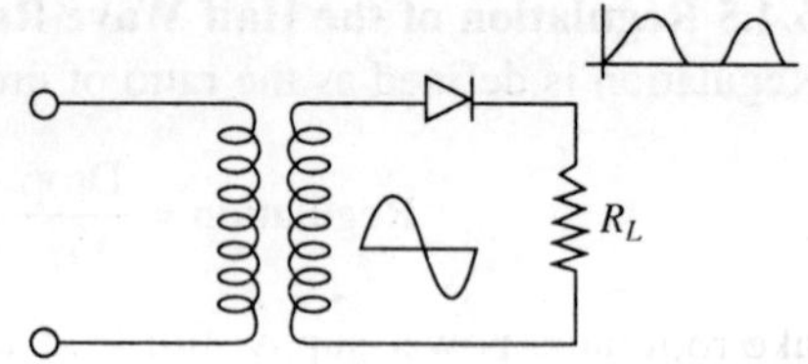

Fig. 6.6 Half wave rectifier.

$$\text{TUF} = \frac{V_d \cdot I_d}{V_{\text{rms}} I_{\text{rms}}} = \frac{\frac{V_m}{\pi} \frac{I_m}{\pi}}{\frac{V_m}{\sqrt{2}} \frac{I_m}{2}} = \frac{2\sqrt{2}}{\pi^2} = 0.286$$

$$\boxed{\text{TUF} = 28.6\%} \tag{6.6}$$

TUF is less since the secondary winding of the transformer is not utilized during the negative half cycles.

6.5 Expressions of Full Wave Rectifier (FWR)

6.5.1 Average Value of Output Current

1. Centre Tapped Type Rectifier

$$I_{\text{av}} = I_d = \frac{\text{Area}}{\text{Base}} = \frac{\int dA}{\text{Base}}$$

The waveform shown in Fig. 6.2(c) repeats with respect to π. Therefore the base is taken as π.

$$I_{\text{av}} = \frac{1}{\pi} \int_0^{\pi} i d\theta = \frac{1}{\pi} \int_0^{\pi} I_m \sin \theta d\theta$$

$$= \frac{I_m}{\pi} \left| -\cos\theta \right|_0^{\pi} = \frac{I_m}{\pi} \left| \cos\theta \right|_{\pi}^{0} = \frac{I_m}{\pi} [1 - (-1)] = \frac{2 I_m}{\pi}$$

$$I_d = I_{av} = \frac{2I_m}{\pi} \tag{6.7}$$

Average value of full wave rectifier is twice that of half wave rectifier.

6.5.2 Rms Value of Output Current of Centre Tap Type Rectifier

The waveform repeats with respect to π. Therefore the base is taken as π.

$$I_{rms} = \sqrt{\frac{1}{\pi}\int_0^{\pi} i^2 d\theta} = \sqrt{\frac{1}{\pi}\int_0^{\pi} I_m^2 \sin^2\theta \, d\theta}$$

$$I_{rms}^2 = \frac{I_m^2}{\pi}\int_0^{\pi}\left(\frac{1-\cos 2\theta}{2}\right)d\theta = \frac{I_m^2}{2\pi}\left|\theta - \frac{\text{Sin}\,2\theta}{2}\right|_0^{\pi}$$

$$= \frac{I_m^2}{2\pi}(\pi - 0)$$

$$I_{rms}^2 = \frac{I_m^2}{2}$$

Take root on both sides

$$I_{rms} = \frac{I_m}{\sqrt{2}} \tag{6.8}$$

6.5.3 Ripple Factor (r) of Centre Tap Type Rectifier

$$r = \sqrt{\left(\frac{I_{rms}}{I_d}\right)^2 - 1} = \sqrt{\left[\left(\frac{I_m}{\sqrt{2}}\right)^2 \Big/ \left(\frac{2I_m}{\pi}\right)^2\right] - 1}$$

$$= \sqrt{\frac{\pi^2}{8} - 1}$$

$$r = 0.48$$

This value is much less than that of half wave rectifier.

6.5.4 Efficiency (η) of Centre Tap Type Rectifier

$$\eta = \frac{\text{D.C. power output}}{\text{A.C. power input}} = \frac{I_d^2 R_L}{I_{rms}^2 (R_f + R_L)} = \frac{I_d^2}{I_{rms}^2\left(1 + \frac{R_f}{R_L}\right)}$$

Substitute (6.7) and (6.8) here

$$\eta = \frac{\left(\frac{2I_m}{\pi}\right)^2}{\left(\frac{I_m}{\sqrt{2}}\right)^2\left(1 + \frac{R_f}{R_L}\right)}$$

$$= \frac{8}{\pi^2\left(1 + \frac{R_f}{R_L}\right)} \tag{6.9}$$

If R_f is neglected, $\eta = 0.81$

$\boxed{\eta \cong 81\%}$ This value is twice that of the efficiency of HWR.

6.5.5 Regulation of Centre Tap Type Rectifier

$$\text{Regulation} = \frac{V_{NL}}{V_{FL}} - 1 = \frac{\dfrac{2V_m}{\pi}}{\dfrac{2V_m}{\pi}\Big/\left(1 + \dfrac{R_f}{R_L}\right)} - 1$$

$$= 1 + \frac{R_f}{R_L} - 1 = \frac{R_f}{R_L}$$

$$\boxed{\%\ \text{Regulation} = \frac{R_f}{R_L} \times 100} \qquad (6.10)$$

6.5.6 TUF of Centre Tap Type Rectifier

Total rms power rating is the sum of rms power rating of the upper half and rms power rating of the lower half of the transformer.

$$V_{\text{rms}} \cdot I_{\text{rms}} = \frac{V_m}{\sqrt{2}} \frac{I_m}{2}$$

$$\text{TUF} = \frac{V_d I_d}{\dfrac{V_m}{\sqrt{2}}\dfrac{I_m}{2} + \dfrac{V_m}{\sqrt{2}}\dfrac{I_m}{2}}$$

$$\text{TUF} = \frac{\dfrac{2V_m}{\pi} \times \dfrac{2I_m}{\pi}}{2\dfrac{V_m I_m}{2\sqrt{2}}} = \frac{4\sqrt{2}}{\pi^2} = 0.573$$

$$\boxed{\text{TUF} = 57.3\%}$$

TUF is not very high since one of the sections of secondary is not utilized.

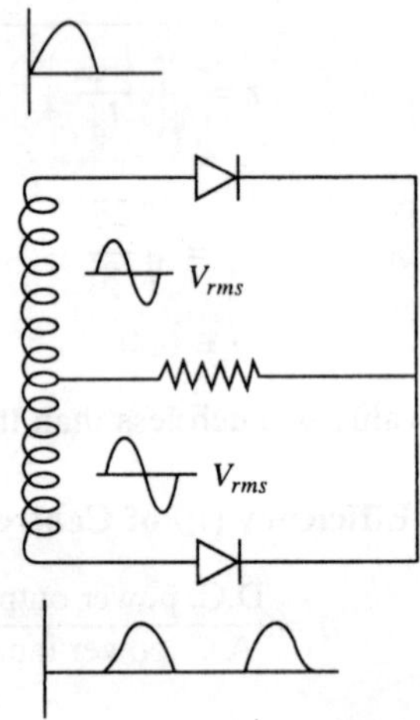

Fig. 6.7 Full wave rectifier.

6.5.7 Bridge Type of FWR

The average value, rms value, ripple factor and efficiency values are similar to those of centre tapped type full wave rectifier.

$$I_d = \frac{2I_m}{\pi}; \quad I_{\text{rms}} = \frac{I_m}{\sqrt{2}}$$

$$\text{Ripple factor} = 0.48; \quad \eta = 81\%$$

6.5.8 Regulation of Bridge Type Rectifier

$$\text{Regulation} = \frac{V_{NL}}{V_{FL}} - 1 = \frac{\dfrac{2V_m}{\pi}}{\dfrac{2V_m}{\pi}\bigg/\left(1 + \dfrac{2R_f}{R_L}\right)} - 1$$

$2R_f$ is taken since two diodes conduct at any time in a bridge type of rectifier.

$$= 1 + 2\frac{R_f}{R_L} - 1 = \frac{2R_f}{R_L}$$

$$\boxed{\%\ \text{Regulation} = \frac{2R_f}{R_L} \times 100} \qquad (6.11)$$

6.5.9 TUF of Bridge Type Rectifier

$$\text{TUF} = \frac{\text{D.C. Power Output}}{\text{rms Rating}} = \frac{V_d \cdot I_d}{V_{\text{rms}} \cdot I_{\text{rms}}}$$

$$= \frac{\dfrac{2V_m}{\pi} \cdot \dfrac{2I_m}{\pi}}{\dfrac{V_m}{\sqrt{2}} \cdot \dfrac{I_m}{\sqrt{2}}} = \frac{8}{\pi^2}$$

$$\boxed{\text{TUF} = 0.81} \quad \boxed{\text{TUF} = 81\%}$$

The transformer utilization factor of bridge rectifier is higher than that of centre tap type. This is because of the utilization of secondary winding in both the half cycles in a bridge rectifier.

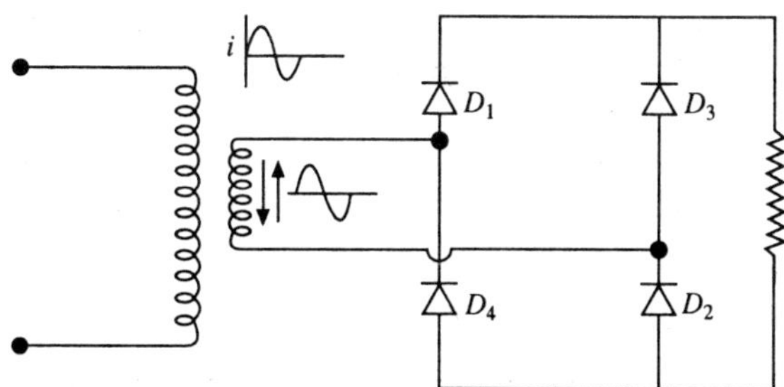

Fig. 6.8a Bridge rectifier.

6.5.10 Polyphase Rectifier

The circuit of 3-phase uncontrolled bridge rectifier is shown in Fig. 6.8(b) and the corresponding waveforms are shown in Fig. 6.8c. The primary of the transformer can be connected in star or delta. The secondary has to be connected in star. Three diodes are connected in the three lines. At any time, one diode conducts. The diode which has largest forward bias conducts. Load is connected between

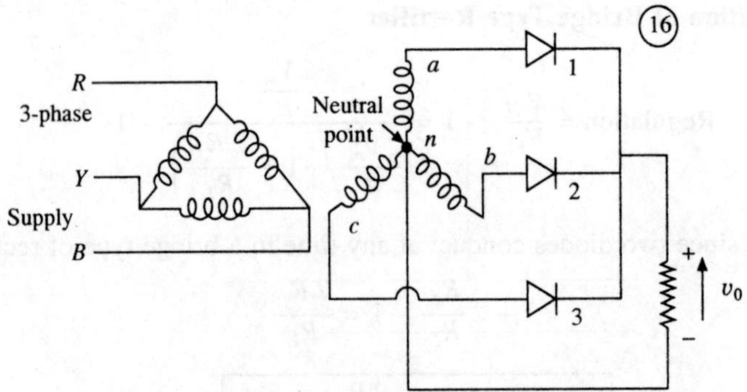

Fig. 6.8b 3 Phase uncontrolled rectifier.

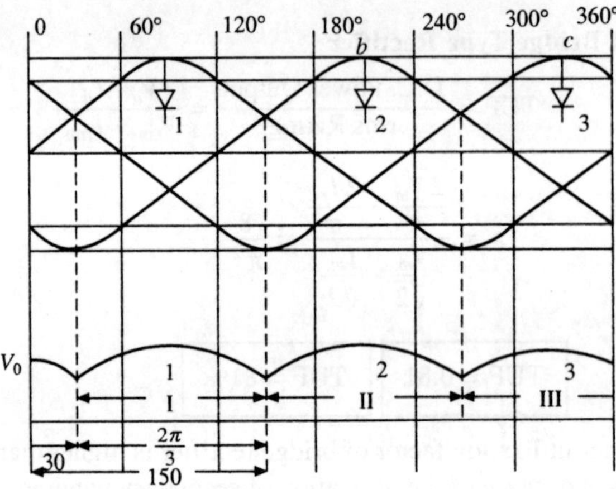

Fig. 6.8c Waveforms.

common cathode point and star point of the transformer secondary winding. When diode D_1 conducts, the voltage across the winding 'a' is applied to the load. From 30° to 150°, diode 1 conducts. From 150° to 270°, diode 2 conducts. Thus each diode conducts for a duration of 120°. Voltage across load is segments of secondary phase voltages. The advantages of 3-phase rectifier are as follows:

1. Average output voltage is higher.
2. Ripple in the output is lesser.

The expressions for output voltage and ripple factor are derived in the following sections.

6.5.11 Average Output of 3-ϕ Rectifier

$$V_d = \frac{\text{Area}}{\text{Base}} = \frac{1}{2\pi/3}\int_{30°}^{150°} v\,d\theta$$

$$= \frac{3}{2\pi}\int_{30}^{150} V_m \sin\theta\,d\theta = \frac{3}{2\pi} V_m\,[-\cos\theta]_{30}^{150}$$

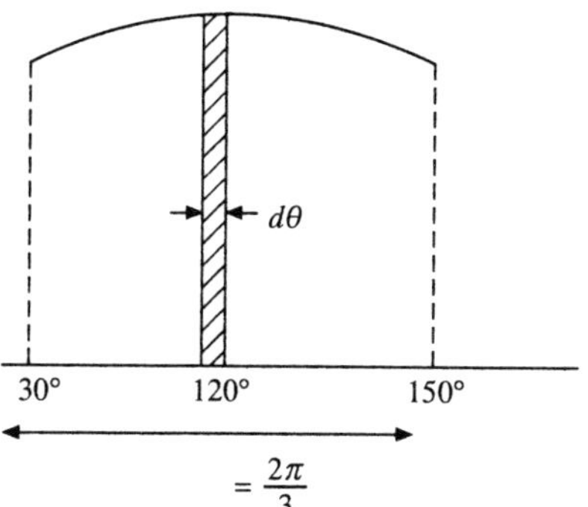

Fig. 6.8d One segment of output voltage

$$= \frac{3V_m}{2\pi}[\cos 30 - \cos 150] = \frac{3\,V_m}{2\pi}[\cos 30 + \cos 30]$$

$$= \frac{3V_m}{2\pi} \times 2\cos 30 = \frac{3\,V_m}{\pi} \cdot \frac{\sqrt{3}}{2}$$

$$V_d = \frac{3\sqrt{3}\,V_m}{2\pi}; \qquad I_d = \frac{V_d}{R_L} = 0.827\,I_m$$

$$V_{rms} = \sqrt{\frac{1}{2\pi/3}\int_{30}^{150} V^2 d\theta}$$

$$V_{rms}^2 = \frac{3}{2\pi}\int_{30}^{150} V_m^2 \sin^2\theta\, d\theta = \frac{3V_m^2}{2\pi}\int_{30}^{150} \frac{1-\cos 2\theta}{2}\, d\theta$$

$$= \frac{3V_m^2}{4\pi}\left[\theta - \frac{\sin 2\theta}{2}\right]_{30}^{150} = \frac{3\,V_m^2}{4\pi}\left[\frac{2\pi}{3} - \left[\frac{\sin 2\theta}{2}\right]_{30}^{150}\right]$$

$$= \frac{3\,V_m^2}{4\pi}\left[2\pi - \left(\frac{\sin 300}{2} - \frac{\sin 60}{2}\right)\right]$$

$$= \frac{3\,V_m^2}{4\pi}[2.09 + 0.866]$$

$$V_{rms}^2 = 0.707\,V_m^2$$

Take root on both sides

$$V_{rms} = 0.84\,V_m$$

Similarly $\quad I_{rms} = 0.84\,I_m$

$$r = \sqrt{\left(\frac{I_{rms}}{I_d}\right)^2 - 1} = \sqrt{\left(\frac{0.84\,I_m}{0.827\,I_m}\right)^2 - 1}$$

$$r = 0.17.$$

The ripple in 3-phase rectifier is much less than that in a 1-phase rectifier.

6.5.12 Summary

	HWR	FWR Centre tap type	FWR Bridge type
I_{av}	$\frac{I_m}{\pi}$	$\frac{2I_m}{\pi}$	$\frac{2I_m}{\pi}$
I_{rms}	$\frac{I_m}{2}$	$\frac{I_m}{\sqrt{2}}$	$\frac{I_m}{\sqrt{2}}$
r	1.21	0.48	0.48
η	0.405	0.81	0.81
TUF	0.286	0.573	0.81
Regulation	$\frac{R_f}{R_L}$	$\frac{R_f}{R_L}$	$\frac{2R_f}{R_L}$
PIV	V_m	$2V_m$	V_m
No. of Diodes	1	2	4

6.6 Filters

Filters filter unwanted A.C. in the output of a rectifier. Our mind has filters to filter unwanted things. Various types of filters are:

(i) Shunt capacitor filter
(ii) Series inductor filter
(iii) L–C filter
(iv) R–C filter
(v) Pi filter

6.6.1 Shunt Capacitor Filter

A rectifier with shunt capacitor filter is shown in Fig. 6.9. This is the simplest and cheapest filter. A large value capacitor C is connected in parallel with load resistor R_L as shown in Fig. 6.9. The capacitance offers low reactance to the A.C. component of current. To the D.C. this acts as an open circuit. All the D.C. current passes through the load. Only a small part of the A.C. component passes through the load producing a small ripple voltage.

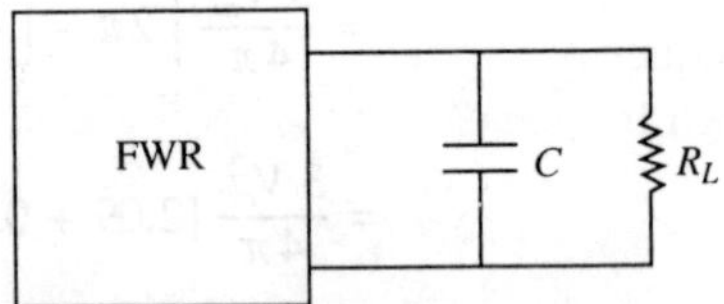

Fig. 6.9 Shunt capacitor filter.

The non-conduction diode disconnects or separates the source from the load. The capacitor discharges through the load. This prevents the load voltage from falling to zero. The capacitor continues to discharge until the source voltage becomes more than the capacitor voltage. When rectifier does not supply any current through the load, the capacitor supplies current through the load. Thus current is maintained through the load all the time. Ripper factor for C-filter is given in section 6.7.

6.6.2 Series Inductor Filter

A rectifier with series inductor filter is shown in Fig. 6.10. An inductor has the fundamental property of opposing any change in current flowing through it. This property is used in the series inductor filter. Whenever, the current through an inductor tends to change, a back e.m.f. is induced in the inductor. This induced emf prevents the current from changing its value. Any sudden change in current that might have occurred in the circuit without an inductor is smoothed out by the presence of the inductor. Since reactance of the inductor increases with frequency, better filtering of the higher harmonic ripples takes place. For D.C., the choke resistance R in series with the load resistance R_L forms a voltage divider circuit. The operation of a series inductor filter depends upon the current through it. This filter can only be used with a full wave rectifier since it requires current to flow at all times. Further more, higher the current flowing through it, better is its filtering action. Therefore, an increase in load current results in reduced ripple.

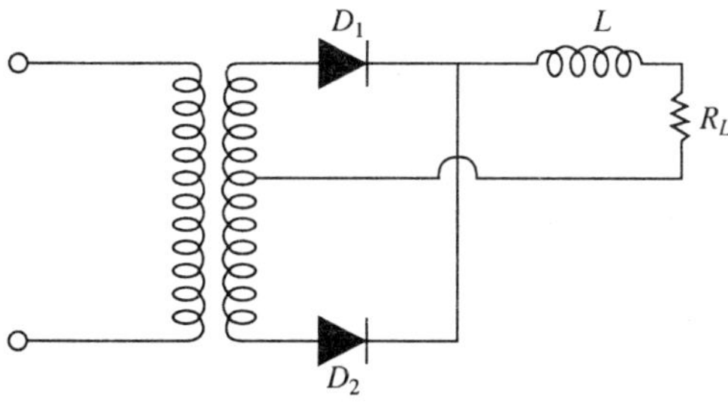

Fig. 6.10 Series inductor filter.

6.6.3 *L-C* Filter

A rectifier with *L-C* filter is shown in Fig. 6.11. *L-C* filter combines the advantages of *L* filter and *C* filter. The series inductor offers zero reactance for D.C. Shunt capacitor offers high reactance for low frequency signals and infinite reactance for D.C. Low frequency voltages are dropped across the series inductor. High frequency currents flow through the capacitor since it offers less reactance to high frequencies.

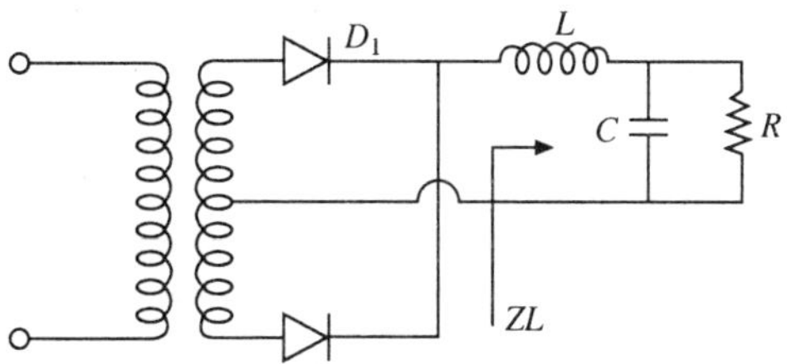

Fig. 6.11 *L–C* filter.

6.6.4 *R-C* Filter

Rectifier with *R-C* filter is shown in Fig. 6.12. The ripple current can be reduced by having a series combination of *R* and *C*. The value of ripple currents can be reduced by increasing the resistance. Thus a high resistance is required to reduce the ripple current. R is taken to be about 10 times X_c. When the rectifier is loaded, D.C. current flows through the resistance R. Large voltage drop and power loss takes place in the resistance R. The regulation is very high since the drop is high. To reduce the regulation, filter resistance must be reduced. To reduce the ripple current, filter resistance must be increased. It is not possible to satisfy both the requirements. Hence R–C filter is not used in practice.

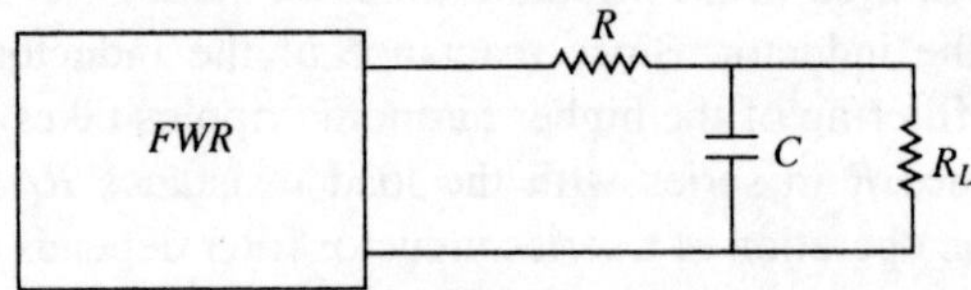

Fig. 6.12 *R-C* filter.

6.6.5 Pi(π)-Filter

π-Filter can be obtained by combining shunt capacitor filter and *L-C* filter. It combines the advantages of *L-C* filter and C-filter. Practically two or three *pi*-filters are cascaded to reduce the ripple to about 0.01 per cent. More details are given in section 6.7.4.

6.7 Expressions for Ripple Factor

6.7.1 Shunt Capacitor Filter

The output voltage of full wave rectifier with shunt capacitor filter is shown in Fig. 6.13.

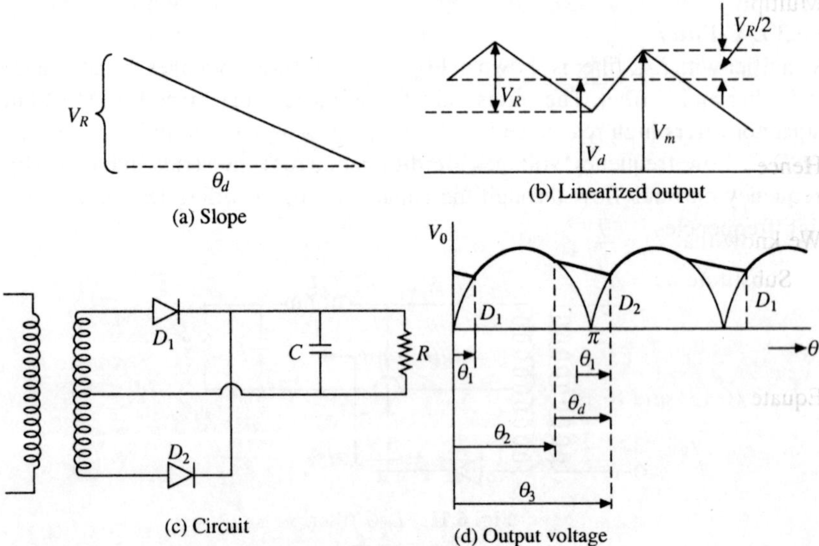

Fig. 6.13 Shunt capacitor filter.

Let V_R —Ripple voltage
θ_c —Charging period
θ_d —Discharge period

V_{rms} —rms Value of ripple
R_L —Load resistance

From Fig. 6.13(d)

$$\theta_d = \theta_3 - \theta_2$$
$$\theta_d = \pi + \theta_1 - \theta_2$$
$$\theta_d = \pi - (-\theta_1 + \theta_2) = \pi - (\theta_2 - \theta_1)$$

Substitute $\theta_2 - \theta_1 = \theta_c$

$$\theta_d = \pi - \theta_c \tag{6.12}$$

From the Fig. 6.3(a)

$$\frac{dv}{d\theta} = \frac{V_R}{\theta_d}$$

$$\frac{dv}{d\theta} = \frac{V_R}{\pi - \theta_c} \simeq \frac{V_R}{\pi} \tag{6.13}$$

$$\pi - \theta_C \simeq \pi \text{ (since } \theta_c \text{ very less)}$$

We know that $q = Cv$ or $v = \dfrac{q}{C}$

differentiate with respect to time

$$\frac{dv}{dt} = \frac{1}{C}\frac{dq}{dt}$$

Multiply with $\dfrac{1}{\omega}$ on both sides

$$\frac{1}{\omega}\frac{dv}{dt} = \frac{1}{\omega C}\frac{dq}{dt}; \text{ substitute } \frac{dq}{dt} = I_d$$

Hence
$$\frac{dv}{d(\omega t)} = \frac{1}{\omega C} I_d$$

We know that $\omega = \dfrac{\theta}{t}$.
Substitute $\omega t = \theta$

$$\frac{dv}{d\theta} = \frac{I_d}{\omega C} \tag{6.14}$$

Equate (6.13) and (6.14)

$$\frac{I_d}{\omega C} = \frac{V_R}{\pi}$$

$$V_R = \frac{\pi I_d}{\omega C} \tag{6.15}$$

From the Fourier series of the capacitor voltage waveform, we know that

$$V'_{rms} = \frac{V_R}{2\sqrt{3}}$$

Substitute equation (6.15) here

$$V'_{rms} = \frac{\pi I_d}{2\sqrt{3}\omega C}$$

$$r = \frac{V'_{rms}}{V_d} = \frac{\pi I_d}{2\sqrt{3}\omega C} \Big/ (I_d R_L)$$

$$\therefore \quad r = \frac{\pi}{2\sqrt{3}\omega C R_L} = \frac{\pi}{2\sqrt{3} \times 2\pi \times f \times C \times R_L}$$

$$\boxed{r = \frac{1}{4\sqrt{3} f C R_L}} \tag{6.16}$$

Larger the value of filter capacitance, lesser the value of ripple factor

From the Fig. 6.13b, $$\boxed{V_d = V_m - \frac{V_R}{2}} \tag{61.7}$$

6.7.2 Series Inductor Filter

Output voltage waveform with series inductor filter is shown in Fig. 6.14.

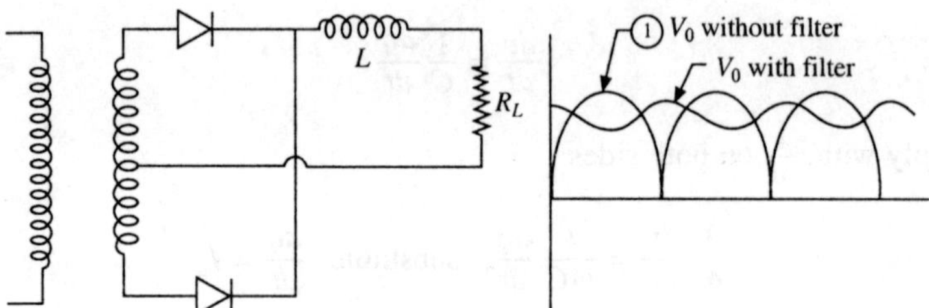

Fig. 6.14 Series inductor filter.

From the Fourier series of output of FWR, we know that

$$V_o = \frac{2V_m}{\pi} - \frac{4V_m}{3\pi}\cos 2\omega t - \ldots$$

Similarly

$$I_o = \frac{2I_m}{\pi} - \frac{4I_m}{3\pi}\cos 2\omega t - \ldots$$

Peak value of 2nd harmonic $I_{p2} = \dfrac{4I_m}{3\pi}$

rms value can be obtained by dividing the above equation by $\sqrt{2}$

$$I'_{rms} = \frac{4I_m}{3\pi\sqrt{2}} = \frac{2\sqrt{2}\, I_m}{3\pi}$$

Substitute $I_m = V_m/Z$

$$= \frac{2\sqrt{2}\left(\frac{V_m}{Z}\right)}{3\pi} = \frac{2\sqrt{2}V_m}{3\pi(\sqrt{R_L^2 + \omega_2^2 L^2})} \tag{6.18}$$

$$r = \frac{I'_{\text{rms}}}{I_d} \tag{6.19}$$

and $$I_d = \frac{2I_m}{\pi} \tag{6.20}$$

Substitute (6.18) and (6.20) in (6.19)

$$r = \frac{2\sqrt{2}\,V_m}{3\pi\sqrt{R_L^2 + \omega_2^2 L^2} \Big/ \frac{2}{\pi}\left(\frac{V_m}{R_L}\right)}$$

$$r = \frac{\sqrt{2}\,R_L}{3\sqrt{R_L^2 + \omega_2^2 L^2}} = \frac{\sqrt{2}R_L}{3R_L\sqrt{1 + \frac{\omega_2^2 L^2}{R_L^2}}}$$

If $$\frac{\omega_2^2 L^2}{R_L^2} >> 1$$

$$r = \frac{\sqrt{2}}{3\,\frac{\omega_2 L}{R_L}} = \frac{\sqrt{2}R_L}{3\omega_2 L} = \frac{\sqrt{2}R_L}{3\,(2\omega)L} = \frac{\sqrt{2}R_L}{6 \times 2\pi f L}$$

Hence $$\boxed{r = \frac{R_L}{6\sqrt{2}\,\pi f L}} \tag{6.21}$$

Larger the value of filter inductance, lesser the ripple factor.

6.7.3 L–C Filter

From the Fourier series of output of FWR we know that

$$I_o = \frac{2V_m}{\pi} - \frac{4I_m}{3\pi}\cos 2\omega t - \ldots$$

$$I'_{\text{rms}} = \frac{4I_m}{3\pi\sqrt{2}} \tag{6.22}$$

From Fig. 6.15

$$Z_L = X_L + X_C \,||\, R_L$$

X_C value is much less since a large filter capacitor is used. Value of R_L is much higher than X_C.

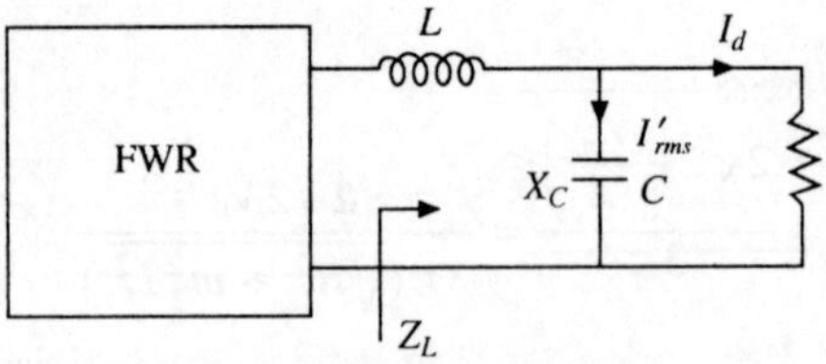

Fig. 6.15 L–C filter.

parallel combination of X_C and R_L is approximately equal to X_c.

$$Z_L \simeq X_L + X_C$$

Z_L is approximately equal to X_L, since X_C is very less.

From equation (6.22)

$$I'_{\text{rms}} = \frac{4\left(\dfrac{V_m}{Z_L}\right)}{3\pi\sqrt{2}}$$

Substitute $Z_L = X_L$

$$I'_{\text{rms}} = \frac{4\,V_m}{3\pi\sqrt{2}\,X_L}$$

$$V'_{\text{rms}} = I'_{\text{rms}}\,X_C = \frac{4\,V_m X_C}{3\pi\sqrt{2}\,X_L}$$

$$r = \frac{V'_{\text{rms}}}{V_d} = \frac{4\,V_m X_C}{3\pi\sqrt{2}\,X_L \dfrac{2\,V_m}{\pi}}$$

$$\boxed{r = \frac{\sqrt{2}}{3}\,\frac{X_C}{X_L}} \qquad (6.23)$$

where $$X_C = \frac{1}{\omega_2 C}$$

$$X_L = \omega_2 L$$

$$\omega_2 = 2\omega = 2(2\pi f)$$

6.7.4 π-Filter

π-filter circuit, voltage across C and equivalent circuit are shown in Fig. 6.16. This circuit can be treated as an L–C filter fed from a rectifier with shunt capacitor filter. The waveform at the input of L–C filter is a triangular waveform with D.C. shift.

The Fourier series of the output waveform across C is required to derive the expression for r.

$$V_C = \frac{2\,V_m}{\pi} - \frac{V_R}{\pi}\left(\cos 2\,\text{wt} - \frac{\cos 4\,\omega t}{2} - \ldots\right)$$

Neglect higher order harmonics.

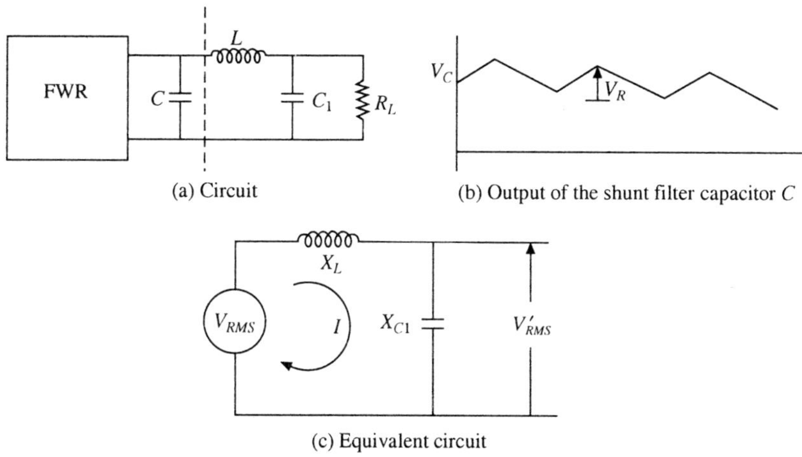

(a) Circuit (b) Output of the shunt filter capacitor C

(c) Equivalent circuit

Fig. 6.16 π-filter.

Peak value of 2nd harmonic $= \dfrac{V_R}{\pi}$

$$\text{rms value} = \frac{V_R}{\pi\sqrt{2}} \tag{6.24}$$

From equation (6.15) of shunt capacitor filter analysis

$$V_R = \frac{\pi I_d}{\omega C}$$

i.e.,

$$V_R = \frac{I_d}{2fC} \tag{6.25}$$

Substitute (6.25) in (6.24)

$$V_{\text{rms}} = \frac{I_d/2fC}{\pi\sqrt{2}} = \frac{I_d}{2\sqrt{2}\,\pi fC} = \frac{I_d}{\sqrt{2}\,\omega C}$$

Multiply numerator and denominator with 2

$$V_{\text{rms}} = \frac{2I_d}{\sqrt{2}\,(2\omega)C} = \frac{\sqrt{2}\,I_d}{\omega_2 C} \quad \text{where} \quad \omega_2 = 2\omega$$

$$= \sqrt{2}\,I_d\left(\frac{1}{\omega_2 C}\right)$$

$$V_{\text{rms}} = \sqrt{2}\,I_d X_C \tag{6.26}$$

But

$$I = \frac{V_{\text{rms}}}{X_L + X_{C1}}$$

$$I \simeq \frac{V_{\text{rms}}}{X_L} \text{ since } X_{C1} \text{ is very small compared to } X_L$$

$$V'_{rms} = I\,X_{C1} = \frac{V_{rms}}{X_L}\,X_{C1}$$

Substitute (6.26) here

$$V'_{rms} = \frac{\sqrt{2}\,I_d X_C}{X_L}\,X_{C1} \tag{6.27}$$

Ripple factor $(r) = \dfrac{V'_{rms}}{V_d} = \dfrac{V'_{rms}}{I_d R_L}$

Substitute (6.27) here

$$r = \frac{\sqrt{2}\,I_d X_C X_{C1}}{X_L I_d R_L}$$

$$\therefore \quad \boxed{r = \frac{\sqrt{2}\,X_C X_{C1}}{X_L R_L}} \tag{6.28}$$

$$= \sqrt{2}\,\frac{1}{\omega_2 C} \times \frac{1}{2\omega\,C_1} \times \frac{1}{\omega_2 L} \times \frac{1}{R_L}$$

$$= \sqrt{2}\,\frac{1}{2\omega C} \times \frac{1}{2\omega\,C_1} \times \frac{1}{2\omega L} \times \frac{1}{R_L}$$

$$r = \frac{\sqrt{2}}{8}\,\frac{1}{\omega^3 C C_1 L R}$$

$$\boxed{r = \frac{1}{4\sqrt{2}\,\omega^3 C C_1 L R_L}} \tag{6.29}$$

Ripple factor value is very less due to the terms ω^3, R_L and L in the denominator of the above equation. Ripple can be reduced by selecting larger values of C, C_1 and L.

6.8 Voltage Regulators

Voltage regulators maintain constant voltage at the output when input voltage and load are fluctuating.

6.8.1 Zener Voltage Regulator

Basic Zener voltage regulator circuit is shown in Fig. 6.17. From the characteristic of a Zener diode, it is seen that the voltage across the device remains constant for

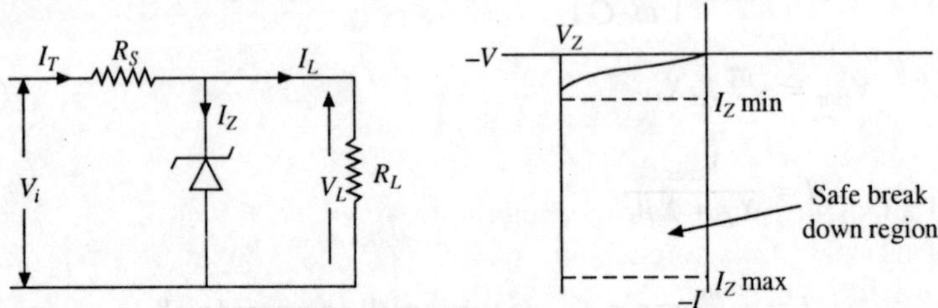

Fig. 6.17 Zener voltage regulators

different values of current. This property is used in the Zener voltage regulator circuit. Consider a fixed load resistance and an increase in the input voltage V_i. When V_i increases, V_L tends to increase. The Zener diode draws more current and the total current increases. The drop in the ballast resistance increases. V_L tends to decrease since the voltage drop is increased.

Consider a constant input voltage and an increase in the load current. This increase in the current is due to the reduction in the load resistance. The voltage V_L tends to decrease and the current drawn by the Zener diode decreases. The total current remains constant and the drop $I_T R_S$ remains constant. Since V_i is constant, V_L also must be a constant. Thus the Zener diode tries to maintain a constant voltage across the load. The resistance R_S is called ballast resistance. This circuit cannot maintain constant voltage if input voltage and load change at the same time.

6.8.2 Basic Transistor Series Regulator

The circuit of basic transistor series regulator is shown in Fig. 6.18. This is called series voltage regulator since the transistor is in series with load resistance. This circuit combines the advantages of Zener regulator and common collector configuration. The power rating of Zener diode is less since Zener diode is connected in base circuit. At the input side, Zener acts as stiff voltage source. Therefore output voltage remains constant. By applying KVL in output loop. $V_{BE} = V_Z - V_L$. Let us consider an increase in load current due to reduction of load resistance value. V_L tends to decrease. When V_L decreases, V_{BE} increases since V_Z is constant. Input current and output current increases, $V_L = I_E R_L$. When R_L decreases I_E increases such that the product remain constant. The disadvantage of this circuit is that the power loss in the transistor is high. This is because the transistor carries load current.

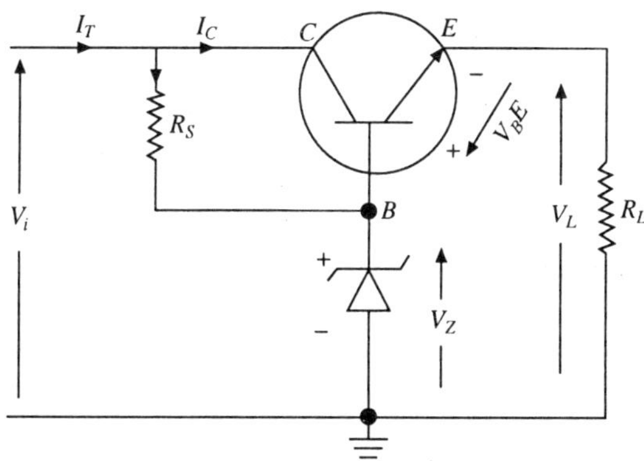

Fig. 6.18 Basic transistor series regulator.

6.8.3 Transistor Shunt Voltage Regulator

Transistor shunt voltage regulator circuit is shown in Fig. 6.19. In this circuit, the transistor is connected in parallel with the load resistance. Let us assume that the load voltage increases. This may be due to an increase in input voltage

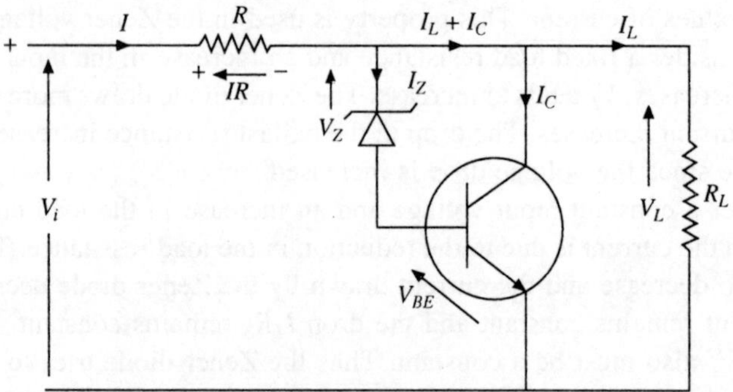

Fig. 6.19 Transistor shunt voltage regulator.

or decrease in the load current. The KVL for the closed loop formed by base emitter junction, Zener diode and load resistance is given by

$$V_{BE} + V_Z - V_L = 0 \quad (6.30)$$

$$V_{BE} = V_L - V_Z \quad (6.31)$$

From equation (6.31), when V_L increases V_{BE} also increases. The base current increases, collector current increases and hence the total current increases. KVL in input loop gives

$$V_i - IR = V_L \quad (6.32)$$

From equation (6.32), when total current increases, the load voltage tends to decrease. In other words when input voltage is increased, the drop IR is also increased to maintain a constant voltage at the output. The power loss in transistor is reduced since the current through the transistor is less when compared to the transistor current in the series regulator circuit.

6.8.4 Series Controlled Transistor Regulator

Closed loop controlled series transistor regulator is shown in Fig. 6.20. The transistor Q_1 is called the pass transistor since most of the load current passes through Q_1. To achieve constant voltage, a closed loop system is employed using controlled transistor series regulator circuit. An additional transistor Q_2 is employed to amplify the error signal. The error signal is the difference between feedback signal and the reference signal of the Zener diode. The closed loop system acts in such a way that the error signal reduces to zero and the output remains always at a constant value. Resistance R_F is of the order of megaohms.

Let us suppose that the load voltage increase due to decrease in the load current. The feed back signal V_2 increases and the base current of Q_2 also increases. I_{C2} increases and I_{B1} decreases for a constant value of I_3 since $I_3 = I_{B1} + I_{C2}$. The collector current of Q_1 decreases and V_{CE1} increases. From the Kirchoff's voltage law between input voltage, output and V_{CE1}, V_L tends to decrease when V_{CE} tends to increase. Thus a constant voltage is maintained at the output.

6.9 Regulated Power Supply (RPS)

The circuits of regulated power supply are shown in Fig. 6.21. A regulated

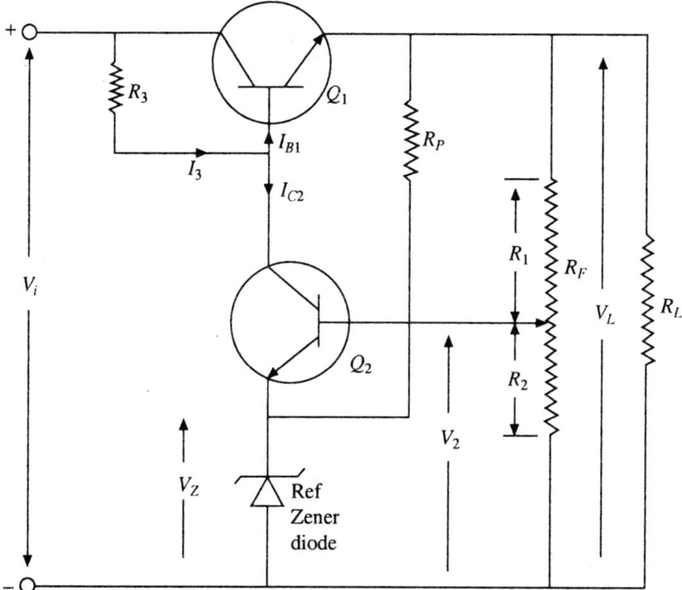

Fig. 6.20 Series controlled transistor regulator.

power supply unit can supply variable D.C. voltage. Commercial A.C. supply is reduced to a lower level by using a step-down transformer. Rectifier converts A.C. voltage into unidirectional voltage which is pulsating D.C. This is filtered by using one or many π-filters. The regulator converts unregulated D.C. voltage into constant D.C. voltage. The Zener diode acts as a voltage source.

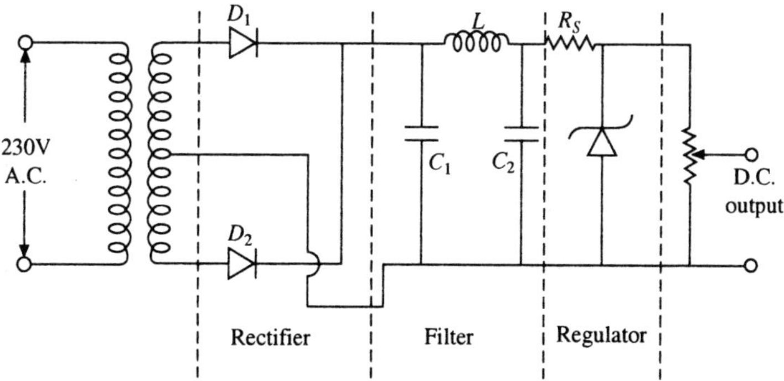

Fig. 6.21 Regulated power supply.

Voltage across Zener diode is applied to a potentiometer. Variable D.C. voltage can be taken from the output of potentiometer. A high resistance is connected in parallel with the filter capacitor. This is called bleeder resistor. The filter capacitor can discharge through the bleeder resistance when A.C. supply is disconnected. Under no-load condition, output current flows through the bleeder resistor and hence output voltage is made available without load. Conventional power supply is bulky. This is due to the large size transformer and filter elements used in this circuit. The size of transformer and filter increases due to low frequency operation. D.C. output voltage can be varied from 0 to 30 V in the power supplies used in laboratories. This is done by using the potentiometer at the output. Potentiometer is not required for a constant power supply.

6.10 Switched Mode Power Supply (SMPS)

Block diagram of SMPS is shown in Fig. 6.22

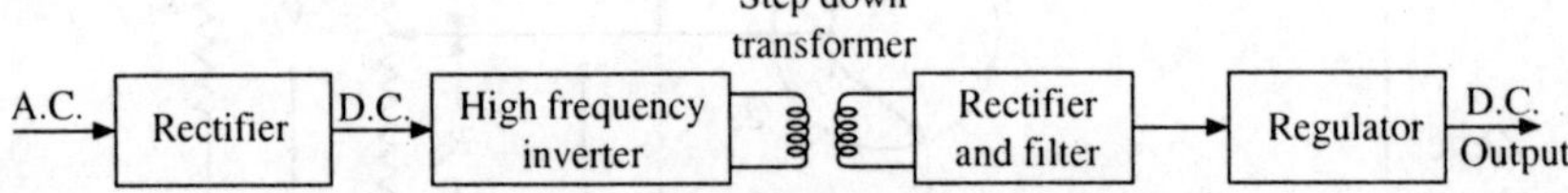

Fig. 6.22 Switched mode power supply.

In this system, A.C. supply is directly rectified using diode bridge rectifiers. Step down transformer is not used in the first stage. A small capacitor filter is connected at the output. Voltage across the filter is the D.C. input voltage to the MOSEFT based inverter. The inverter converts D.C. voltage into high frequency A.C. It is better to use transformer at high frequency since flux and area are reduced at high frequency. The size of transformer is very much reduced due to the high frequency operation. High frequency A.C. is converted into dc using full wave rectifier having fast recovery diodes. Then the output is filtered using π-filter. We know that ripple factor of π-filter is inversely proportional to L, C_1, C_2 and ω^3. For a specified value of ripple factor, size of the filter is very much reduced since the ripple frequency is very high. D.C. output voltage is taken through a regulator. In a transformer, emf depends on flux and frequency. For a given value of e.m.f. the flux value gets reduced if the transformer operates at very high frequency. Therefore the size of the core gets reduced. Overall size of the SMPS is very much reduced due to the reduction in the size of transformer and filter elements. This circuit requires fast recovery diodes and a ferrite core transformer.

It is popularly used in television system, computers and spacecrafts since it is very compact. Power density of SMPS is higher since the volume is lesser.

6.11 Selection of Ballast Resistance

Load current varies from $I_{L\min}$ to $I_{L\max}$ when R_L is varied from maximum to minimum value. The input voltage fluctuates between $V_{i\min}$ and $V_{i\max}$.

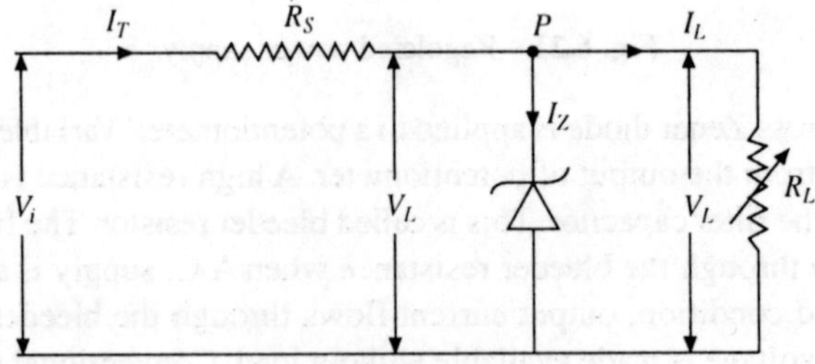

Fig. 6.23 Zener regulater.

K_{CL} at p gives

$$I_T = I_L + I_Z$$

$$I_Z = I_T - I_L$$

$$I_Z = \frac{V_i - V_L}{R_S} - I_L$$

Zener diode draws maximum current when input voltage is maximum and load current is minimum, therefore

$$I_{Z\max} = \frac{V_{i\max} - V_L}{R_S} - I_{L\min} \tag{6.33}$$

Zener diode draws minimum current when input voltage is minimum and load current is maximum

$$I_{Z\min} = \frac{V_{i\min} - V_L}{R_S} - I_{L\max} \tag{6.34}$$

The value of Rs can be designed using (6.33) or (6.34). If R_L is fixed, $I_{L\min} = I_{L\max} = I_L$. If input voltage is fixed $V_{i\max} = V_{i\min} = V_i$.

Worked Problems

Ex. 6.1 A centre tapped transformer has a 220 V primary and secondary rated at 12–0–12 V and is used with a rectifier circuit having a load of 100 Ω. What is the D.C. output voltage, D.C. load current and PIV rating?

$$V_{\text{rms}} = 12 \text{ V}$$

$$R_L = 100\ \Omega; \quad V_{\text{rms}} = \frac{V_m}{\sqrt{2}}$$

$$V_m = \sqrt{2}\, V_{\text{rms}}$$

$$V_m = \sqrt{2}\ 12$$

$$V_d = \frac{2\,V_m}{\pi} = \frac{2 \times 12\sqrt{2}}{\pi} = 10.8 \text{ V}$$

$$I_d = \frac{V_d}{R_L}; \quad I_d = \frac{10.8}{100} = 0.108 \text{ A}$$

$$\text{PIV} = 2\,V_m$$

$$\text{PIV} = 2 \times 12 \times \sqrt{2} = 33.9 \text{ V}$$

Ex. 6.2 The rms secondary voltage of a rectifier circuit is 117 V at 60 Hz. Find the turns ratio of the transformer for an output of 200V D.C. Also find the frequency of output wave form.

$$V = V_m \sin \omega t; \qquad V_1 = 117$$

$$V_d = \frac{2\,V_m}{\pi}; \quad 200 = \frac{2\,V_{m2}}{\pi}$$

$$V_{m2} = 314 \text{ V}$$

V_{m2} is peak value of secondary voltage

$$V_2\,(\text{rms}) = \frac{V_{m2}}{\sqrt{2}} = \frac{314}{\sqrt{2}} = 222 \text{ V}$$

$$\text{Turns ratio } (K) = \frac{V_2}{V_1} = \frac{222}{117} = 1.89$$

Frequency of the output waveform

$$f_o = 2f = 2 \times 60 = 120 \text{ Hz}$$

Ex. 6.3 A full wave rectifier uses 2 diodes. Forward resistance of each diode is 20 Ω. The transformer secondary voltage from centre tap to each end of secondary is 50 V and R_L = 980 Ω. Find (i) Average current and (ii) rms value of load current.

R_f = 20 Ω; R_L = 980 Ω; V = 50; I_d = ? I_{rms} = ?

$$I_m = \frac{V_m}{R_f + R_L}; \qquad V_m = V_{\text{rms}}\sqrt{2} = 50\sqrt{2}$$

$$I_m = \frac{50\sqrt{2}}{20 + 980} = 0.0707 \text{ A}$$

$$I_d = \frac{2\,I_m}{\pi} = \frac{2 \times 0.0707}{\pi} = 0.04 \text{ A}$$

$$I_{\text{rms}} = \frac{I_m}{\sqrt{2}} = \frac{0.0707}{\sqrt{2}} = 0.05 \text{ A}$$

Ex. 6.4 A FWR using centre tapped transformer of 200-0-200 V has 2 diodes. Each of them is rated at a maximum current of 700 mA. Determine (a) R_L that can be connected to give largest D.C. output power without exceeding diode rating, (b) V_d and I_d for the condition of (a) and (c) PIV of each diode. Take operating current as 80% of maximum current. Neglect R_f.

Given

$$I_m = 0.8 \times \text{Rated current}$$

$$I_m = 0.8 \times 700 \times 10^{-3} \text{ A} = 0.56 \text{ A}$$

$$I_m = \frac{V_m}{R_L}$$

(a) $R_L = \dfrac{V_m}{I_m} = \dfrac{200\sqrt{2}}{0.56} = 505\ \Omega$

(b) $V_d = \dfrac{2\,V_m}{\pi} = \dfrac{2 \times 200\sqrt{2}}{\pi} = 180 \text{ V}$

$$I_d = \frac{2\,I_m}{\pi} = \frac{2 \times 0.56}{\pi} = 0.35 \text{ A}$$

(or)

$$I_d = \frac{V_d}{R_L} = \frac{180}{505} = 0.35 \text{ A}$$

(c) PIV = $2 \times V_m = 2 \times 200\sqrt{2} = 565.6$ V

Ex. 6.5 Design a capacitor filter that meets the following specifications. $V_d = 20$ V; $I_d = 100$ mA; $r = 0.01$. Use FWR operating at a supply frequency of 60 Hz.

Given $$V_d = 20 \text{ V} \qquad I_d = 100 \text{ mA}$$

$$r = 0.01 \qquad f = 60 \text{ Hz}$$

$$R_L = \frac{V_d}{I_D} = \frac{20}{100\,\text{mA}} = 200\ \Omega$$

$$C = \frac{1}{4\sqrt{3}\, r f R_L} = \frac{1}{4\sqrt{3} \times 0.01 \times 60 \times 200} = 1203\ \mu\text{F}$$

Ex. 6.6 Calculate the Ripple factor and D.C. voltage at the output of FWR with a capacitor filter of 525 μF, $R_L = 200\ \Omega$. Waveform shows that $V_m = 8.8$ V and $V_R = 0.8$ V. The supply frequency is 50 Hz.

$$R_L = 200\ \Omega; \quad V_m = 8.8 \text{ V}; \quad f = 50 \text{ Hz}$$

$$C = 525\ \mu\text{F}; \quad V_R = 0.8 \text{ V}; \quad r = ? \quad V_d = ?$$

$$r = \frac{1}{4\sqrt{3}\, f C R_L} = \frac{1}{4\sqrt{3} \times 50 \times 525 \times 10^{-6} \times 200} = 0.027$$

$$V_d = V_m - \frac{V_R}{2} = 8.8 - \frac{0.8}{2}; \; V_d = 8.8 - 0.4 = 8.4 \text{ V}$$

Ex. 6.7 A FWR uses inductive filter with $L = 20$ H. A sinusoidal voltage 300 sin 314 t is applied at the output. Assuming that the rectified output has only 2nd harmonic, find out (i) D.C. current and (ii) Ripple factor. The load resistance is 20 Ω.

$$R_L = 20\ \text{k}\Omega; \quad \omega = 314; \quad I_d = ?$$

$$L = 20\text{H}; \quad 2\pi f = 314; \quad r = ?$$

$$V_m = 300 \quad f = 50$$

(i) $$I_d = \frac{2\,V_m}{\pi R_L} = \frac{2 \times 300}{\pi \times 20} = 9.6 \text{ A}$$

(ii) $$r = \frac{R_L}{6\sqrt{2}\,\pi f L} = \frac{20}{6\sqrt{2}\,\pi \times 50 \times 20} = 0.75 \times 10^{-3}$$

Ex. 6.8 An L–C filter with $L = 2$H and $C = 40\ \mu$F is used in the output of a single phase full wave rectifier that is fed from 40-0-40 V transformer. Load current is 0.2A. Calculate ripple factor and D.C. output voltage. The frequency is 50 Hz.

$$L = 2\text{H}; \quad C = 40\ \mu\text{F}; \quad V_m = 40\sqrt{2}; \quad I_d = 0.2; \; r = ?; \quad V_d = ?$$

$$X_L = \omega_2 L = 2\omega L = 2(2\pi f)L = 4\pi f L = 4\pi \times 50 \times 2 = 1256\ \Omega$$

$$X_C = \frac{1}{\omega_2 C} = \frac{1}{2\omega C} = \frac{1}{2 \times 2\pi f C} = \frac{1}{4\pi f C}$$

$$= \frac{1}{4\pi \times 50 \times 40 \times 10^{-6}} = 39.8\ \Omega$$

$$r = \frac{\sqrt{2}}{3} \cdot \frac{X_C}{X_L} = \frac{\sqrt{2}}{3} \times \frac{39.8}{1256} = 0.014$$

$$V_d = \frac{2V_m}{\pi} = \frac{2 \times 40 \times \sqrt{2}}{\pi} = 36\ \text{V}$$

Ex. 6.9 The current supplied by a single phase bridge rectifier is 0.25 A at 20 V D.C. Find the ripple factor when two 100 μF capacitors each and 5H inductor are used in a π-filter with 50 Hz power supply.

Given $C_1 = C$

$$R_L = \frac{V_d}{I_d} = \frac{20}{0.25} = 80\ \Omega$$

$$r = \frac{1}{4\sqrt{2}\,\omega^3 (C^2) LR_L} = \frac{1}{4\sqrt{2}\,(314)^3 (100 \times 10^{-6})^2 \times 5 \times 80}$$

$$= 1.42 \times 10^{-3} = 0.14\%$$

Ex. 6.10 A single phase FWR uses π section filter rms with two capacitors of 10 μF and a choke of 10 H. Secondary voltage is 280 V RMS with respect to centre tap. If the load current is 10 μA, determine D.C. output voltage and % ripple in the output. Supply frequency is 50 Hz.

$$V_d = V_m - \frac{V_R}{2} = V_m - \frac{1}{2} \times \frac{I_d}{2fc} = V_m - \frac{I_d}{4fc}$$

$$= 280\sqrt{2} - \frac{10 \times 10^{-6}}{4 \times 50 \times 10 \times 10^{-6}}$$

$$V_d = 395.9\ \text{V}$$

$$R_L = \frac{V_d}{I_d} = \frac{395.9}{10 \times 10^{-6}} = 39.5\ \text{M}\Omega$$

$$r = \frac{1}{4\sqrt{2}\,\omega^3 C^2 LR_L} = \frac{1}{4\sqrt{2}\,(314)^3 \times (10 \times 10^{-6})^2 \times 10 \times 39.5 \times 10^6}$$

$$= 1.4 \times 10^{-7}$$

Ex. 6.11 Design a power supply using a π section filter to give D.C. output of 30 V, 100 mA with a ripple factor not exceeding 0.01%. Use a FWR operating with a supply of 50 Hz. Assume the choke to have an inductance of 20 H.

$$R_L = \frac{V_d}{I_d} = \frac{30}{0.1} = 300\ \Omega$$

$$r = 0.01\% = 10^{-4}$$

$$r = \frac{1}{4\sqrt{2}\,\omega^3 C^2 LR_L}$$

$$10^{-4} = \frac{1}{4\sqrt{2}\,(314)^3\,C^2 \times 20 \times 300}$$

$$C = 97.5\ \mu\text{F}$$

$$V_m = V_d + \frac{V_R}{2} = V_d + \frac{I_d}{2fC \times 2} \quad \text{since } V_R = \frac{I_d}{2fC}$$

$$= 30 + \frac{0.1}{4 \times 50 \times 975 \times 10^{-6}} = 30 + 5.1 = 35.1\ \text{V}$$

$$V_{\text{RMS}} = \frac{35.1}{\sqrt{2}} = 24.8\ \text{V} \simeq 25\ \text{V}$$

Use a transformer of 230/25–0–25 V.

Ex. 6.12 A Zener diode with Zener voltage of 30 V is used in a regulator circuit with a series resistance of 200 Ω connected across the Zener diode. Over what range of input voltage will the circuit operate? Maximum Zener current is 25 μA. Minimum Zener current is zero.

$$V_z = 30\ \text{V}; \qquad R_s = 200\ \Omega$$

$$R_L = 2000\ \Omega; \qquad I_{z\max} = 25\ \text{mA}$$

$$I_{z\min} = 0; \quad I_L = \frac{V_L}{R_L}; \quad = \frac{30}{2000} = 15\ \text{mA}.$$

$$I_{z\max} = \frac{V_{i\max} - V_L}{R_S} - I_L$$

$$25 \times 10^{-3} = \frac{V_{i\max} - 30}{200} - 15 \times 10^{-3}$$

$$(40 \times 10^{-3}) \times 200 = V_{i\max} - 30$$

$$V_{i\max} = 38\ \text{Volts}$$

$$I_{z\min} = \frac{V_{i\max} - V_L}{R_S} - I_L$$

$$0 = \frac{V_{i\min} - 30}{200} - 15 \times 10^{-3}$$

$$15 \times 10^{-3} = \frac{V_{i\min} - 30}{200}$$

$$15 \times 200 \times 10^{-3} = V_{i\min} - 30$$

Hence $V_{i\min} = 33$ V

33 V < V_i < 38 V. The range of input voltage is 33 to 38 V.

Ex. 6.13 Design a Zener regulator to provide an output voltage of 15 volts over a load current of 10 to 20 mA. Input voltage is 20 V and $V_z = 15$ V. Take maximum Zener current as 90% of the load current.

Since load current is 10–20 mA, R_L should be variable.

$$R_{L\min} = \frac{V_L}{I_{L\max}} = \frac{15}{20\text{ mA}} = 750\ \Omega$$

$$R_{L\max} = \frac{V_L}{I_{L\min}} = \frac{15}{10\text{ mA}} = 1500\ \Omega$$

R_L varies from 750 to 1500 Ω

$$I_{z\max} = 90\% \text{ of } I_{L\max} = 0.9 \times 20 = 18\text{ mA}$$

$$I_{z\min} = 10\% \text{ of } I_{L\max} = 1 \times 20 = 2\text{ mA}$$

$$I_{z\max} = \frac{V_{i\max} - V_L}{R_S} - I_{L\min}$$

$$18 \times 10^{-3} = \frac{20 - 15}{R_S} - 10 \times 10^{-3}$$

$$\therefore \quad R_{s\min} = 178\ \Omega$$

$$I_{z\min} = \frac{V_{i\min} - V_L}{R_S} - I_{L\max}$$

$$2 \times 10^{-3} = \frac{20 - 15}{R_S} - 20 \times 10^{-3}$$

$$R_{s\max} = 227\ \Omega$$

$178 < R_S < 227\ \Omega$

The ballast resistance lies between 178 and 227 Ω.

Ex. 6.14 Calculate the ripple voltage of a full-wave rectifier with a 100 μF filter capacitor connected to a load drawing 50 mA,

Ripple voltage $$V_R = \frac{\pi I_d}{\omega C} = \frac{\pi \times I_d}{2\pi f C}$$

$$= \frac{\pi \times 50 \times 10^{-3}}{2\pi\, 50 \times 100 \times 10^{-6}} = 5\text{ V}$$

Short Questions and Answers

Q. 1. Explain the function of a capacitor filter of a full-wave rectifier.

Ans. The shunt capacitor offers infinite reactance to D.C. It offers less reactance to A.C. The A.C. component in the output of the rectifier flows through capacitor. It reduces the value of ripple factor.

Q. 2. What is the peak inverse voltage?

Ans. Maximum reverse voltage that appears across a non-conducting diode is called peak inverse voltage (PIV).

Q. 3. What are the advantages of bridge rectifiers?

Ans.
- PIV is V_m
- Centre tapped transformer is not required.
- It is cheaper

Q. 4. Define ripple factor.

Ans. It is the ratio of rms value of A.C. component in the output to the D.C. component.

Q. 5. What is meant by transformer utilization factor?

Ans. It is the ratio of D.C. power delivered to the rms A.C. rating of transformer.

Q. 6. Mention the relative merits of centre-tapped type rectifier.

Ans.
- Less conduction loss
- It requires only two diodes
- Less voltage drop

Q. 7. Define voltage regulation.

Ans. It is the ratio of drop in terminal voltage of the rectifier with load to the full load voltage.

$$\text{Reg} = \frac{V_{\text{dc}}\,(\text{No load}) - V_{\text{dc}}\,(\text{Full load})}{V_{\text{dc}}\,(\text{Full load})}$$

Q. 8. What is the ripple factor for half-wave and full-wave rectifiers?

Ans. r for half-wave rectifier is 1.21
r for full-wave rectifier is 0.48

Q. 9. Give the maximum efficiency values of half-wave and full-wave rectifiers.

Ans. Efficiency of half-wave rectifier is 40.5%
Efficiency of full-wave rectifier is 81%

Q. 10. What are transformer utilization factor values for various types of rectifiers?

Ans. TUF of half-wave rectifier = 0.286
TUF of centre tapped rectifier = 0.573
TUF of bridge rectifier = 0.8

Q. 11. List some advantages of full wave choke-input filter.

Ans.
- It smoothens the D.C. output current
- It does better filtering of higher order harmonics
- Ripple factor reduces at higher currents

Q. 12. List important features of IC voltage regulators.

Ans.
- They produce constant D.C. voltage and output voltage does not change with change in load.
- Output voltage does not change when input voltage fluctuates.

Q. 13. Give the PIV of diodes for each type of rectifier.

Ans. PIV for half rectifier = V_m
PIV for centre tapped rectifier = $2V_m$
PIV for Bridge type = V_m

Q. 14. Primary utilization factor of a centre tapped full wave rectified circuit is 2 times that a secondary utilization factor. (True/False)

Ans. True

Q. 15. What is the PIV of a half wave rectifier feeding a capacitance filter?

Ans. 2 V_m.

Q. 16. How do you ensure continuous load current in an inductive load supplied by a controlled rectifier?

Ans. Using a free wheeling diode, continuous load current can be maintained. When the load current reduces, the polarity of the voltage across the load inductance gets reversed. This polarity forward biases the free wheeling diode. The diode conducts and maintains continuity of current. Thus the energy stored in load inductance maintains continuous load current.

Q. 17. What is the function of bleeder resistance?

Ans.
- It acts as dummy load
- Filter capacitor can discharge through the bleeder after A.C. is switched off.

Q. 18. What is the ripple factor of a full-wave rectifier feeding capacitor filter?

Ans.

$$r = \frac{1}{4\sqrt{3}\, f\, C R_L}$$

Q. 19. What are the advantages of π-filter?

Ans.
- Ripple factor can be reduced to 0.01%
- Higher and lower order harmonics can be filtered.
- Average output voltage is increased

Q. 20. A centre tap fullwave 1 – ϕ rectifier uses 2 diodes. The PIV of each diode is

Ans.

$$\text{PIV} = 2\, V_m$$
$$= 2 \times 280\, \sqrt{2} = 791 \text{ volts}$$

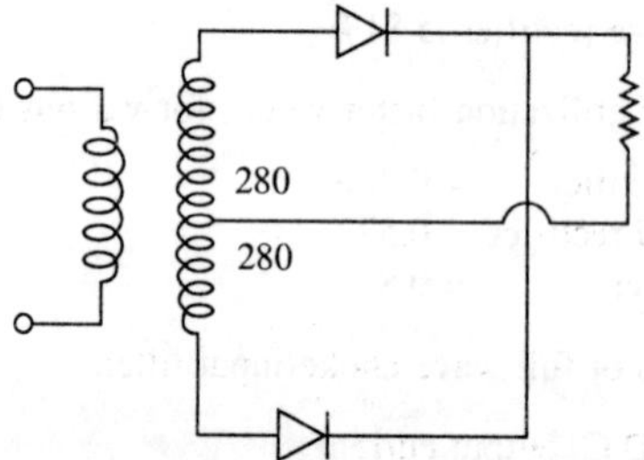

Q. 21. A voltage source $V_{AB} = 4 \sin \omega t$ is applied to A and B. The diodes are assumed to be ideal. Resistance offered by the circuit between A and B is . . . .

Ans. During positive half cycle D_1 conducts. During negative half cycle D_2 conducts. Only one diode conducts at a time. Therefore the resistance looking from AB is 10 kΩ.

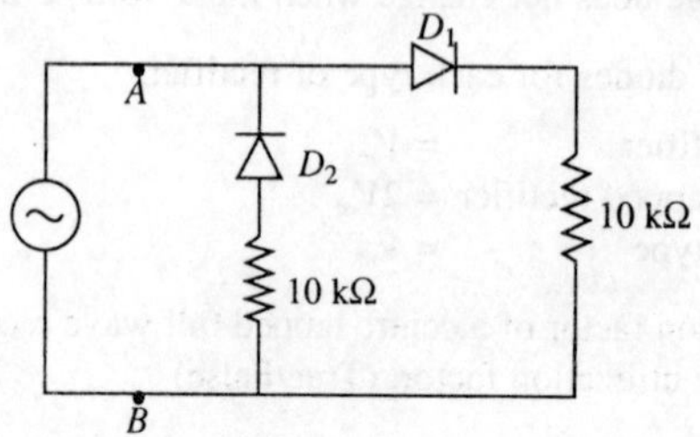

Q. 22. What would be the current in ammeter if it is measured by MI ammeter?

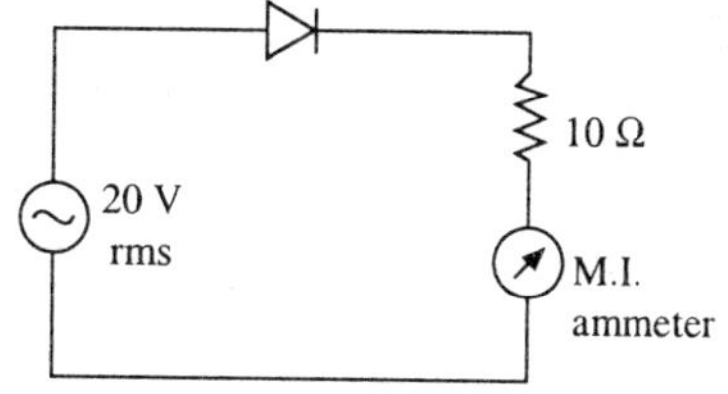

Ans. MI meter indicates rms value.

$$I_o = \frac{I_m}{2} = \frac{2\sqrt{2}}{2}$$

$$I_o = \sqrt{2}$$

Q. 23. A bleeder resistor in a power supply filter performs which function.?

Ans. Ensures discharge of capacitors when power supply is turned off.

Q. 24. A voltage of 200 cos 100 t is applied to a half wave rectifier with R_L = 5 k. Rectifier circuit has a diode in series with a resistance of 1 kΩ. I_m, I_0 and I_{rms} are . . . . . .

Ans.

$$I_m = \frac{200}{6\text{ k}} = 33.3\text{ mA}$$

$$I_0 = \frac{I_m}{\pi} = 10.6\text{ mA}$$

$$I_{rms} = \frac{I_m}{2} = 16.67\text{ mA}$$

Q. 25. The Zener diode shown in the circuit has a reverse breakdown voltage of 10 V. The power dissipation in R_S would be . . . . . .

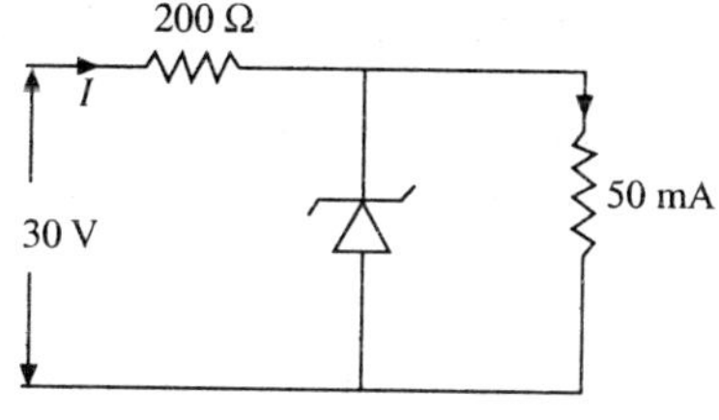

Ans. $I = \frac{30 - 10}{200} = \frac{20}{200} = 0.1\text{ A}$

$= 0.1^2 \times 200 = 2w$

Q. 26. A 5 V reference is drawn from the circuit shown below. If the Zener diode is of 5 mW and 5 V, then R_z is . . . . . .

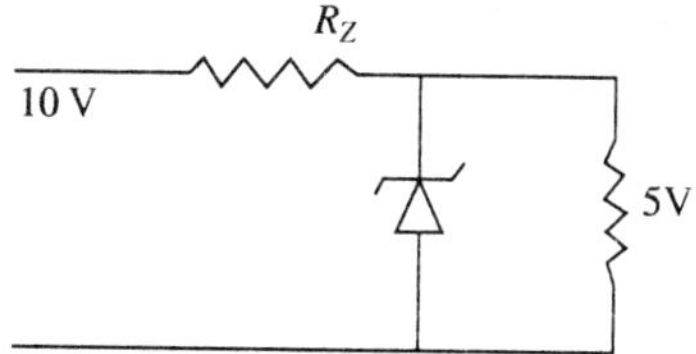

Ans. $I_z = \frac{5\text{ mW}}{5\text{ V}}$

$$I = 1\text{ mA}$$

$$R_z = \frac{10 - 5}{1\text{ mA}} = 5\text{ k}\Omega$$

7

7. Biasing Methods of Transistor

7.1 Biasing

Biasing means to fix a particular value of V_{CE} and I_C so that the transistor operates in the active region. A transistor must be properly biased before it is used as an amplifier. Base emitter junction must be forward biased and collector base junction must be reverse biased.

7.1.1 Requirements of Biasing

1. To select the Q-point at the centre of the load line.
2. To make the collector current independent of temperature.
3. To make the performance of the amplifier independent of the changes in β value (collector current to be independent of β).

7.2 Relation between the Leakage Currents

The total collector current from common base characteristics is the sum of the current due to transistor action (αI_E) and the leakage current I_{CBO}.

$$I_C = \alpha I_E + I_{CBO} \tag{7.1}$$

The collector current from common emitter characteristic is the sum of the current due to transistor action and the leakage current.

$$I_C = \beta I_B + I_{CEO} \tag{7.2}$$

From (7.1), $$I_C = \alpha I_E + I_{CBO}$$

Substitute $$I_E = I_C + I_B$$

Therefore, $$I_C = \alpha(I_C + I_B) + I_{CBO}$$

$$I_C = \alpha I_C + \alpha I_B + I_{CBO}$$

$$I_C - \alpha I_C = \alpha I_B + I_{CBO}$$

$$I_C(1 - \alpha) = \alpha I_B + I_{CBO}$$

Divide throughout by $(1 - \alpha)$

$$I_C = \frac{\alpha}{1 - \alpha} I_B + \frac{I_{CBO}}{1 - \alpha}$$

Substitute $\frac{\alpha}{1 - \alpha} = \beta$

$$I_C = \beta I_B + \frac{I_{CBO}}{1-\alpha} \tag{7.3}$$

Consider the relation

$$\beta = \frac{\alpha}{1-\alpha}$$

Add 1 on both sides,

$$1+\beta = 1 + \frac{\alpha}{1-\alpha}$$

$$1+\beta = \frac{1-\alpha+\alpha}{1-\alpha}$$

$$1+\beta = \frac{1}{1-\alpha} \tag{7.4}$$

Substitute (7.4) in (7.3),

$$I_C = \beta I_B + (1+\beta)\, I_{CBO} \tag{7.5}$$

Equate (7.2) and (7.5),

$$\beta I_B + I_{CEO} = \beta I_B + (1+\beta)\, I_{CBO}$$

βI_B gets cancelled on both sides

$$\boxed{I_{CEO} = (1+\beta)\, I_{CBO}} \tag{7.6a}$$

7.3 Thermal Runaway

Consider an increase in temperature. The leakage current I_{CEO} increases since more covalent bonds are broken.

From equation (7.2), when I_{CEO} increases, collector current I_C increases.

With the increase in collector current, the temperature of the transistor increases and the leakage current increases and the process continues until the transistor is damaged due to the excessive heat produced. This effect is known as thermal runaway. This is one of the drawbacks of BJT. This is given in Fig. 7.1(a).

7.4 Stability Factor

Stability factor is defined as the rate of change of collector current with respect to the leakage current for a constant value of β.

$$S_f = \frac{dI_C}{dI_{CO}}$$

where $I_{CBO} = I_{CO}$

From equation (7.2),

$$I_C = \beta I_B + I_{CEO}$$

Substitute (7.6a) in this equation.

$$I_C = \beta I_B + (1+\beta)\, I_{CBO}$$

$$I_C = \beta I_B + (1+\beta)\, I_{CO}$$

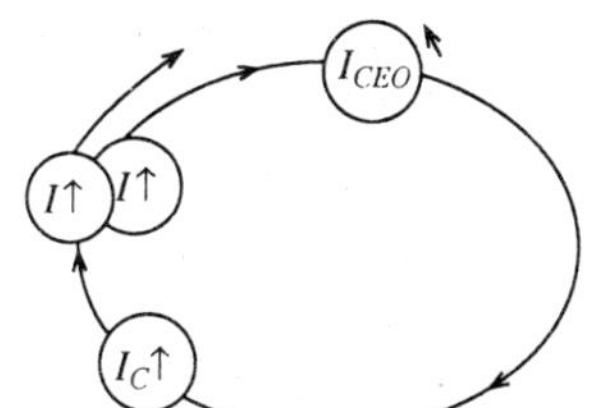

Fig. 7.1(a) Thermal Runaway

Differentiate with respect to I_C,

$$1 = \beta \frac{dI_B}{dI_C} + (1 + \beta) \frac{dI_{CO}}{dI_C}$$

$$\left(1 - \beta \frac{dI_B}{dI_C}\right) = (1 + \beta) \frac{dI_{CO}}{dI_C}$$

$$\frac{dI_C}{dI_{CO}} = \frac{(1 + \beta)}{\left(1 - \beta \frac{dI_B}{dI_C}\right)}$$

But $$S_f = \frac{dI_C}{dI_{CO}}$$

Therefore, $$S_f = \frac{(1 + \beta)}{\left(1 - \beta \frac{dI_B}{dI_C}\right)} \qquad (7.6b)$$

Practically S_f should be less than 5.

$dI_C = S_f \, dI_{CO}$. If S_f is less, change in collector is lesser. Therefore it is better to have lesser value of S_f.

7.5 Biasing Methods

Various methods of biasing a BJT are

1. Fixed bias
2. Collector to base bias
3. Emitter bias (self bias)
4. Voltage divider bias (or) universal bias

7.5.1 Fixed Bias

Fixed bias circuit is shown in Fig. 7.1(b). Common wire used at the bottom of an electronic circuit is called ground wire. It is shown at the bottom of Fig. 7.1(b). The meaning of + V_{ce} Fig. 7.1(b) is that there is a battery connected between the top line and ground. In this method, the voltage required at the base is obtained from the battery used in collector circuit through a resistance R_B. KVL in the base loop gives

$$V_{CC} - I_B R_B - V_{BE} = 0$$

$$I_B R_B = V_{CC} - V_{BE}$$

$$I_B = \frac{V_{CC} - V_{BE}}{R_B} \qquad (7.7)$$

$$I_C = \beta I_B$$

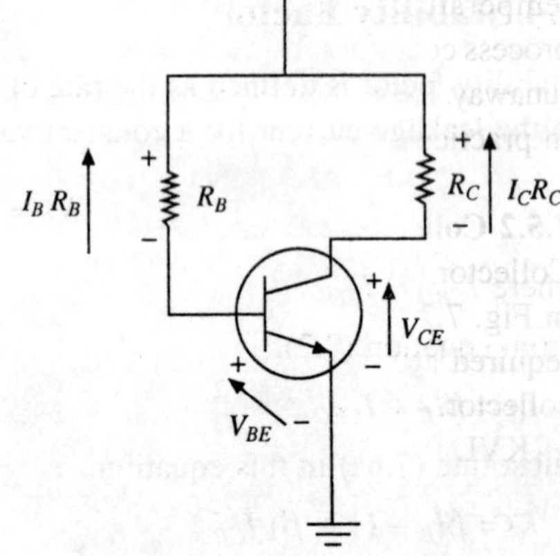

Fig. 7.1(b) Fixed bias.

Applying KVL in output loop,

$$V_{CE} + I_C R_C - V_{CC} = 0$$

$$V_{CE} = V_{CC} - I_C R_C$$

$$I_C = \frac{V_{CC} - V_{BE}}{R_C}$$

Differentiating equation (7.7) with respect to I_C

$$\frac{dI_B}{dI_C} = \frac{d}{dI_C} \frac{(V_{CC} - V_{BE})}{R_B} = 0$$

$$S_F = S = \frac{dI_C}{dI_{CO}} = \frac{1 + \beta}{1 - \beta\left(\frac{dI_B}{dI_C}\right)}$$

Substitute $\frac{dI_B}{dI_C} = 0$

$$= \frac{1 + \beta}{1 - 0} = 1 + \beta$$

$$S = \frac{dI_C}{dI_{CO}} = 1 + \beta \quad (7.8)$$

Assuming β value of 100,

$$1 + \beta = 1 + 100 = 101$$

$$\frac{dI_C}{dI_{CO}} = 101$$

Therefore, $dI_C = 101\ dI_{CO}$

When the temperature increases, the leakage current increases and the collector current increases by 101 times the increase in the leakage current. Hence the collector current increases and the temperature further increases. This process continues till it leads to thermal runaway. Hence fixed biasing is not used in practice.

7.5.2 Collector to Base Bias

Collector to base bias circuit is shown in Fig. 7.2. In this circuit, the voltage required at the base is taken from the collector.

KVL for the input loop gives

$$V_{CC} - (I_C + I_B)R_C - I_B R_B - V_{BE} = 0$$

$$I_B R_B = V_{CC} - I_C R_C - I_B R_C - V_{BE}$$

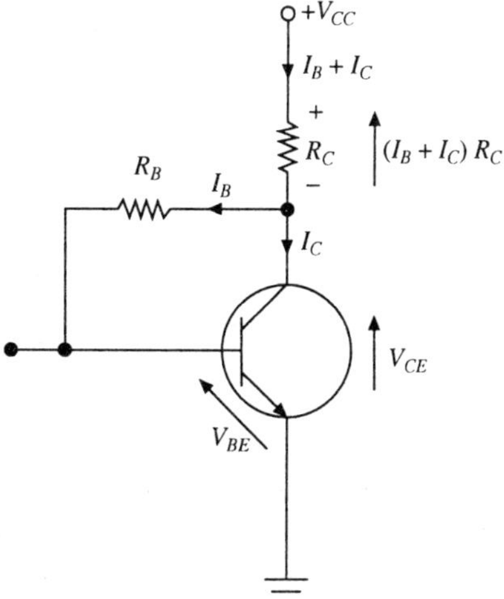

Fig. 7.2 Collector to base bias

Neglect the term I_BR_C

$$I_B = \frac{(V_{CC} - I_C R_C) - V_{BE}}{R_B} \tag{7.9}$$

KVL in the output loop gives

$$V_{CE} + (I_C + I_B)R_C - V_{CC} = 0$$

$$V_{CE} = V_{CC} - R_C(I_C + I_B)$$

Neglect the term I_BR_C

$$V_{CE} = V_{CC} - I_CR_C \tag{7.10}$$

KVL for the input loop gives

$$V_{CC} = (I_C + I_B)\ R_C + I_BR_B + V_{BE}$$

$$V_{CC} = I_CR_C + I_BR_C + I_BR_B + V_{BE}$$

$$V_{CC} - V_{BE} = I_CR_C + I_B(R_B + R_C)$$

$$(V_{CC} - V_{BE}) - I_CR_C = I_B\ (R_B + R_C)$$

$$\frac{(V_{CC} - V_{BE}) - I_C R_C}{(R_B + R_C)} = I_B$$

Divide term by term

$$\frac{(V_{CC} - V_{BE})}{(R_B + R_C)} - \frac{I_C R_C}{R_B + R_C} = I_B$$

Differentiate I_B with respect to I_C,

$$0 - \frac{R_C}{R_B + R_C} = \frac{dI_B}{dI_C}$$

Substitute $\frac{dI_B}{dI_C}$ in the stability factor equation (7.6)

$$S = \frac{1 + \beta}{1 - \beta \frac{dI_B}{dI_C}} = \frac{1 + \beta}{1 + \beta \frac{R_C}{R_B + R_C}}$$

$$S = \frac{1 + \beta}{1 + \beta \frac{R_C}{R_B + R_C}} \tag{7.11}$$

If $\beta = 100$, $R_C = 5\text{K}$, $R_B = 500$ K.

$$S = \frac{1 + 100}{1 + 100 * \frac{5}{500 + 5}} \approx 60$$

This value is less than that of fixed bias.

This biasing is not used in practice due to the following reasons.

1. The resistance R_B is connected between the output and input terminals. This resistance feeds part of the output voltage to the input. This negative feedback reduces the amplification. The biasing is provided to have a good amplification.

2. The stability factor is approximately equal to $\frac{1+\beta}{2}$. Thus the stability factor is very high.

3. This circuit can check the raising tendency of the collector current. When temperature increases, leakage current increases and the collector current increases. From equation (7.10), when collector current increases, V_{CE} decreases. When V_{CE} decreases, base current decreases from equation (7.9). The collector current and the temperature get reduced. Thus the raising tendency of temperature is checked.

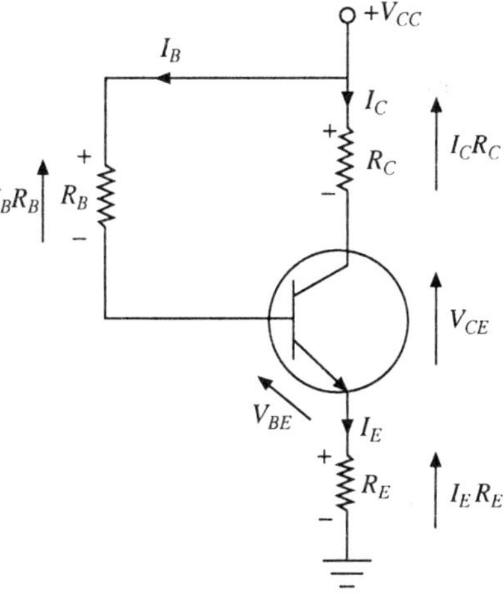

Fig. 7.3 Emitter bias.

7.5.3 Emitter Bias

Emitter bias circuit is shown in Fig. 7.3. In this method, a resistance R_E is connected between emitter and ground. KVL in the input loop gives

$$V_{CC} = I_BR_B + V_{BE} + I_ER_E \tag{7.12}$$

$$(V_{CC} - V_{BE}) - I_ER_E = I_BR_B$$

$$I_B = \frac{(V_{CC} - V_{BE}) - I_ER_E}{R_B}$$

$$I_E = I_B + I_C = I_B + \beta I_B = I_B(1 + \beta)$$

In Equation (7.12), substitute $I_E = I_B + I_C$

$$V_{CC} = I_BR_B + V_{BE} + (I_B + I_C)R_E$$

Therefore, $V_{CC} - V_{BE} = I_BR_B + I_BR_E + I_CR_E$ (7.13)

$$(V_{CC} - V_{BE}) - I_CR_E = I_B(R_B + R_E)$$

$$I_B = \frac{(V_{CC} - V_{BE})}{R_B + R_E} - \frac{I_CR_E}{(R_B + R_E)}$$

Differentiating I_B with respect to I_C

$$\frac{dI_B}{dI_C} = 0 - \frac{R_E}{(R_B + R_E)}$$

We know that

$$S = \frac{1+\beta}{1 - \beta\frac{dI_B}{dI_C}}$$

Substitute $\frac{dI_B}{dI_C}$ here

$$S = \frac{1 + \beta}{1 + \beta \frac{R_E}{R_B + R_E}} \quad (7.14)$$

From equation (7.13)

$$V_{CC} - V_{BE} = I_B R_B + I_B R_E + I_C R_E$$

neglect $I_B R_E$

$$V_{CC} - V_{BE} = I_B R_B + 0 + \beta I_B R_E$$

$$V_{CC} - V_{BE} = I_B (R_B + \beta R_E)$$

$$I_B = \frac{(V_{CC} - V_{BE})}{R_B + \beta R_E} \quad (7.15)$$

$$I_C = \beta I_B$$

$$= \beta \frac{(V_{CC} - V_{BE})}{R_B + \beta R_E}$$

Dividing both numerator and denominator by β

$$I_C = \frac{V_{CC} - V_{BE}}{\frac{R_B}{\beta} + R_E} \quad (7.16)$$

From equation (7.16), if collector current has to be independent of β, R_E must be much higher than R_B/β. A large value of the resistance R_E requires a large voltage battery V_{CC} since the drop $I_E R_E$ increases for a given value of collector current. Because of this drawback, emitter bias is not used. In the amplifier circuit, input A.C. source is connected between base and ground, some amount of A.C. voltage gets dropped across R_E. The voltage applied between base and emitter gets reduced, output voltage of amplifier also gets reduced. This can be eliminated by connecting a capacitor across R_E.

7.5.4 Voltage Divider Bias

This is a very popular method of biasing. The resistors R_1 and R_2 act as a voltage divider network. Voltage divider biasing circuit is shown in Fig. 7.4.

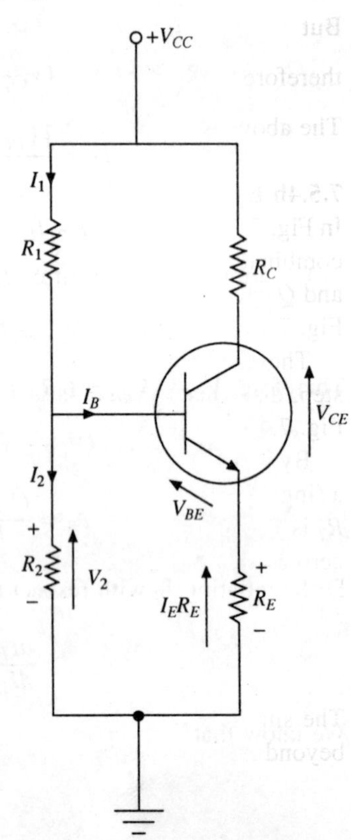

Fig. 7.4 Voltage divider bias.

7.5.4a Approximate Method

In approximate method, I_1 is assumed to be equal to I_2 since base current is very less.

$$I_1 = I_2 = \left(\frac{V_{CC}}{R_1 + R_2} \right) R_2 \tag{7.17}$$

Apply KVL in the closed loop formed by base, emitter, R_2 and R_E

$$V_2 - V_{BE} - V_E = 0$$

$$I_E R_E = V_E = V_2 - V_{BE}$$

$$V_E = I_E R_E \qquad I_E = \frac{V_E}{R_E}$$

$$I_B = \frac{I_E}{I + \beta}; \quad I_C = \beta I_B$$

Apply KVL in the collector loop,

$$V_{CC} - I_C R_C - V_C = 0$$

$$V_C = V_{CC} - I_C R_C \tag{7.18}$$

But $\quad V_C = V_{CE} + V_E$

therefore, $\quad V_{CE} + V_E = V_{CC} - I_C R_C$

The above equations can be used to design R_1, R_2, R_E and R_C.

7.5.4b Exact Analysis of Voltage Divider Bias

In Fig. 7.4(a), single source V_{CC} in the biasing circuit is represented as a parallel combination of two voltage sources of same value i.e. V_{CC}. The wire between P and Q can be removed since P and Q are at the same potential. This is shown in Fig. 7.4(b).

Thevenin equivalent circuit is obtained by looking from A and B. As a first step, disconnect the circuit beyond AB. The left part of the circuit is shown in Fig. 7.4(c).

By using Thevenin's theorem, the circuit between A and B can be reduced to a single source V_T in series with resistance R_T. Where V_T is Thevenin voltage and R_T is Thevenin resistance. To find Thevenin resistance, reduce the source V_{CC} to zero as shown in Fig. 7.4(e). R_T is parallel combination of R_1 and R_2.

$$\boxed{R_T = \frac{R_1 R_2}{R_1 + R_2}}$$

The simplified voltage divider bias circuit is shown in Fig. 7.4(f). The circuit beyond AB is reconnected. Apply KVL in input loop.

$$V_T - I_B R_T - V_{BE} - I_E R_E = 0$$

$$V_T - V_{BE} - I_B R_T - (I_B + I_C) R_E = 0$$

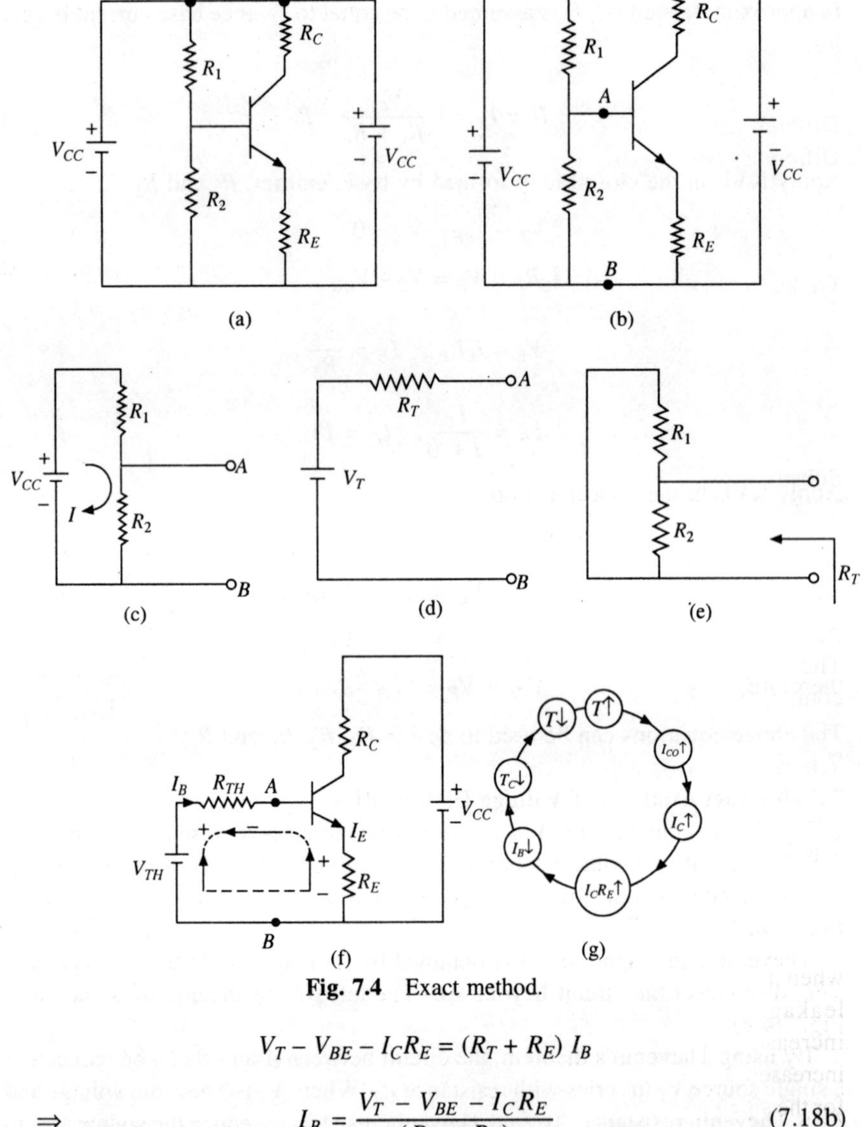

Fig. 7.4 Exact method.

$$V_T - V_{BE} - I_C R_E = (R_T + R_E)\, I_B$$

$$\Rightarrow \qquad I_B = \frac{V_T - V_{BE} - I_C R_E}{(R_T + R_I)} \qquad (7.18b)$$

when temperature increases, collector current increases, $I_C R_E$ increases. From (7.18b), when $I_C R_E$ increases, I_B decreases. When I_B decreases thes is shown in Fig. 7.4(g). temperature tends to decrease. Therefore thermal runaway is eliminated. Apply KVL in the input loop of Fig. 7.4(f).

$$V_{TH} = I_B R_{TH} + V_{BE} + I_E R_E$$

$$V_{TH} = I_B R_{TH} + V_{BE} + (I_B + I_C) R_E$$

$$V_{TH} - V_{BE} = I_B R_{TH} + I_B R_E + I_C R_E$$

$$(V_{TH} - V_{BE}) - I_C R_E = I_B(R_{TH} + R_E)$$

$$\frac{V_{TH} - V_{BE}}{R_{TH} + R_E} - \frac{I_C R_E}{R_{TH} + R_E} = I_B \qquad (7.19)$$

Differentiate I_B with respect to I_C
Differentiation of first term is zero, since it is independent of I_C.

$$\frac{dI_B}{dI_C} = \frac{-R_E}{R_{TH} + R_E}$$

We know that stability factor

$$S = \frac{1 + \beta}{1 - \beta \frac{dI_B}{dI_C}}$$

Substitute $\frac{dI_B}{dI_C}$ here

$$S = \frac{1 + \beta}{1 + \beta \frac{R_E}{R_{TH} + R_E}} \qquad (7.20)$$

The stability factor gets reduced if R_{TH} is much less than R_E. R_{TH} is parallel combination of R_1 and R_2. R_{TH} gets reduced since R_2 is much less than R_1.

7.6 Bias Compensation

7.6.1 Compensation for I_{CO}

The circuit is shown in Fig. 7.5. By applying KCL in Fig. 7.5.

$$I_1 = I_B + I_r$$

$$I_B = I_1 - I_r \qquad (7.21)$$

when the temperature increases, the leakage current of the transistor increases and the collector current increases. To provide the compensation for the silicon transistor, a silicon diode is connected between the base and ground in reverse biased condition.

When temperature increases, the reverse current I_r through the diode also increases. From equation (7.21), when I_r increases, the base current I_B decreases and hence the collector current tends to decrease. Thus the thermal runaway is eliminated.

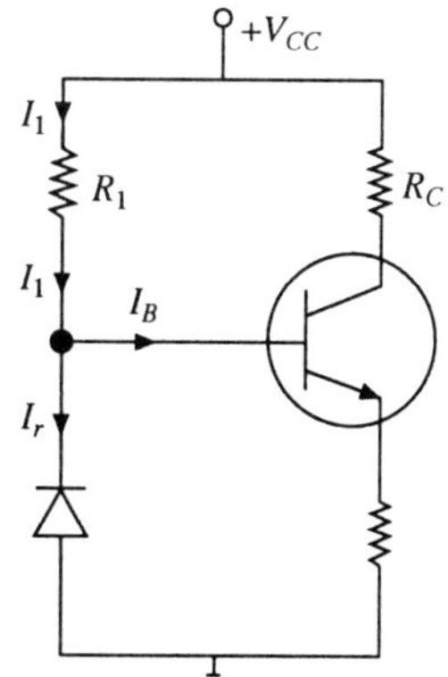

Fig. 7.5 Compensation for I_{CO}.

7.6.2 Thermistor Compensation

Thermistor compensation circuit is shown in Fig. 7.6. A thermistor is connected

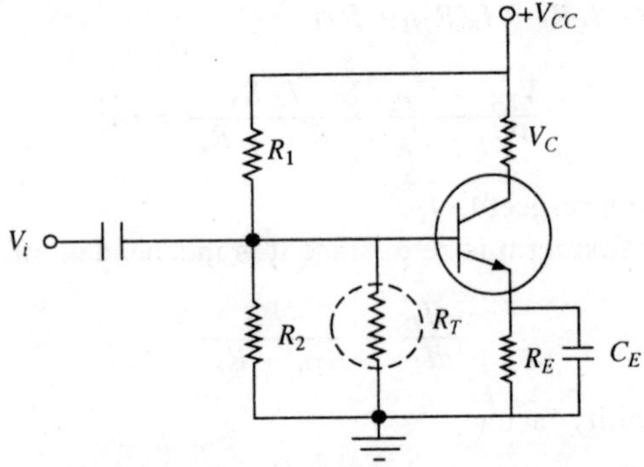

Fig. 7.6 Thermistor compensation.

in parallel with R_2 as shown in the above figure. The thermistor has negative temperature coefficient i.e. its resistance decreases with increase in temperature. Whenever temperature increases, the resistance R_T decreases and voltage applied between base and emitter decreases. Therefore base current and collector current decreases. Thus the rising tendency of collector current is eliminated by using a thermistor.

7.6.3 Compensation for V_{BE}

A reverse biased diode is connected in the emitter circuit as shown in Fig. 7.7. Transistor and diode must be made of similar material so that V_{BE} of transistor and V_f of diode will have same value. For silicon devices, the value is around 0.7V and for germanium devices, the value is around 0.3V. From the circuit, it can be seen that V_{BE} and V_f are in series opposition. Whenever there is a change in temperature, V_{BE} and V_f change equally. The change in V_{BE} due to the change in temperature is compensated by the change in V_f due to same change in temperature of the diode.

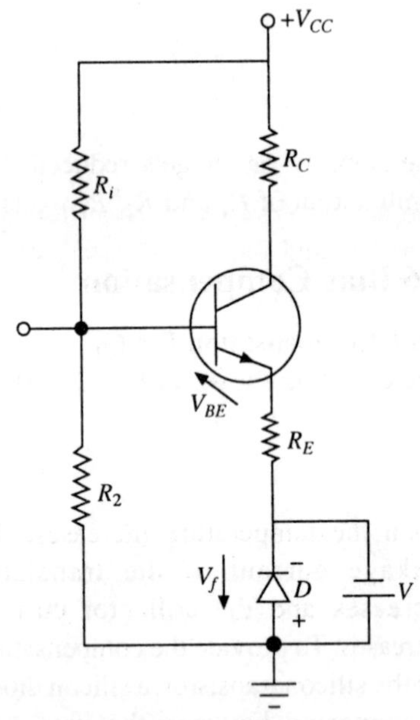

Fig. 7.7 Compensation for V_{BE}.

7.7 Comparison of Biasing Methods

Sl. No	Fixed bias	Collector to bias	Emitter bias	Voltage divider bias
Thermal runaway	Yes	No	No	No
Stability factor	Very high	High	High	< 5
S_r	$1+\beta$	$\dfrac{1+\beta}{1+\beta\dfrac{R_C}{R_B+R_C}}$	$\dfrac{1+\beta}{1+\beta\dfrac{R_E}{R_E+R_C}}$	$\dfrac{1+\beta}{1+\beta\dfrac{R_E}{R_{TH}+R_E}}$
Dis-advantage	(i) Rising tendency of temperature (ii) Stability factor is very high	– ve feedback reduces gain	requires large D.C. power supply	Nil

7.8 Heat Sink

Heat sink is a large metal sheet attached to the body of the transistor to dissipate the heat effectively. Without the heat sink, thc thermal resistance between the junction and the ambient will be high. By providing a heat sink the thermal resistance decreases. Thermal resistance (θ) is the resistance offered to the flow of heat. The unit of θ is °C/W

$$P = \frac{T_j - T_a}{\theta} \tag{7.21}$$

T_j = Junction temperature
T_a = Ambient temperature
θ = Thermal resistance
P = Power disipation in watts

If ambient temperature, thermal resistance and power loss are given, maximum permissible junction temperature can be obtained using the equation (7.21).

Worked Problems

Ex. 7.1 Find the Q point of a silicon transistor when $V_{CC} = 8\ V$, $R_C = 2$ k, $R_B = 500\ k$, $\beta = 100$. Determine the stability factor of fixed bias circuit shown in Fig. 7.8.

$$I_B = \frac{V_{CC} - V_{BE}}{R_B} = \frac{8 - 0.7}{500 \times 10^3} = 14.6\ \mu\text{A}$$

$$I_C = \beta I_B = 100 \times 14.6\ \mu A = 1460\ \mu A = 1.46\ \text{mA}$$

$$V_{CE} = V_{CC} - I_C R_C = 8 - 1.46 \times 2 = 8 - 2.92 = 5.08\ \text{V}$$

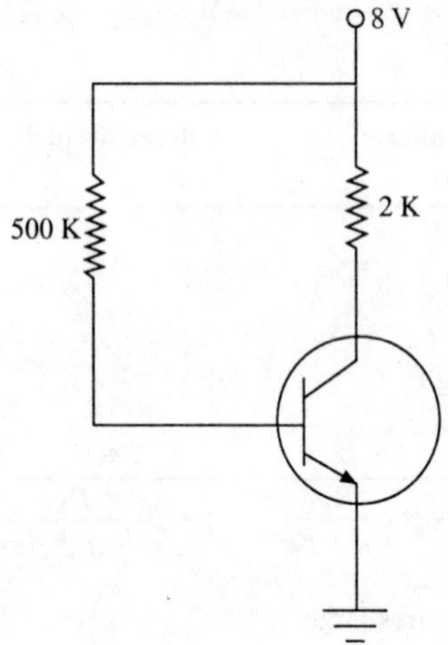

Fig. 7.8 Fixed bias circuit.

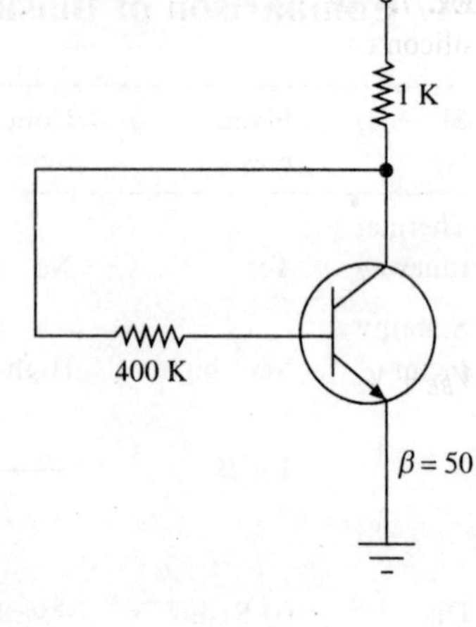

Fig. 7.9 Fixed bias circuit

Stability factor $S_f = 1 + \beta = 1 + 100 = 101$

Ex. 7.2 Determine the collector current for the given circuit shown in Fig. 7.9. Also determine the stability factor. Given V_{CE} = 10V. Assume silicon transistor.

$$I_B = \frac{V_{CE} - V_{BE}}{R_B + R_C}$$

$$= \frac{10 - 0.7}{(400 + 1) \times 10^3} = 0.0231 \text{ mA} = 23.19 \text{ μA}$$

$$I_C = \beta I_B = 50 \times 20.6 = 1030 \, \mu\text{A} = 1.03 \text{ mA}$$

Stability factor $S_f = \dfrac{1 + \beta}{1 + \beta \dfrac{R_C}{R_C + R_B}} = \dfrac{1 + 50}{1 + 50 \dfrac{1 \times 10^3}{(400 + 1) \times 10^3}} = 45.35$

Ex. 7.3 Determine various currents in the following circuit. Also determine the stability factor if the β value of the silicon transistor is 50.

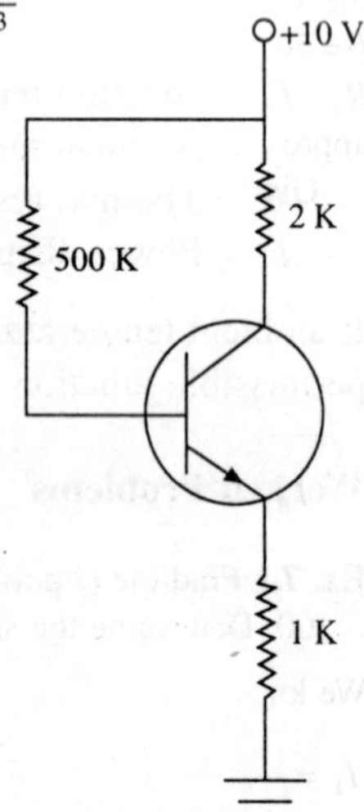

Fig. 7.10 Emitter bias circuit.

$$I_B = \frac{V_{CC} - V_{BE}}{R_B + \beta R_E} = \frac{10 - 0.7}{(500 + 50 \times 1)10^3} = 16.9 \, \mu\text{A}$$

$$I_C = \beta I_B = 50 \times 16.9 \times 10^{-6} = 845 \text{ μA}$$

$$S_f = \frac{1 + \beta}{1 + \beta \dfrac{R_E}{R_B + R_E}} = \frac{1 + 50}{1 + 50 \times \dfrac{10^3}{(500 + 1) \times 10^3}} = 46.37$$

$$V_{CE} = V_{CC} - I_C R_C - I_E R_E$$

$$= 10 - 845 \times 10^{-6} \times 2 \times 10^3 - 0.8619 \times 10^{-3} \times 1 \times 10^3$$

$$= 7.4481 \text{ V}$$

Ex. 7.4 Determine the operating point of the circuit shown below. The circuit uses a silicon transistor with $\beta = 100$. Use approximate method.

$$I = \frac{V_{CC}}{R_1 + R_2} = \frac{10}{(10+5)10^3}$$

$$= \frac{10}{15000} = 0.66 \text{ mA}$$

$$V_2 = I\,R_2 = 0.666 \times 10^{-3} \times 5 \times 10^3 = 3.33$$

$$V_{BE} + V_E = V_2$$

$$V_E = V_2 - V_{BE} = 3.33 - 0.7 = 2.63 \text{ V}$$

$$V_E = I_E R_E$$

$$I_E = \frac{V_E}{R_E} = \frac{2.63}{05 \times 10^{-3}} = 5.26 \text{ mA}$$

$$I_B = \frac{I_E}{1+\beta} = \frac{5.26 \times 10^{-3}}{1 + 100} = 52.08 \ \mu\text{A}$$

$$I_C = \beta I_B = 100 \times 52.08 = 5208 \ \mu A = 5.208 \text{ mA}$$

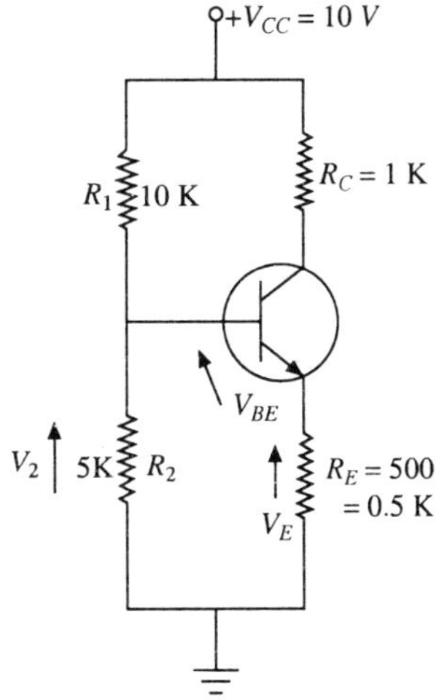

Fig. 7.11 Voltage divider bias circuit.

$$V_{CC} = I_C R_C + V_{CE} + I_E R_E$$

$$V_{CE} = V_{CC} - I_C R_C - I_E R_E$$

$$= 10 - 5.208 \times 10^{-3} \times 1 \times 10^{+3} - 5.26 \times 10^{-3} \times 0.5 \times 10^{+3}$$

$$= 2.162 \text{ V}$$

Ex. 7.5 The operating point is selected such that I_C = 5 mA, V_{CE} = 5 volts, If $R_C = 2\ k$, and V_{CC} = 20 volts, determine the values of R_1, R_2 and R_E. Given that $\beta = 50$, $V_{BE} = 0.7\ V$ and the current through R_1 is 10 times base current. Use approximate method.

Given

$$I_C = 5 \text{ mA} \quad R_C = 2 \text{ k}\Omega$$

$$V_{CC} = 20 \text{ V} \quad \beta = 50$$

$$V_{BE} = 0.7 \quad I_1 = 10\, I_B$$

$$I_B = \frac{I_C}{\beta} = \frac{5 \times 10^{-3}}{50} = 0.1 \text{ mA}$$

$$I_E = I_C + I_B = (5 + 0.1) \times 10^{-3}$$
$$= 5.1 \text{ mA}$$

We know

$$I_1 = \frac{V_{CC}}{R_1 + R_2};\quad 1 \times 10^{-3} = \frac{20}{R_1 + R_2}$$

$$R_1 + R_2 = \frac{20}{10^{-3}} = 2 \times 10^4$$

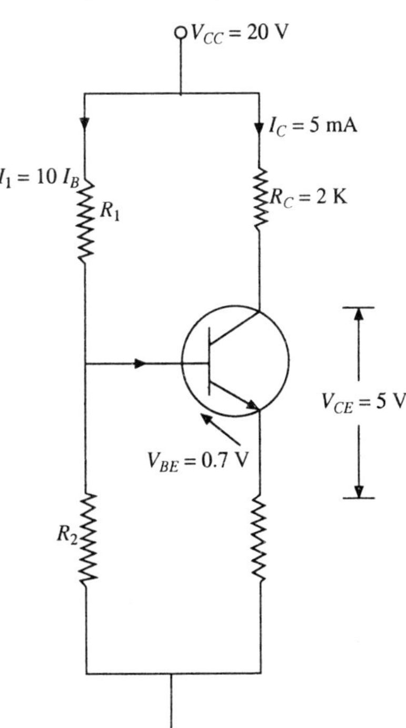

Fig. 7.12 Voltage divider bias circuit.

$$R_1 + R_2 = 20\ k$$

Apply KVL in the output

$$V_{CC} = I_C R_C + V_{CE} + V_E$$

$$20 = 5 \times 10^{-3} \times 2 \times 10^3 + 5 + V_E$$

$$10 + 10 = 10 + 5 + V_E$$

$$V_E = +5$$

$$R_E = \frac{V_E}{I_E} = \frac{5}{5.1 \times 10^{-3}} = 0.98\ \text{k}\Omega$$

Applying KVL in the base emitter loop

$$V_2 - V_{BE} - V_E = 0$$

$$V_2 = V_E + V_{BE} = 5 + 0.7 = 5.7$$

But

$$I_1 \simeq I_2$$

$$V_2 = I_2 R_2 = 5.7 = 1 \times 10^{-3}\ R_2$$

$$R_2 = 5.7\ \text{k}\Omega$$

But

$$R_1 + R_2 = 20\ \text{k}$$

$$R_1 = 20 - 5.7 = 14.3\ \text{k}\Omega$$

Ex. 7.6 A silicon transistor with $\beta = 50$, has $V_{BE} = 0.7V$ and $V_{CC} = 22.5$ V and $R_C = 5.6$ k. It is designed to establish Q-point at $V_{CE} = 12V$ and $I_C = 1.5$ mA and stability factor = 3. Find the values of R_1, R_2 and R_E for voltage divider biasing network. Use exact analysis.

$$\beta = 50$$

$$V_{BE} = 0.7\ \text{V}$$

$$V_{CC} = 22.5\ \text{V}$$

$$R_C = 5.6\ \text{K}$$

$$V_{CE} = 12\ \text{V}$$

$$I_C = 1.5\ \text{mA}$$

$$S_F = 3$$

Step 1: Calculate R_E from the KVL in output loop of Fig. 7.4(b).

Apply KVL in output loop.

$$V_{CC} = I_C R_C + V_{CE} + I_E R_E$$

$$I_E = (1 + \beta) I_B$$

$$I_C = 1.5\ \text{mA}: \quad \beta = 50$$

Therefore,

$$I_B = \frac{1.5 \times 10^{-3}}{50} = 0.03\ \text{mA}$$

$$I_E = I_C + I_B = (1.5 + 0.03)\ \text{mA} = 1.53\ \text{mA}$$

$$V_{CC} = 22.5\ \text{V}$$

Therefore, $22.5 = 1.5 \times 10^{-3} \times 5.6 \times 10^3 + 12 + 1.53 \times 10^{-3} R_E$

$$R_E = 1.37 \text{ k}\Omega$$

Step 2: Calculate R_{TH} from the expression for stability factor.

$$S_f = \frac{1+\beta}{1+\beta \dfrac{R_E}{R_E + R_{TH}}}$$

$$3 = \frac{1+50}{1+50 \dfrac{1.37 \times 10^3}{R_{TH} + 1.37 \times 10^3}}$$

$$3 = \frac{51(R_{TH} + 1.37 + 10^3)}{1.37 \times 10^3 + R_{TH} + 685 \times 10^3}$$

$$R_{TH} = 291 \text{ K}$$

Step 3: Solve for R_1 and R_2 from V_{TH} and R_{TH} expressions.
Apply KVL in input loop to Fig. 7.4(b).

$$V_{TH} = I_B R_{TH} + V_{BE} + I_E R_E$$

$$= 0.03 \times 10^{-3} \times 2.91 \times 10^3 + 0.7 + 1.53 \times 10^{-3} \times 1.37 \times 10^3$$

$$V_{TH} = 2.88 \text{ V}$$

$$V_{TH} = \frac{V_{CC}}{R_1 + R_2} = 2.88$$

$$\frac{R_2}{R_1 + R_2} = \frac{2.88}{V_{CC}} = \frac{2.88}{22.5}$$

But $$R_{TH} = \frac{R_1 R_2}{R_1 + R_2} = R_1 \frac{R_2}{(R_1 + R_2)}$$

$$R_{TH} = R_1 \frac{R_2}{R_1 + R_2}$$

$$2.91 \times 10^3 = R_1 \frac{2.88}{22.5}; \; R_1 = 22.73 \text{ k}\Omega$$

$$\frac{R_1 + R_2}{R_2} = \frac{22.5}{2.88}$$

$$\frac{R_1}{R_2} = \frac{22.5}{2.88} - 1$$

$$\frac{R_1}{R_2} = 6.8125$$

$$\frac{22.73 \text{ K}}{6.8125} = R_2$$

$$R_2 = 3.33 \text{ k}\Omega$$

Note: Follow the three steps in the sequence given above.

Ex. 7.7 A power transistor has a thermal resistance of 0.8°C/w and dissipates 4w. If maximum junction temperature is 90°C, what is the maximum permissible ambient temperature?

$$R_T = \frac{\theta}{P}$$

$$\theta = R_T \times P = 0.8 \times 4 = 3.2\ ^\circ\text{C}$$

$$\theta = \theta_j - \theta_A = 3.2$$

$$\theta_A = \theta_j - 3.2 = 90 - 3.2$$

Maximum ambient temperature = 86.8 °C

Ex. 7.8 Figure shows bipolar junction transistor amplifier using silicon transistor with V_{CC} = 20 V, h_{fe} = 400 and V_{BE} = 0.65 V. Bipolar junction transistor should be biased at V_{CE} = 10 V, I_C = 0.6 mA. Find the values of R_E, R_1 and R_2 such that it meets the following specification. I_{CO} at 25 °C = 5nA. I_{CO} at 145 °C = 3 μA. Assume change in collector current to be 5%.

$$h_{fe} = \beta = 400$$

$$S_f = \frac{dI_C}{dI_{CO}} = \frac{5\%\ \text{of}\ 0.6\ \text{mA}}{3 \times 10^{-6} - 5 \times 10^{-9}}$$

$$S_f = 10$$

$$R_C = 2\ k\Omega$$

$$I_E = I_B + I_C = \frac{I_C}{\beta} + I_C = \frac{0.6}{400} + 0.6$$

$$I_E = 0.6015\ \text{mA}$$

Step 1: KVL in output loop

$$V_{CC} = I_C R_C + V_{CE} + I_E R_E$$

$$20 = 0.6 \times 2 \times 10^3 + 10 + 0.6015 \times R_E$$

$$R_E = 14.6\ k\Omega$$

Step 2:

$$S_f = \frac{1 + \beta}{1 + \beta \dfrac{R_E}{R_E + R_{TH}}}$$

$$10 = \frac{1 + 400}{1 + 400 \dfrac{14.6}{14.6 + R_{TH}}}$$

$$R_{TH} = 137.4\ k\Omega$$

$$\frac{R_1 R_2}{R_1 + R_2} = 137.4\ k\Omega \qquad (7.22)$$

Step 3: KVL in input loop

$$V_{TH} = I_B R_{TH} + V_{BE} + I_E R_E$$

$$V_{TH} = 9.65.$$

Again $$V_{TH} = V_{CC} \cdot \frac{R_2}{R_1 + R_2} = 9.65$$

$$9.65 = 20 \frac{R_2}{R_1 + R_2}$$

$$\frac{R_2}{R_1 + R_2} = 0.4825 \qquad (7.23)$$

From (7.22) and (7.23) we get

$$R_1 = 279 \text{ k}\Omega$$

$$R_2 = 299.2 \text{ k}\Omega$$

Ex. 7.9 A transistor having $\alpha = 0.99$ and $V_{BE} = 0.7$ V is used in the circuit shown. What is the value of collector current?

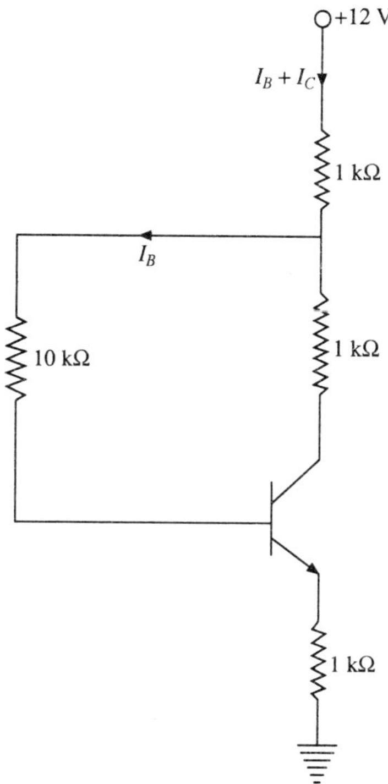

$V_{BE} = 0.7$ V; $\alpha = 0.99$

$$\beta = \frac{\alpha}{1 - \alpha}$$

$$\beta = \frac{0.99}{0.01} = 99$$

Apply KVL in input loop

$$V_{CC} = (I_B + I_C)\, R_C + I_B R_B + V_{BE} + I_E R_E$$

$$V_{CC} = \left(\frac{I_C}{\beta} + I_C\right) 1 + \frac{I_C}{\beta} \times 10 + V_{BE} + \frac{(\beta)}{1 + \beta} I_C R_E$$

$$V_{CC} = \left(\frac{I_C}{99} + I_C\right) + \frac{I_C}{99} 10 + 0.7 + \left(\frac{99}{1+99}\right) I_C$$

$$I_C = 5.33 \text{ mA.}$$

Short Questions and Answers

Q. 1. Define stability factors S, S' and S''.

Ans. $$S = \frac{\partial I_C}{\partial I_{CO}};\quad S' = \frac{\partial I_C}{\partial V_{BE}};\quad S'' = \frac{\partial I_C}{\partial \beta}$$

Q. 2. Why is fixed bias not used in practice?
Ans. It has high stability factor.

Q. 3. What a is the advantage of voltage divider bias?
Ans. It has a very low stability factor.

Q. 4. What is the disadvantage of collector to base bias?
Ans. It introduces negative feedback.

Q. 5. What is the disadvantage of emitter bias?
Ans. It requires a battery of large voltage rating.

Q. 6. Which is the best method of biasing?
Ans. Voltage divider bias.

Q. 7. What is meant by bias stability. What factors affect *BJT* biasing?
Ans. Aim of bias stability is to keep the operating point stable. Change in the values of temperature, beta and V_{BE} affect *BJT* biasing.

Q. 8. In the circuit shown in figure, the transistor is in Q-active mode. The approximate value of V_C is . . . . . .

Ans.

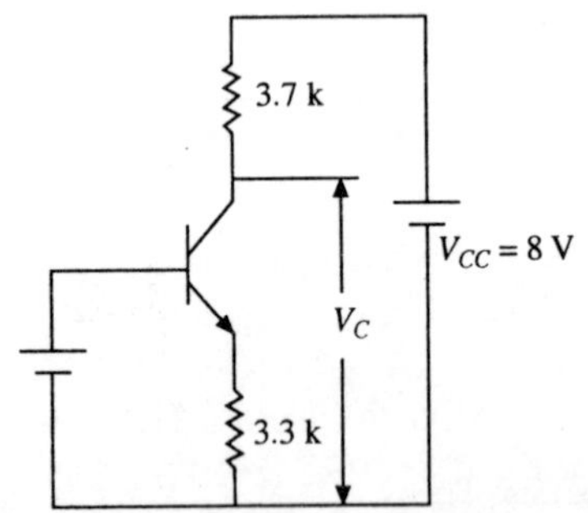

$$I_C = \frac{8-1}{3.7+3.3} = 1 \text{ mA}$$

$$V_C = 8 - 1 \times 3.7 = 8 - 3.7$$

$$\boxed{V_C = 4.3 \text{ V}}$$

Q. 9. If $\alpha = 0.98$, $I_{CO} = 6\ \mu$A and $I_B = 100\ \mu$A for a transistor, then the value of I_C will be . . . . . .

Ans. $$\beta = \frac{1}{0.02} = 50$$

$$I_{CEO} = (1 + \beta)\, I_{CO} = 51 \times 6 = 306\ \mu A$$

$$I_C = \beta I_B + I_{CEO} = 50 \times 100 + 306$$

$$= 5306\ \mu A$$

$$\boxed{I_C = 5.306\ mA}$$

Q. 10. Find the value of V_{CE} for the figure shown below?

Ans.

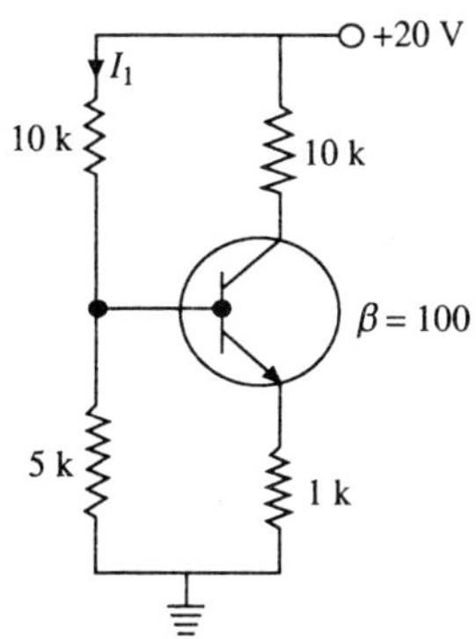

$$I_1 = \frac{20}{15} = 1.33\ mA$$

$$I_B = \frac{I_1}{100} = 0.0133\ mA$$

$$I_C = \beta\, I_B$$

$$I_C = 1.33\ mA$$

$$V_{CE} = V_{CC} - I_C(R_C + R_E) = 20 - 1.33\ (10 + 1) = 5.37\ V$$

Q. 11. Given $V_{BE} = 0.7$, $\beta = 50$ for the circuit shown below. For $V_{CE} = 2V$, the value of R_B is . . . . . .

Ans.

$$I_C = \frac{V_{CC} - V_{CE}}{R_C} = \frac{12 - 2}{5} = \frac{10}{5} = 2\ mA$$

$$I_B = \frac{I_C}{\beta} = \frac{2}{50} = 0.04\ mA$$

$$R_B = \frac{V_{CC} - V_{BE}}{I_B} = \frac{12 - 0.7}{0.04} = 282.5\ \Omega$$

8

Amplifiers

8.1 Introduction

An amplifier converts a weak signal into a strong signal. Block diagram of a public address (PA) system is shown in Fig. 8.1(a). The most important block of a PA is amplifier. The microphone at the input of PA system converts sound energy into electrical energy. The amplifier amplifies weak signal into strong signal. The speaker at the output converts electrical energy into sound energy. Based on the connection, the amplifiers are classified as (i) common base (ii) common emitter and (iii) common collector amplifier. In common base amplifier, the base of transistor is directly connected to ground. There should not be any resistance between base and ground. In common emitter amplifier, emitter is directly connected to the ground. In common collector amplifier, collector is directly connected to the ground.

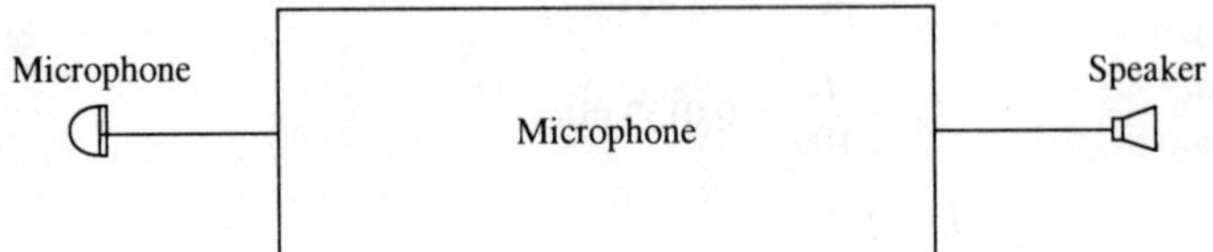

Fig. 8.1(a) Public address system.

8.2 Common Base Amplifier

Comon base amplifier circuit is shown in Fig. 8.1(b). The total current in the collector is the sum of A.C. current and the D.C. current. These currents can be obtained separately using A.C. and D.C. equivalent circuits. The total current can be obtained by using superposition theorem.

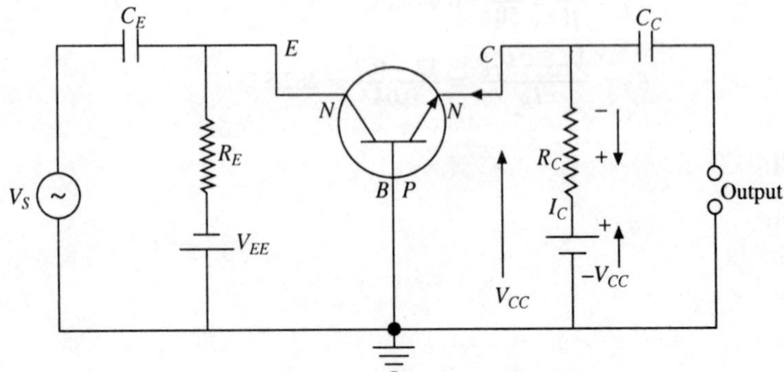

Fig. 8.1(b) Common base amplifier.

The battery V_{EE} forward biases emitter junction and the battery V_{CC} reverse biases the collector junction. The coupling capacitors acts as short circuit for A.C. and open circuit for D.C. The input voltage applied to the amplifier is V_s. During the positive half cycle, the top terminal assumes positive and this positive voltage is applied to the emitter. Emitter is already forward biased by the battery V_{EE}. Because of the positive voltage at the *N*-terminal, the forward bias decreases and the emitter current decreases. The collector current is also decreased. Thus the output voltage increases when the input voltage increases. The output voltage and the input voltage are in phase. The disadvantage of this amplifier is that it has very less input impedance and very high output impedance. Analysis of this amplifier is given in section 8.7.

8.3 Common Emitter Amplifier

Common emitter amplifier circuit is shown in Fig. 8.2. It can be seen that emitter is directly connected to the ground. During the positive half cycle, the top terminal assumes positive polarity. This positive voltage is applied to the *P*-terminal of the transistor. The transistor emitter junction is more forward biased. The base current increases, and collector current also increases. When collector current increases, the output voltage V_{CE} decreases. Thus a negative half cycle is produced at the output. The voltage at output and input have a phase difference of 180°. The current gain of the circuit is very high when compared to the current gain of common base amplifier. This is because the collector current is beta times base current. The input impedance is about 1 KΩ and the output impdedance is about 40 KΩ. The power gain is high since the current gain is high.

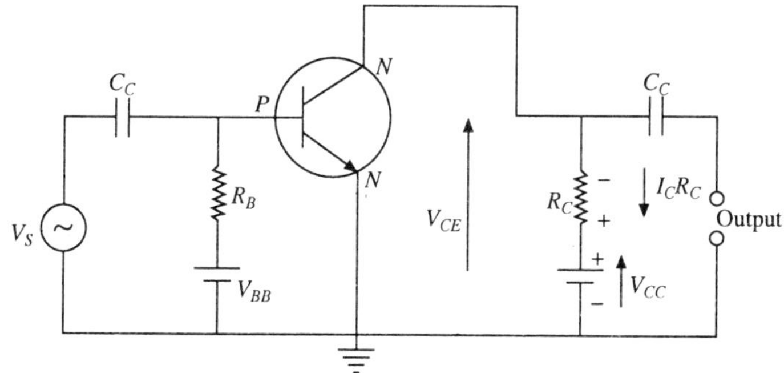

Fig. 8.2 Common emitter amplifier.

8.4 Classification of Amplifiers

Operating point of an amplifier is also called *Q*-point. Based on the location of *Q*-point, the amplifiers are classified as class *A*, class *B*, class *AB* and class *C*. The waveforms are shown in Fig. 8.3.

Class A: The operating point is selected at the middle of the load line. The input signal produces a fluctuating current. The peak value of this current has to be less than the saturation current. The device conducts throughout the cycle i.e. 360°.

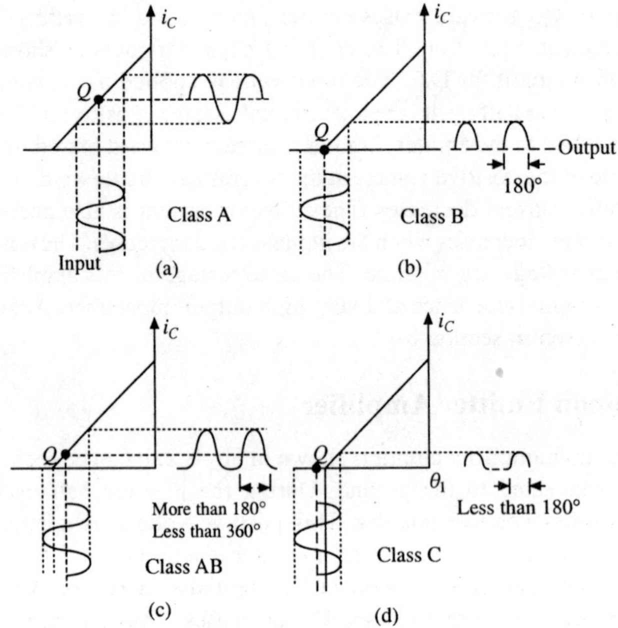

Fig. 8.3 Classification of amplifier.

Class B: The operating point is selected at the cut-off point. For half a cycle, the device conducts and for the other half cycle, the device does not conduct since the device is in the cut-off region. The output waveform is similar to the half wave rectified output.

Class AB: The operating point is selected slightly before the cut-off point. The device is biased such that it conducts for more than 180° and less than 360° during one cycle. The conduction period is in between the periods of Class *A* and Class *B* (and hence it is called Class *AB*).

Class C: The operating point is selected beyond the cut-off point. The transistor conducts for a period less than 180° during one cycle. The output waveform has a train of positive pulses.

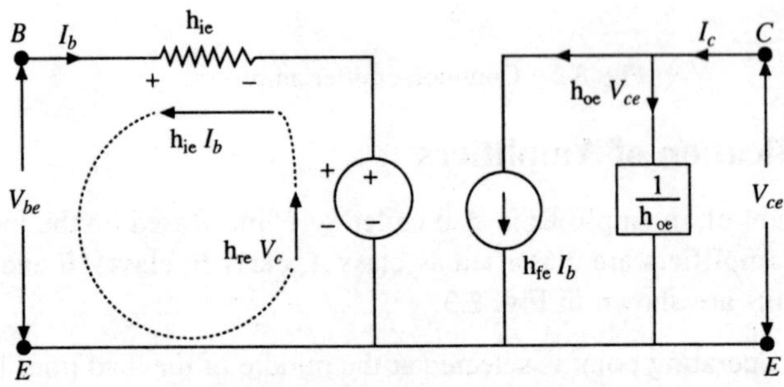

Fig. 8.4 Linear model of BJT.

8.5 Linear Model of BJT

The low frequency model of BJT is required for analysing amplifiers. Hybrid parameters are defined as follows:

Let $h_{11} = h_i$ (i stands for input)
$h_{21} = h_f$ (f stands for forward)
$h_{22} = h_o$ (o stands for output)
$h_{12} = h_r$ (r stands for reverse)

Let
$$\begin{pmatrix} h_{11} & h_{12} \\ h_{21} & h_{22} \end{pmatrix} = \begin{pmatrix} h_i & h_r \\ h_f & h_o \end{pmatrix}$$

suffixes are positioned as follows:

$$\begin{array}{|cc|} i & r \uparrow \\ f & o \\ \hline \end{array}$$

When you read the suffix in the matrix in anti clockwise direction, you get *if or*.

Let $V_1 = V_b \qquad V_2 = V_c$

Replacing, $V_1 = V_b$, $V_2 = V_c$, $I_1 = I_b$ and $I_2 = I_c$

$$V_b = f_1(i_b, v_c)$$

$$i_c = f_2(i_b, v_c)$$

$$\begin{pmatrix} V_b \\ I_c \end{pmatrix} = \begin{pmatrix} h_i & h_r \\ h_f & h_o \end{pmatrix} \begin{pmatrix} I_b \\ V_c \end{pmatrix}$$

$$V_b = h_i I_b + h_r V_c \qquad (8.1)$$

$$I_c = h_f I_b + h_o V_c \qquad (8.2)$$

$$h_i = \frac{V_b}{I_b} \text{ when } V_c = 0$$

$$h_f = \frac{I_c}{I_b} \text{ when } V_c = 0$$

$$h_r = \frac{V_b}{V_c} \text{ when } I_b = 0$$

$$h_o = \frac{I_c}{V_c} \text{ when } I_b = 0.$$

All the parameters are not impedances or admittances. All parameters are mixed.
Therefore they are called hybrid parameters.

By synthesizing equations, we get exact model for BJT. This is shown in Fig. 8.4. By applying KVL in the input loop, we get (8.1). By applying KCL at *P* we get (8.2).

h_i—input impedance
h_f—forward current gain
h_o—output admittance
h_r—reverse voltage ratio

8.6 Approximate Analysis of CE Amplifier

Approximate analysis uses approximate equivalent circuit of BJT. It is obtained by neglecting h_{oe} and h_{re} since they are of the order of 10^{-4}. In common emitter circuit, the emitter terminal is directly connected to the A.C. ground. i.e., there is no resistance between the emitter and ground in the A.C. equivalent circuit. Equivalent circuit of CE amplifier is shown in Fig. 8.5.

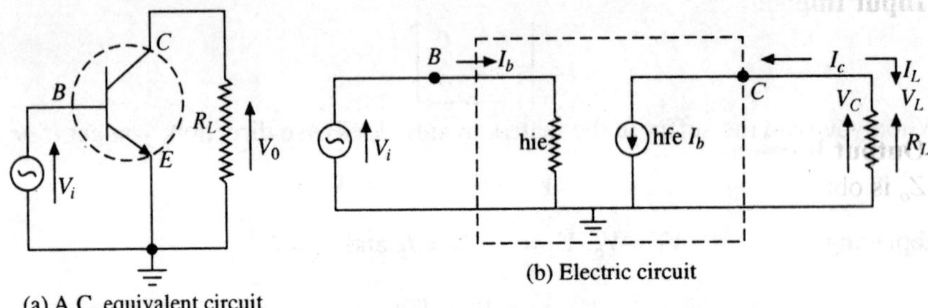

Fig. 8.5 Equivalent circuit of CE amplifier.

The value of $h_{re} = 10^{-4}$. Therefore $h_{re}V_{ce}$ is negligible $1/h_{oe} = 100$ kΩ. This can be treated as open circuit. By short circuiting the source $h_{re}\,V_{ce}$ and open circuiting the branch $1/h_{oe}$ we get approximate model for BJT. This is shown in Fig. 8.5(b).

Current Gain : In Fig. 8.5(b), BJT is replaced by its approximate equivalent circuit.

$$\text{Current gain} = \frac{\text{Load current}}{\text{Input current}}$$

$$A_I = \frac{I_L}{I_b}$$

$$A_I = \frac{-I_c}{I_b} = \frac{-h_{fe}I_b}{I_b}$$

$$\boxed{A_I = -h_{fe}} \tag{8.3}$$

Voltage Gain :

$$A_v = \frac{V_L}{V_i}$$

$$A_v = \frac{I_L R_L}{V_i} = \frac{-I_c R_L}{V_i}$$

substitute $V_i = h_{ie} I_B$

$$= \frac{-I_c R_L}{h_{ie} I_B} = \frac{-h_{fe} I_B R_L}{h_{ie} I_B}$$

$$\boxed{A_v = \frac{-h_{fe} R_L}{h_{ie}}} \tag{8.4}$$

$$= \left| \frac{h_{fe} R_L}{h_{ie}} \right| \angle 180^\circ$$

Here the negative sign indicates about the action of CE amplifier, i.e. phase inversion of 180°.

Input Impedance: $$Z_i = \frac{V_L}{I_b} = \frac{h_{ie} I_b}{I_b}$$

$$\boxed{Z_i = h_{ie}} \tag{8.5}$$

Output Impedance:
Z_o is obtained by shorting the voltage sources and opening the current sources.

$$Z_o = \left. \frac{V_L}{I_c} \right|_{v_i = 0}$$

Z_o is the impedance without considering R_L

$$Z_o = \propto$$

Z_o' is the output impedance considering R_L.

$$\frac{1}{Z_o'} = \frac{1}{\infty} + \frac{1}{R_L}$$

8.7 Approximate Analysis of Common Base Amplifier

In common base amplifier, base is directly connected to A.C. ground.

A.C. equivalent circuit of common base amplifier is shown in Fig. 8.6(a). In Fig. 8.6(b), the BJT is represented by its approximate equivalent circuit. applying KCL at P, we get

$$I_e + I_b + I_c = 0$$

$$I_e = -(I_b + I_c) = -(I_b + h_{fe} I_b)$$

$$I_e = -(1 + h_{fe}) I_b \tag{8.6}$$

$$A_I = \frac{I_L}{I_e} = \frac{-I_c}{I_e}$$

$$A_I = \frac{-h_{fe} I_b}{-(1 + h_{fe}) I_b} = \frac{h_{fe}}{1 + h_{fe}}$$

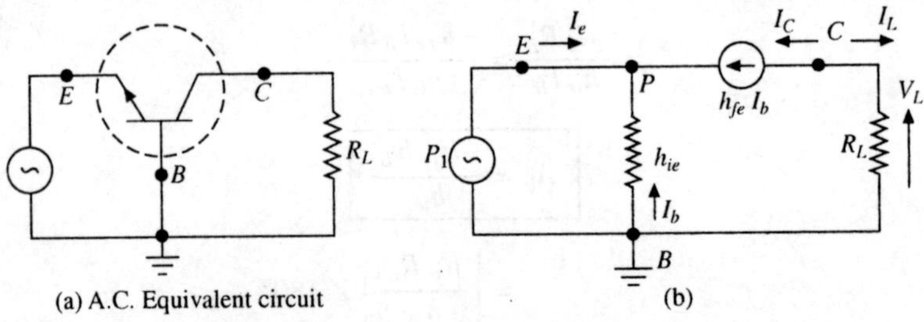

Fig. 8.6 Common base amplifier circuit.

$$\boxed{A_I = \frac{h_{fe}}{1 + h_{fe}}} \tag{8.7}$$

Voltage Gain

$$A_v = \frac{V_L}{V_i} = \frac{I_L R_L}{V_i}$$

By KVL in input loop, $V_i + h_{ie}I_b = 0$.

Thus $$V_i = -h_{ie}I_b$$

Therefore, $$A_v = -\frac{I_c R_L}{V_i} = \frac{-h_{fe} I_b R_L}{-h_{ie} I_b}$$

$$\boxed{A_v = \frac{h_{fe} R_L}{h_{ie}}} \tag{8.8}$$

Input impedance

$$Z_i = \frac{V_i}{I_e}$$

Hence $$Z_i = \frac{+h_{ie} I_b}{+(1 + h_{fe}) I_b}$$

$$\boxed{Z_i = \frac{h_{ie}}{1 + h_{fe}}} = \frac{1\ \text{k}\Omega}{1 + 100} = 10\ \Omega \tag{8.9}$$

This is very low value. Common base amplifier is not used in practice due to low input impedance.

8.8 Analysis of Common Collector Amplifier

A.C. equivalent circuit is shown in Fig. 8.7. Apply KCL at point P

$$I_b + I_c + I_e = 0 \tag{8.10}$$

$$I_e = -(I_b + I_c) = -(I_b + h_{fe} I_b)$$

$$I_e = -(1 + h_{fe})I_b \tag{8.11}$$

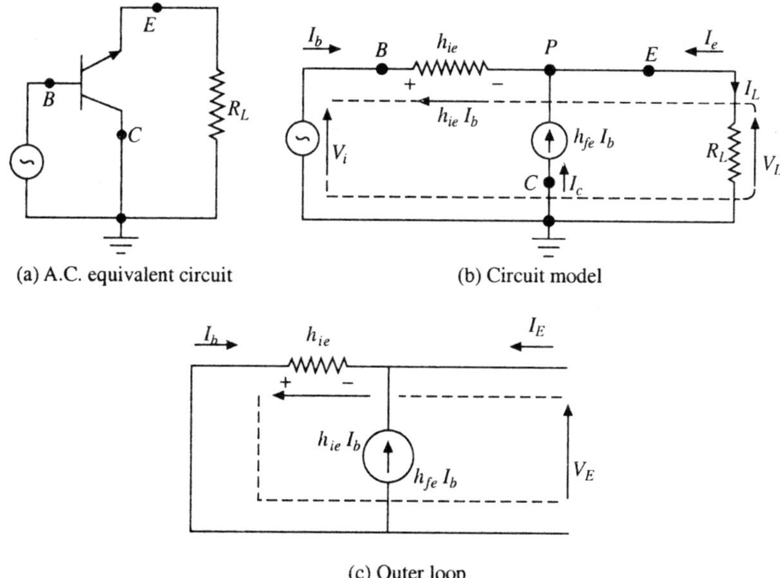

Fig. 8.7 Common collector amplifier circuit.

Apply KVL in the closed loop shown in Fig. 8.7.(b)

$$V_i - h_{ie}I_b - V_L = 0 \tag{8.12}$$

$$V_i = h_{ie}I_b + V_L = h_{ie}I_b + I_L R_L$$

$$V_i = h_{ie}I_b + (-I_e)\, R_L \tag{8.13}$$

Substituting (8.11) in (8.13)

$$V_i = h_{ie}I_b + (1 + h_{fe})I_b R_L$$

$$V_i = [h_{ie} + (1 + h_{fe})R_L]I_b$$

Input Impedance

$$Z_i = \frac{V_i}{I_b} = h_{ie} + (1 + h_{fe})R_L \tag{8.14}$$

$$h_{ie} = 1k$$

Hence $\quad Z_i = 1k + (1 + 100)\ 1k = 102\ k$

Highest input impedance can be obtained using this configuration. The input impedance can be increased by increasing the value of R_L.

Current Gain $\quad \dfrac{I_L}{I_b} = -\dfrac{I_e}{I_b} = \dfrac{I_b(1 + h_{fe})}{I_b}$

$$\boxed{A_I = 1 + h_{fe}} \tag{8.15}$$

Voltage Gain $$A_v = \frac{V_L}{V_i} = \frac{I_L R_L}{V_i} = -\frac{I_e R_L}{V_i} \tag{8.16}$$

$$= \frac{I_b(1 + h_{fe})R_L}{[h_{ie} + (1 + h_{fe})R_L]I_b}$$

Divide both numerator and denominator by $(1 + h_{fe})$

$$A_v = \frac{R_L}{\dfrac{h_{ie}}{1 + h_{fe}} + R_L}$$

if $$R_L >> \frac{h_{ie}}{1 + h_{fe}}$$

$$A_v = \frac{R_L}{R_L} = 1$$

$$\boxed{A_v \cong 1}$$

$$A_v = \frac{V_E}{V_i} = 1$$

$$V_E = V_i$$

Output voltage is equal to input voltage. Since output voltage follows the emitter voltage, this circuit is named as emitter follower. This is suitable as a buffer since the output voltage of this circuit is approximately equal to input voltage. A buffer should have high input impedance and low output impedance. Buffer is connected between processor and output device of a microprocessor or microcomputer system. Because of the buffer, current drawn from the processor is reduced. Thus the buffer prevents loading of processor.

Output Impedance
Apply KVL in the loop shown in Fig. 8.7c.

$$V_E + h_{ie} I_b = 0$$

$$V_E = -h_{ie} I_b$$

To find the Thevenin impedance of a circuit having dependant and independant sources, apply a voltage V_E, find the current I_E. Only independant sources must be reduced to zero.
Output impedance is the ratio of voltage applied to the current I_E when V_s is reduced to zero.

$$Z_o = \left.\frac{V_E}{I_E}\right|_{V_s=0} = \frac{-I_b h_{ie}}{-I_b(1 + h_{fe})}$$

$$Z_o = \frac{h_{ie}}{1 + h_{fe}} = \frac{1k}{1 + 100} = 10\ \Omega \tag{8.17}$$

The circuit has high input impedance and low output impedance, but voltage gain is 1.

Summary

	CB	*CE*	*CC*
A_i	less than 1	50 to 250	51 to 251
	$\frac{h_{fe}}{1+h_{fe}}$	$-h_{fe}$	$1 + h_{fe}$
A_v	50 to 250	50 to 250	$\simeq 1$
	$\frac{h_{fe} R_L}{h_{ie}}$	$-\frac{h_{fe} R_L}{h_{ie}}$	
Z_i	$\frac{h_{ie}}{1+h_{fe}}$	h_{ie}	$h_{ie} + (1 + h_{fe})R_L$
	less	Medium	High
Z_o	∞	∞	$\frac{h_{ie}}{1+h_{fe}}$

8.9 Exact Analysis of CE Amplifier

In exact analysis, h_{re} and h_{oe} are considered. The AC equivalent circuit of common emitter amplifier with exact equivalent circuit is shown in Fig. 8.8.

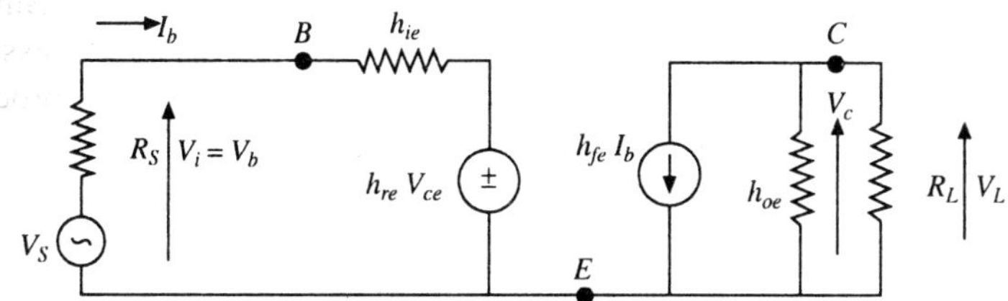

Fig. 8.8 CE amplifier circuit.

Current Gain (A_I)

Current gain is defined as load current by input current.

$$A_I = \frac{I_L}{I_b} = -\frac{-I_c}{I_b}$$

$$I_c = h_{fe} I_b + h_{oe} V_c$$

Divide throughout by I_b,

$$\frac{I_c}{I_b} = h_{fe} + h_{oe}\frac{V_c}{I_b} = h_{fe} + h_{oe}\frac{V_L}{I_b}$$

$$\frac{I_c}{I_b} = h_{fe} + h_{oe}\,\frac{I_L R_L}{I_b}$$

$$\frac{I_c}{I_b} = h_{fe} - h_{oe}\,\frac{I_C R_L}{I_b}$$

$$\frac{I_c}{I_b}(1 + h_{oe} R_L) = h_{fe}$$

$$\frac{I_c}{I_b} = \frac{h_{fe}}{1 + h_{oe} R_L}$$

Multiply with (–1) on either sides

$$-\frac{I_c}{I_b} = -\frac{h_{fe}}{1 + h_{oe} R_L}$$

Hence

$$\boxed{A_I = \frac{-h_{fe}}{1 + h_{oe} R_L}} \tag{8.18}$$

Input Impedance (Z_i)

$$Z_i = \frac{V_i}{I_b} = \frac{V_b}{I_b}; \quad V_b = h_{ie} I_b + h_{re} V_c$$

Divide throughout by I_b; $\dfrac{V_b}{I_b} = h_{ie} + \dfrac{h_{re} V_c}{I_b}$

$$Z_i = h_{ie} + h_{re}\,\frac{V_L}{I_b} = h_{ie} + h_{re}\,\frac{I_L}{I_b} R_L$$

Substitute $A_I = I_L/I_b$

$$\boxed{Z_i = h_{ie} + h_{re} \cdot A_I R_L} \tag{8.19}$$

$$Z_i = h_{ie} + \frac{h_{re} h_{fe}}{1 + h_{oe} R_L} R_L$$

Voltage Gain without Considering R_s

$$A_v = \frac{V_L}{V_i} = \frac{I_L R_L}{V_b}$$

Multiply and divide with

$$= \frac{-I_c R_L}{V_b} \times \frac{I_b}{I_b} = -\frac{I_c}{I_b} \times \frac{I_b}{V_b} \times R_L$$

Substitute $A_I = -I_C/I_b$ and $Z_i = V_b/I_b = \dfrac{A_I R_L}{Z_i}$

$$\boxed{A_v = \frac{A_I R_L}{Z_i}} \tag{8.20}$$

Output Admittance

$$Y_o = \left.\frac{I_c}{V_c}\right|_{v_s=0}$$

Applying KCL, $I_c = h_{fe}I_b + h_{oe}V_c$

Divide by V_c on both sides, $\frac{I_c}{V_c} = h_{fe}\frac{I_b}{V_c} + h_{oe}$ (8.21)

Applying KVL in input loop with $V_s = 0$,

$$h_{re}V_c + h_{ie}I_b + R_sI_b = 0$$

$$h_{re}V_c + (h_{ie} + R_s)\,I_b = 0$$

$$(h_{ie} + R_s)\,I_b = -\,h_{re}V_C$$

$$\frac{I_b}{V_c} = \frac{-h_{re}}{(h_{ie} + R_s)} \tag{8.22}$$

Substituting (8.22) in (8.21), $Y_o = \frac{I_c}{V_c} = \frac{-h_{fe}h_{re}}{(h_{ie} + R_s)} + h_{oe}$

$$Y_o = h_{oe} - \frac{h_{fe}h_{re}}{h_{ie} + R_s}$$

$$Z_o = \frac{1}{Y_o}$$

Voltage Gain Considering R_s (A_{vs})

$$A_{vs} = \frac{V_L}{V_s}$$

Multiply and divide with V_b

$$A_{vs} = \frac{V_L}{V_b} \times \frac{V_b}{V_s} \tag{8.23}$$

$$V_b = IZ_i = \frac{V_sZ_i}{(R_s + Z_i)} \tag{8.24}$$

Substituting (8.24) in (8.23),

$$A_{vs} = \frac{A_v}{V_s}\frac{V_s \cdot Z_i}{(Z_i + R_s)}$$

$$\boxed{A_{vs} = \frac{A_vZ_i}{Z_i + R_s}}$$

When $R_s = 0$, $A_{vs} = A_v$

Because of source resistance, the voltage gain is reduced.

Current Gain Considering $R_s(A_{Is})$

By current division rule

$$I_b = \frac{I_s R_s}{R_s + Z_i}$$

$$A_{IS} = I_L/I_s$$

Multiply and divide with I_b

$$A_{Is} = \frac{I_L}{I_S} \times \frac{I_b}{I_b}$$

On rearranging the terms

$$= \frac{I_L}{I_b} \times \frac{I_b}{I_s} = A_I \frac{I_b}{I_s} = A_I \frac{R_s}{R_s + z_i}$$

$$A_{IS} = A_I \frac{R_s}{R_s + z_i}$$

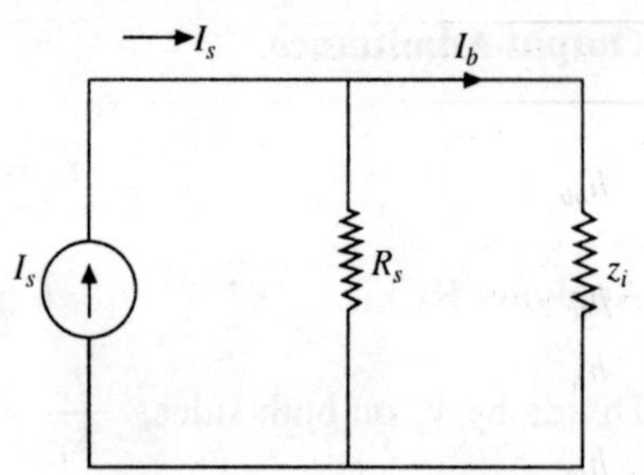

Fig. 8.9 Equivalent circuit.

8.9.1 Typical *h*-Parameters

The *h*-parameters for various configurations are shown below:

	CE	*CC*	*CB*
h_i	1100 ohm	1100 ohm	21 ohm
h_r	2.5×10^{-4}	$\simeq 1$	2.9×10^{-4}
h_f	50	–51	– 0.98
$1/h_o$	40 *k*	40 *k*	2 mega ohm

8.9.2 Conversion of *h*-Parameters

Conversion formulae are summarized here.

	CE	*CB*
h_{ie}		$h_{ib}/(1 + h_{fb})$
h_{re}		$\frac{h_{ib} h_{ob}}{1 + h_{fb}} - h_{rb}$
h_{fe}		$-h_{fb}/(1 + h_{fb})$
h_{ob}		$\frac{h_{ob}}{1 + h_{fb}}$
h_{ib}	$\frac{h_{ie}}{1 + h_{fe}}$	
h_{rb}	$\frac{h_{ie} h_{oe}}{1 + h_{fe}} - h_{re}$	
h_{fb}	$\frac{-h_{fe}}{1 + h_{fe}}$	

	CE	CB
h_{ob}	$\frac{h_{oe}}{1+h_{fe}}$	
h_{ic}		$\frac{h_{ib}}{1+h_{fb}}$
h_{rc}		1
h_{fc}		$\frac{1}{1+h_{fb}}$
h_{oc}		$\frac{h_{ob}}{1+h_{fb}}$

These formulae can be used to convert the h-parameters.

8.9.3 Small Signal Amplifier

In power amplifier, we use power transistor whereas in voltage amplifier, the transistor used is a signal transistor. R_1, R_2, R_c and R_E represent biasing elements.

C_{C1} is the coupling capacitor, it blocks $D.C.$
C_{C2} is also coupling capacitor.
R_E provides necessary voltage.
C_E is called emitter bypass capacitor. It bypasses $A.C.$ Without C_E the drop across R_E is increased if C_E is used, then AC current flows through C_E.

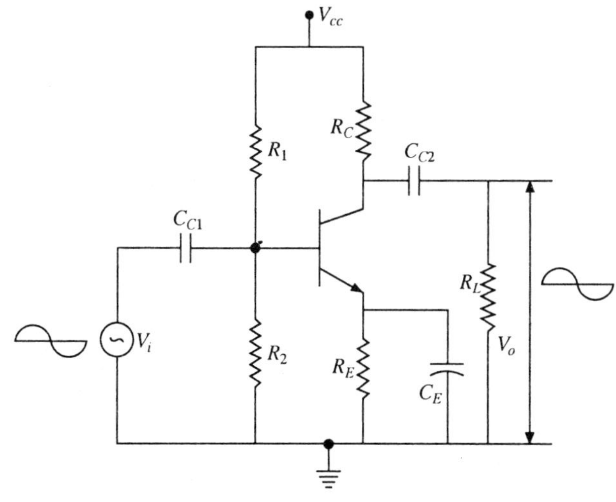

Fig. 8.9.1

8.9.4 Rules to Get D.C. Equivalent Circuit

(i) Short circuit $A.C.$ sources.
(ii) Consider capacitor as open since it blocks D.C.
(iii) Inductors are short circuited.

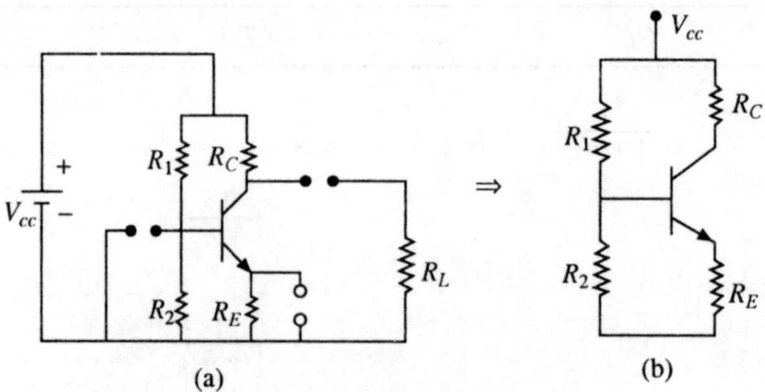

Fig. 8.9.2 D.C. equivalent circuit

Fig. 8.92(b) is nothing but voltage divider biasing network.

8.9.5 Rules to Get A.C. Equivalent Circuit

(i) Short circuit D.C. source.

(ii) Capacitor is considered as a short circuit.

A.C. Analysis

Fig. 8.9.3(b) is equivalent to Fig. 8.9.3(a) . Parallel combination of R_1 and R_2 is equal to R_b. Parallel combination of R_c and R_L is equal to R_{AC}. In Fig. 8.9.3(c), R_b is in parallel with an ideal voltage source V_i. The resistance R_b can be removed.

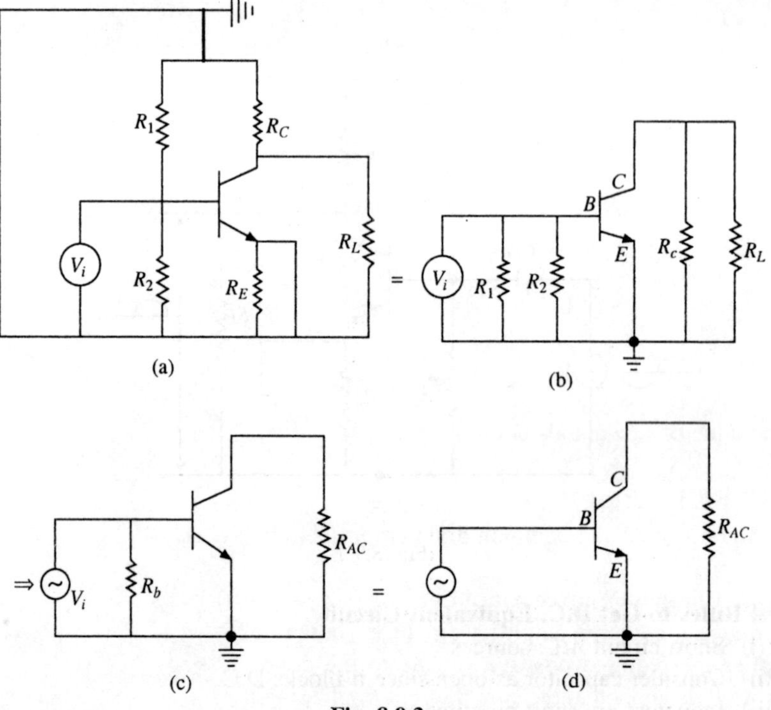

Fig. 8.9.3

where $$R_b = \frac{R_1 R_2}{R_1 + R_2}$$

and $$R_{AC} = \frac{R_C R_L}{R_C + R_L}$$

8.9.6 Design of Coupling Capacitor

For low frequency analysis C_c cannot be treated as short circuit.

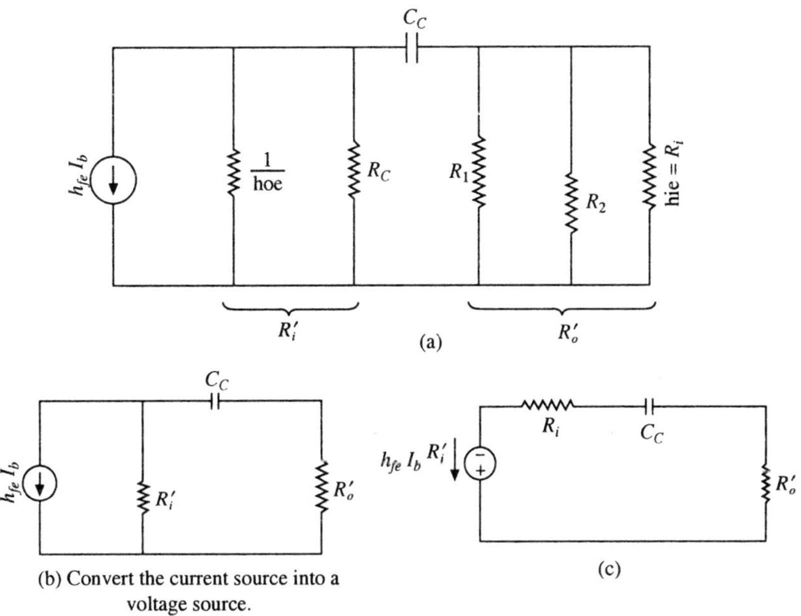

Fig. 8.9.4 Low frequency circuit

At half power frequency f_2, $X_c = R$; $\frac{1}{2\pi f_L C} = R; f_L = \frac{1}{2\pi RC}$

$$f_L = \frac{1}{2\pi (R_i' + R_o') C_c}$$

or $$C_c = \frac{1}{2\pi f_L (R_i' + R_o')} \quad (8.25)$$

The value of C_c defends upon f_L.

8.10 FET Amplifiers

Common source and common drain amplifiers are discussed here.

8.10.1 Common Source Amplifier

The common source amplifier circuit is shown in Fig. 8.10. The battery V_{DD} applies the required positive voltage to the drain. The negative voltage drop across R_s is applied at the gate through R_G. During the positive half cycle of the input voltage, the negative bias of the gate is reduced. The width of the depletion layer decreases and the width of the conducting channel of FET increases. The drain current increases and V_{DS} decreases. This indicates that the output voltage is increasing in the negative direction. Thus there is a phase shift of 180° between the input and output. The fluctuating voltage applied at the input causes the drain current to fluctuate. Large fluctuations in drain current produce higher output voltage. Thus the output voltage is much higher than the input voltage.

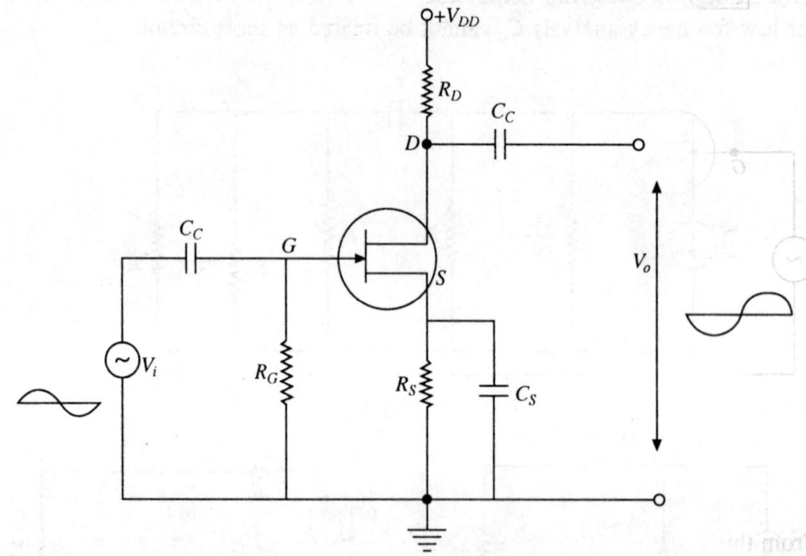

Fig. 8.10 Common source amplifier.

Analysis of Common Source Amplifier

For obtaining A.C. equivalent circuit of Fig. 8.10, the capacitors must be treated as short circuit. D.C. sources must be reduced to zero by short circuiting them. In Fig. 8.11(a), the D.C. source is represented as a short circuit. The capacitor C_s across R_s is shown as a short circuit. Parallel combination of R_s and short circuit is a short circuit. It is shown in Fig. 8.11(b). The high resistance R_G can be treated as open circuit. In Fig. 8.11(c), R_G is shown as open circuit.

Substitute (8.26) in (8.27)

$$V_o = \frac{-\mu V_{GS}}{r_D + R_D} R_D$$

$$\frac{V_o}{V_{GS}} = \frac{-\mu R_D}{r_D + R_D}$$

From the above circuit $V_{GS} = V_i$

$$\frac{V_o}{V_i} = -\frac{\mu R_D}{r_D + R_D}$$

$$\boxed{A_v = \frac{-\mu R_D}{r_D + R_D}} \quad (8.28)$$

$$\frac{V_o}{V_i} = A_v = |A_v| \angle 180°$$

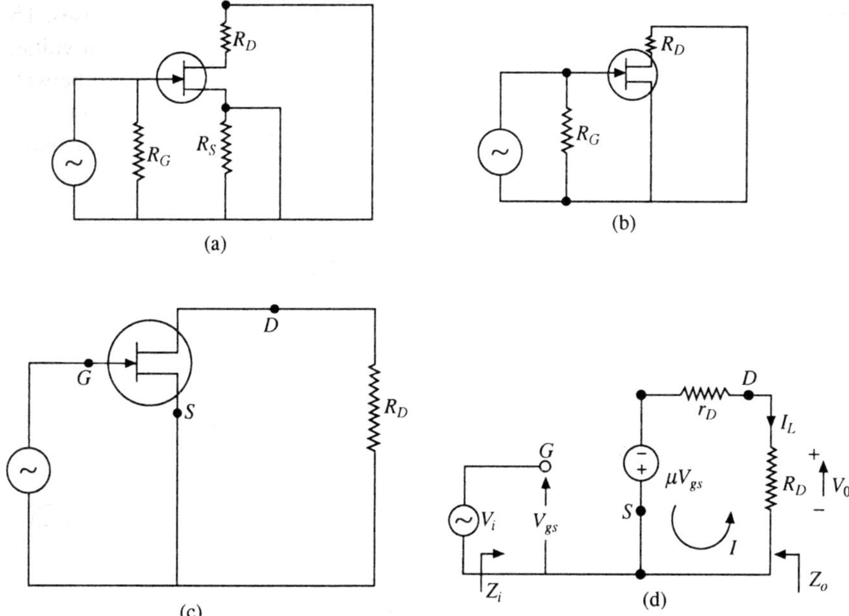

Fig. 8.11 A.C. equivalent circuit of common source amplifier.

$$V_o = V_i \mid A_v \mid \angle 180°$$

From this equation, it can be seen that output voltage is displaced by 180° with respect to input voltage.

8.10.2 Common Drain Amplifier (Source Follower)

Common drain amplifier circuit is shown in Fig. 8.12. During the positive half cycle of the input voltage, the negative gate bias reduces and the drain current increases. When drain current increases, the output voltage increases. Thus the input and output voltages are in phase. The voltage gain is approximately equal to unity. This circuit is used where a high input impedance is required. In the A.C. equivalent circuit, the drain terminal is nothing but the A.C. ground. Thus

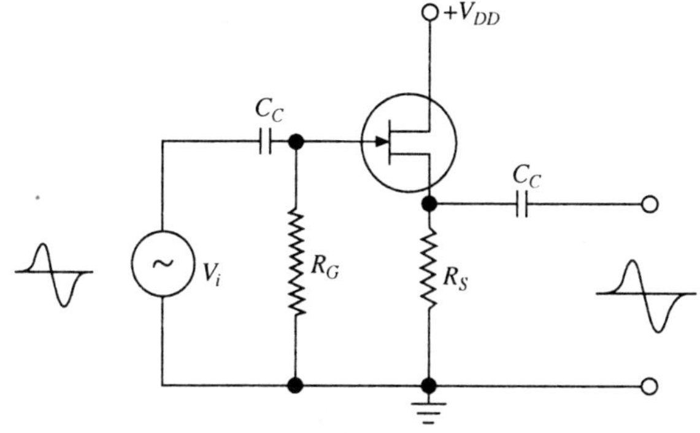

Fig. 8.12 Common drain amplifier.

the amplifier is a common drain amplifier since the drain terminal is directly grounded. The output voltage or source voltage is equal to the input voltage as the gain is unity. This circuit is called source follower since the output voltage or source voltage follows the input voltage.

Analysis of Common Drain Amplifier

In Fig. 8.13(a), capacitors and D.C. source are represented as short circuit. In Fig. 8.13(b), FET is represented by its Norton model. In Fig. 8.13(c), the resistance r_D is treated as open circuit since its value is very high.

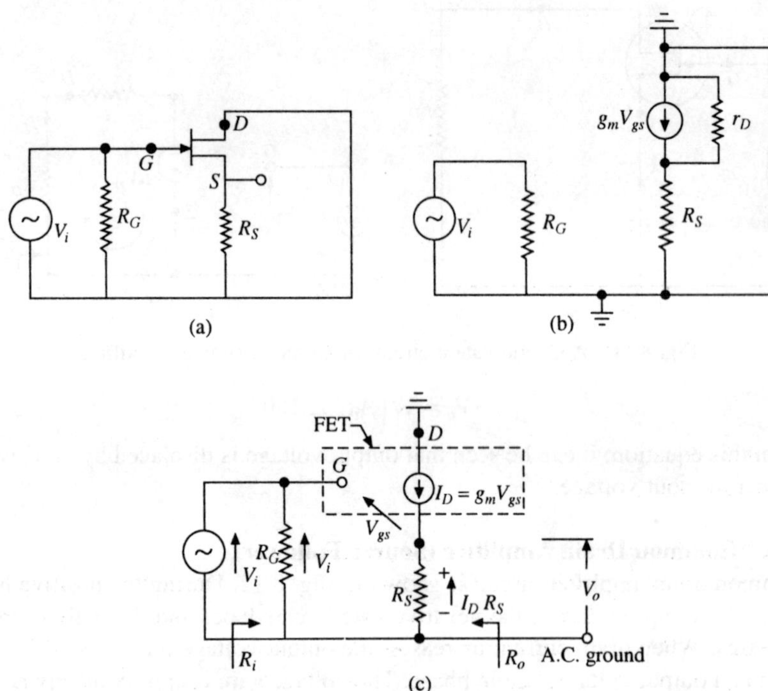

Fig. 8.13 A.C. equivalent circuit of common drain amplifier.

D.C. voltage sources and capacitors have to be shorted for A.C. analysis. The resistance r_D is very high and hence it can be treated as open circuit. Apply KVL in the loop formed by V_i, V_{gs} and $I_D R_s$ (Fig. 8.13).

$$V_i - V_{gs} - I_D R_s = 0$$

$$V_i = V_{gs} + I_D R_s = V_{gs} + g_m V_{gs} R_s = (1 + g_m R_s) V_{gs} \tag{8.29}$$

$V_o = I_D R_s$ from Fig. 8.13

$$= g_m V_{gs} R_s \tag{8.30}$$

On dividing (8.30) and (8.29) we get

$$\frac{V_o}{V_i} = \frac{g_m V_{gs} R_s}{(1 + g_m R_s) V_{gs}}$$

$$A_v = \frac{g_m R_s}{1 + g_m R_s}$$

If $g_m R_s >> 1$

$$\frac{V_o}{V_i} = A_v \simeq 1$$

$$g_m = 3 \times 10^{-3}\ R_s = 10k\Omega \text{ then } A_v = \frac{30}{1 + 30} = \frac{30}{31} \simeq 1.$$

8.11 Cascode Amplifier

The circuit of cascode amplifier is shown in Fig. 8.14. In cascode amplifier, a common source amplifier drives a common gate amplifier. The common source amplifier gives unity voltage gain. Common gate amplifier gives a voltage gain of $g_m r_d$. This circuit has high input impedance since FET is used. This circuit has low input capacitance since input junction of FET is reverse biased. This low value of capacitance reduces the Miller effect. The high frequency response of this amplifier is improved due to reduction in input capacitance. There is a good isolation between input and output due to high input impedance.

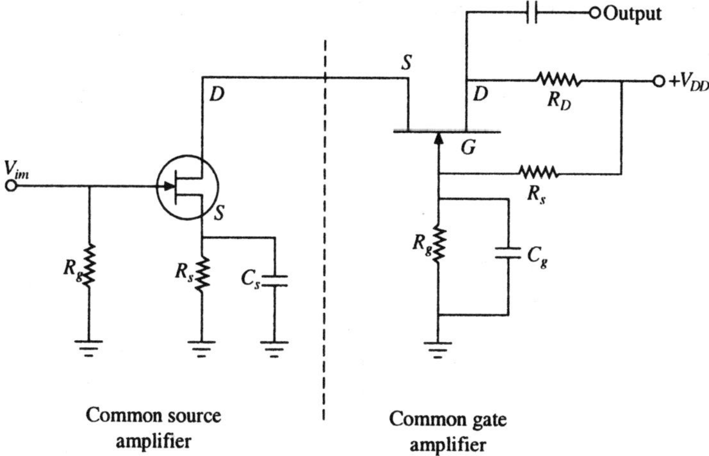

Fig. 8.14 Cascode amplifier.

8.12 Wide Band Amplifier

The bandwidth can be increased by providing low frequency compensation and high frequency compensation.

Low Frequency compensation

Amplifier with low frequency compensation is shown in Fig. 8.15. The bandwidth of R–C-coupled amplifier can be increased by decreasing the value of lower cut-off frequency and increasing the value of upper cut-off frequency.

The lower cut-off frequency can be decreased by using low frequency

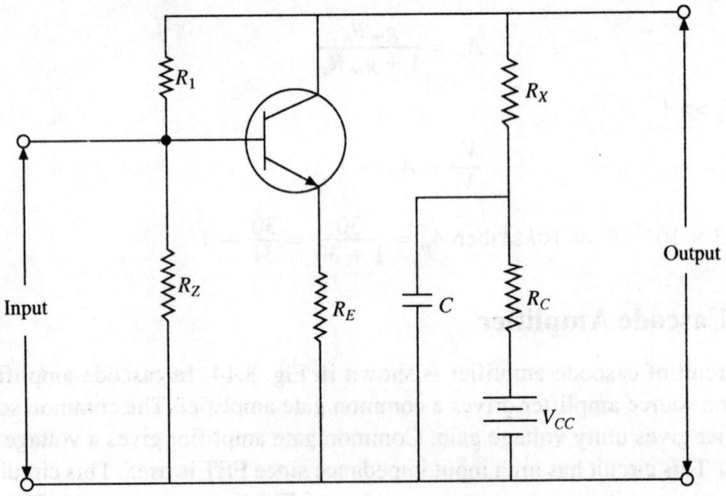

Fig. 8.15 Wide band amplifier.

compensation technique. In the above circuit, the resistance across the output under high frequency operation is R_X since R_c is bypassed by the shunt capacitor 'C'. Under low frequency operation, this capacitor acts as open circuit and the resistance across the output increases to $(R_x + R_c)$. Hence the output voltage increases.

Note that there is no input coupling capacitor and output coupling capacitor in this circuit.

The capacitors are removed since they produce more voltage drop at low frequencies.

High Frequency compensation

Amplifier with high frequency compensation is shown in Fig. 8.16. Because of the stray capacitance or the wiring capacitance, the output voltage is reduced in the high frequency range. This is because the shunt capacitance offers very low reactance. The high frequency compensation is done by connecting an inductor in series with R_c. In the D.C. equivalent circuit, the L and C will be in parallel and they offer very high impedance to the D.C. Hence D.C. components cannot be bypassed to the ground.

Thus the bandwidth can be increased by two or three times. Wideband amplifiers are used in applications where the range of input frequencies is very large like oscilloscopes and television receivers. Wideband amplifier is also known as video amplifier.

8.13 Darlington Configuration

Darlington configuration is shown in Fig. 8.17. Two transistors in common collector configuration are cascaded and embedded in one casing. This connection is the cascading of two CC transistors. This circuit has very high current gain and very high input impedance. But the voltage gain of this circuit is approximately equal to 1.

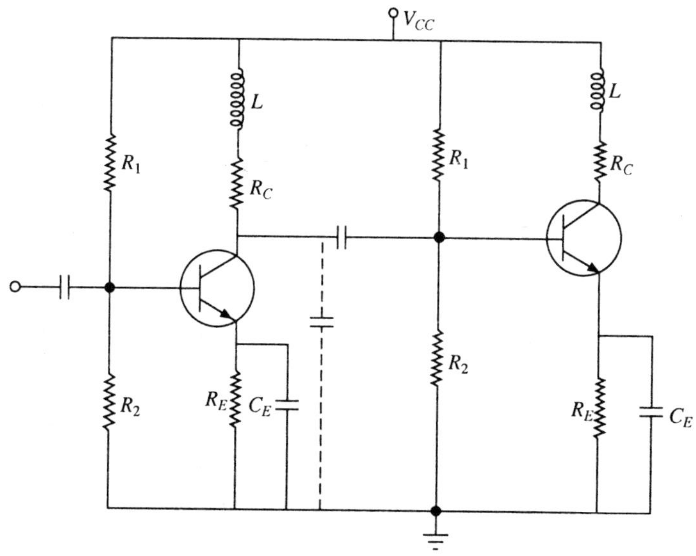

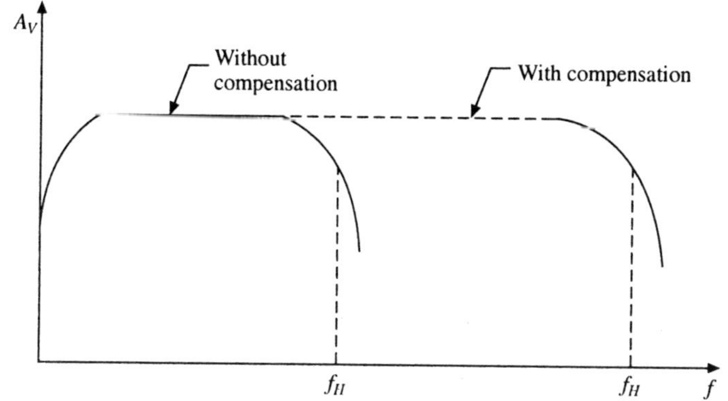

Fig. 8.16 High frequency compensation.

The approximate analysis is done by assuming that I_c is equal to I_E. The overall current gain $\approx \beta^2$. Overall voltage gain is the product of voltage gains of each stage. Each stage gives a voltage gain of unity. Therefore the overall voltage gain is unity.

Current Gain $$A_I = \frac{I_o}{I_1}$$

Multiply and divide with I_2

$$= \left(\frac{I_o}{I_2}\right) \cdot \left(\frac{I_2}{I_1}\right) = \beta \cdot \beta = \beta^2$$

$$\boxed{A_1 = h_{fe}^2 = \beta^2} \text{ approximately.} \qquad (8.31)$$

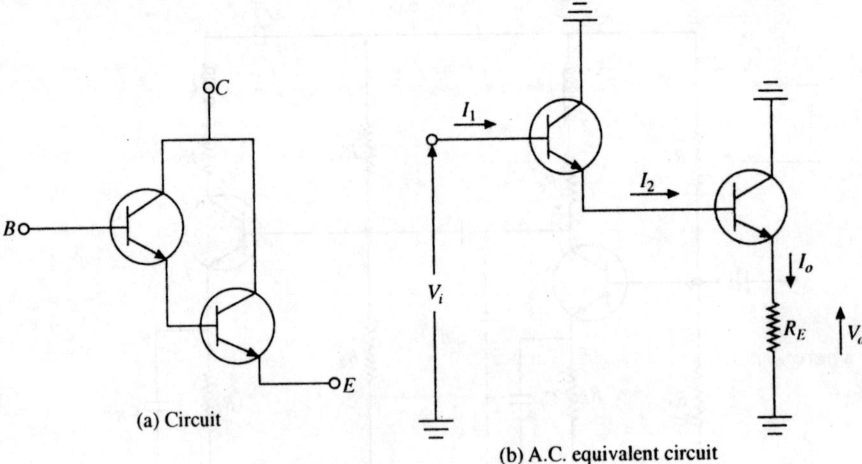

Fig. 8.17 Darlington configuration.

Hence it is called high β transistor

$$\frac{I_o}{I_2} = \frac{I_E}{I_2} \simeq \frac{I_C}{I_B} = h_{fe}$$

$$I_o = h_{fe} I_2$$

Multiply with R_E on both sides

$$I_o R_E = h_{fe} I_2 R_E; \qquad h_{fe} = \frac{I_2}{I_1}; \quad I_2 = h_{fe} I_1$$

Substitute

$$I_2 = h_{fe} I_1$$

$$V_o = h_{fe} \cdot h_{fe} I_1 R_E$$

$$V_o = h_{fe}^2 I_1 R_E$$

In common collector configuration $V_o = V_i$

Hence

$$V_i = h_{fe}^2 I_1 R_E$$

$$\frac{V_i}{I_1} = h_{fe}^2 R_E$$

$$\boxed{\text{Input impedance } Z_i = h_{fe}^2 R_E} \qquad (8.32)$$

Thus the input impedance is very high.

8.14 Miller's Capacitance (Miller's Theorem)

Statement: When an impedance Z connects the input and the output terminals of an inverting amplifier, this can be replaced by two impedances, one at the input and one at the output.

$$Z_{in} = \frac{Z}{1 - A}$$

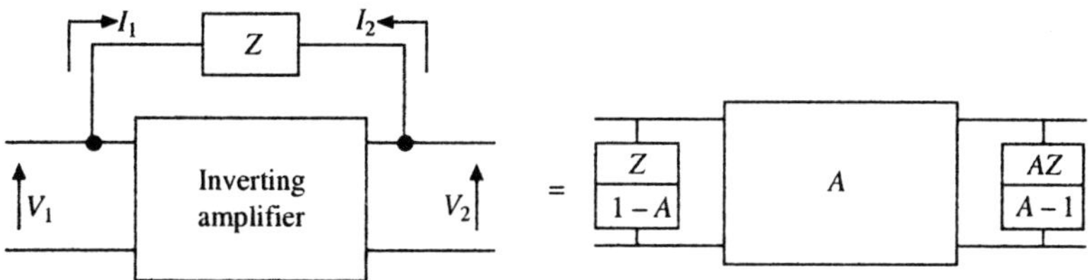

Fig. 8.18 Miller's theorem.

$$Z_{out} = \frac{AZ}{A-1}$$

where Z is the reactance of capacitor

$$\boxed{Z = X_C = \frac{1}{\omega C}} \qquad (8.33)$$

Apply Ohm's law at the input junction

$$I_1 = \frac{V_1 - V_2}{Z} = \frac{V_1\left(1 - \dfrac{V_2}{V_1}\right)}{Z}$$

Substitute $A = V_2/V_1$

$$I_1 = \frac{V_1(1-A)}{Z}$$

Rearrange the terms

$$\frac{Z}{(1-A)} = \frac{V_1}{I_1} = Z_{in}$$

Hence

$$\boxed{Z_{in} = \frac{Z}{(1-A)}} \qquad (8.34)$$

Apply Ohm's law at the output junction

$$I_2 = \frac{V_2 - V_1}{Z} = \frac{V_2\left(1 - \dfrac{V_1}{V_2}\right)}{Z}$$

Substitute $\dfrac{V_1}{V_2} = 1/A = \dfrac{V_2\left(1 - \dfrac{1}{A}\right)}{Z} = \dfrac{V_2(A-1)}{AZ}$

$$\frac{AZ}{(A-1)} = \frac{V_2}{I_2} = Z_{out}$$

$$\boxed{Z_{out} = \frac{AZ}{A-1}} \qquad (8.35)$$

Worked Problems

Ex. 8.1 A common source *FET* amplifier has $R_D = 500\ k\Omega$. If r_D and μ are 100 $k\Omega$ and 24, calculate the voltage gain.

$$A_v = \frac{-\mu R_D}{r_D + R_D} = \frac{-24 \times 500}{500 + 100}$$

$$A_v = -20$$

Ex. 8.2 A common source *FET* amplifier should have a gain of 29. If $\mu = 50$, $r_D = 5\ k\Omega$, find the value of R_D to be used in the drain circuit.

$$-29 = -\frac{50 \times R_D}{5k + R_D}$$

$$R_D = 6.9\ k\Omega$$

Ex. 8.3 A certain transistor has I_{CO} of 5 μA. When it is used in *CE* circuit, a base current of 30 μA is required to give a collector current of $1mA$. What is beta value of transistor?

$$I_C = \beta I_B + (1 + \beta) I_{CO}$$

$$10^{-3} = \beta \times 30 \times 10^{-6} + 5 \times 10^{-6} + 5 \times 10^{-6}\beta$$

$$10^{-3} - 5 \times 10^{-6} = \beta(30 + 5)10^{-6}$$

$$9.95 \times 10^{-4} = 35 \times 10^{-6}\beta$$

$$\beta = 28.4$$

Ex. 8.4 An emitter follower circuit is having input impedance of 500 $k\Omega$ and output impedance of 20Ω. The circuit uses a transistor with $h_{fe} = 50$, $h_{ie} = 1K$. Find the current gain and voltage gain for this circuit.

$$Z_i = 500\ k\Omega \quad h_{fe} = 50 \quad h_{ie} = 1K \quad Z_o = 20\ \Omega$$

$$Z_i = h_{ie} + (1 + h_{fe})R_L$$

$$500\ K = 1 + (1 + 50)\ R_L$$

$$R_L = 9.78\ k\Omega$$

$$A_I = 1 + h_{fe} = 1 + 50 = 51$$

$$A_v = \frac{R_L}{\frac{h_{ie}}{1 + h_{fe}} + R_L} = \frac{9.78 \times 10^3}{\frac{1 \times 10^3}{1 + 50} + 9.78 \times 10^3} = 0.998 \simeq 1$$

Ex. 8.5 Calculate A_I and Z_i of Darlington configuration if

$$\alpha_1 = \alpha_2 = 0.98 \quad h_{ie} = 1\ k\Omega \text{ and } R_L = 100\Omega$$

$$R_L = 100\ \Omega = 0.1\ k\Omega$$

$$\beta = \frac{\alpha}{1 - \alpha} = \frac{0.98}{0.02}$$

$$h_{fe} = 49$$

$$A_I = h_{fe}^2 = 49^2 = 2401$$

$$Z_i = h_{fe}^2 R_L = 240.1\ k\Omega$$

Ex-8.6 Draw the approximate model for the circuit shown in fig. 8.19(a) Given $h_{ie} = 1\ k\Omega$, $h_{fe} = 50$ what is the lower 3-*db* frequency f_L of this circuit?

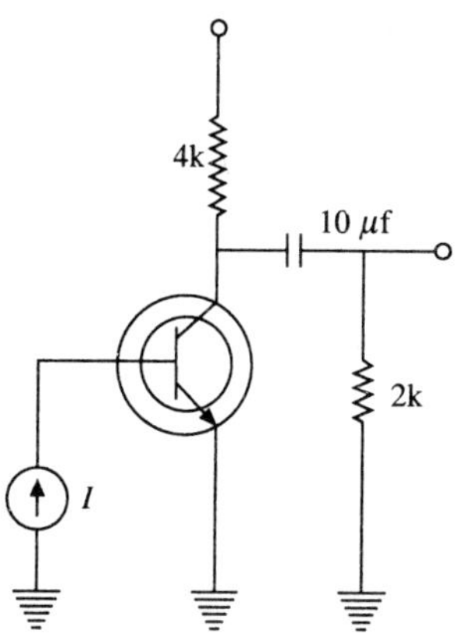

Fig. 8.19(a) CE Amplifier

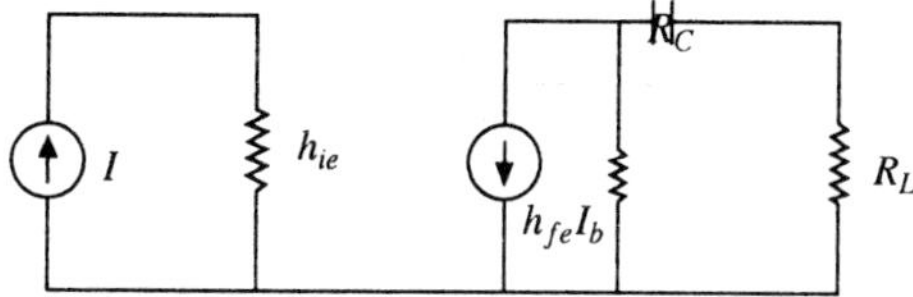

Fig. 8.19(b) Low frequency model

we know that

At $\quad f = f_2$

$$X_c = R$$

$$\frac{1}{2\pi f_2 C} = R_C + R_1$$

$$f_L = \frac{1}{2\pi(R_C + R_L)C}$$

$$f_L = \frac{1}{2\pi(4+2)10^3 \times 10 \times 10^{-6}}$$

$$f_L = 2.6\ \text{Hz}.$$

Ex-8.7 The FET shown in Figure 8.20 has $\mu = 50$, $r_d = 10\ \text{k}\Omega$. Draw a small signal equivalent circuit for the amplifier and calculate midband voltage gian.

$$V_S - V_{gs} - I_D R_S = 0$$

$$V_{gs} = V_S - I_D R_S = V_S - \frac{\mu V_{gs} R_S}{r_D + R_S + R_L'}$$

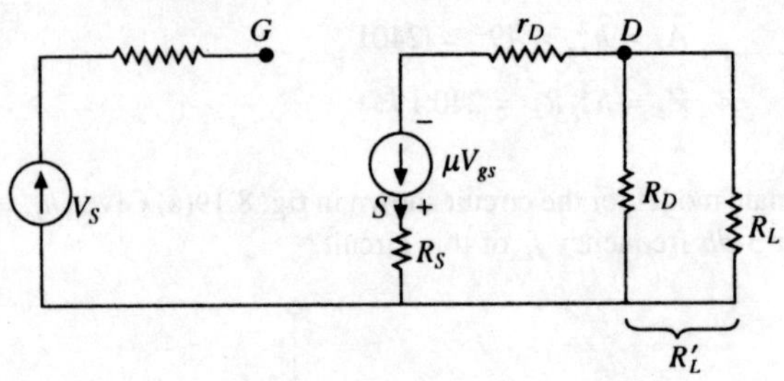

⇓

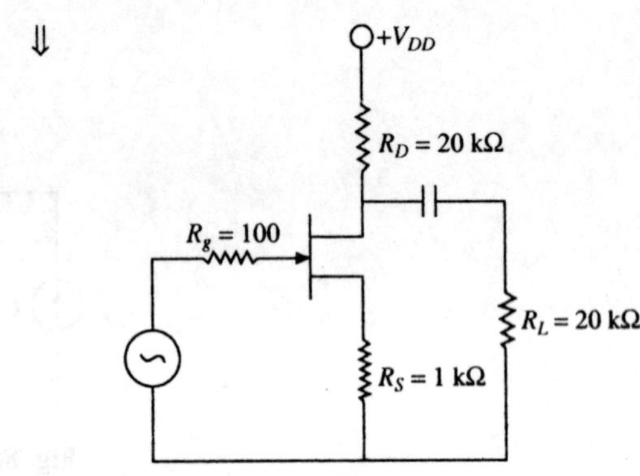

Fig. 8.20 FET amplifier

$$V_{gs}\left(1 + \frac{\mu R_S}{r_D + R_S + R'_L}\right) = V_S$$

$$V_{gs}(1 + R_S k) = V_S \tag{1}$$

$$V_O = -I_D R'_L$$

$$V_O = -\frac{\mu V_{gs}}{r_D + R_S + R'_L} = -kV_{gs}$$

$$A = \frac{V_O}{V_S} = \frac{-kV_{gs}}{V_{gs}(1 + R_S k)}$$

$$A = \frac{-k}{1 + R_S k}$$

substitute $K = \dfrac{\mu}{R_T}$ where $P_T = r_D + R_s + R'_L$

$$A = \frac{-\mu/R_T}{1 + R_S \dfrac{\mu}{R_T}}$$

$$A = \frac{-\mu/R_T}{\dfrac{R_T + R_S\mu}{R_T}}$$

$$A = \frac{-\mu}{r_D + R_S + R_L^1 + R_S\mu}$$

$$A = \frac{-50}{20 \times 10^3 + 1 \times 10^3 + 20 \times 10^3 + 1 \times 10^3 \times 50}$$

$$A = -7$$

Ex. 8.8 In the Fig. 8.21 $R_d = 12\ k\Omega$, $R_g = 1\ M\Omega$, $R_S = 2\ k\Omega$, $V_{DD} = 25$ V $I_{Dss} = 4\ mA$, $V_{gs}(\text{off}) = -4$, $g_m = 1\ m\Omega$, $r_d = R_d$, $V_{gs} = -1$
Determine I_d, and A_v.

$$I_d = I_{dss}\left(1 - \frac{V_{gs}}{V_{gs(\text{off})}}\right)^2$$

$$I_d = 2.25 \text{ mA}$$

$$I_d = I_{dss}$$

$$I_d = 2.25 \text{ mA}$$

$$\mu = g_m\, r_d$$

$$\mu = 10^{-3} \times 12\ k$$

$$\mu = 12$$

$$A = \frac{-\mu R_D}{r_D + R_D}$$

$$A = \frac{-12 \times 12}{2 \times 12}$$

$$A = -6$$

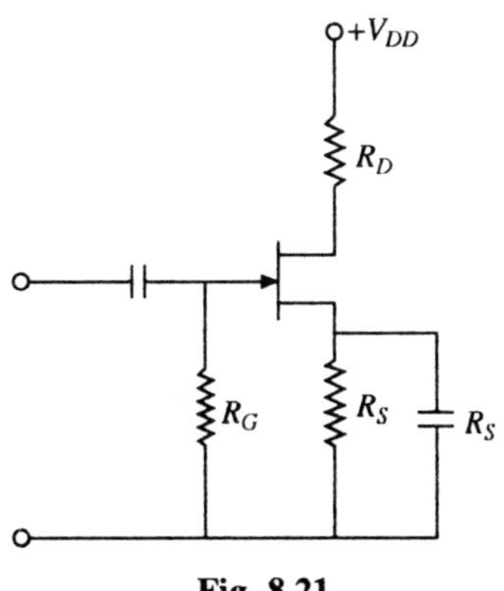

Fig. 8.21

Ex. 8.9 If *FET* used has $g_m = 0.004$, $r_d = 40$ $k\Omega$, what is the output voltage, and gain? The circuit shown in Fig. 8.22.

$$A = \frac{g_m R_S}{1 + g_m R_S}$$

$$A = \frac{0.004 \times 2500}{1 + 0.004 \times 2500} = 0.909$$

$$V_o = AV_i = 0.909 \times 200 = 181.8\ mV$$

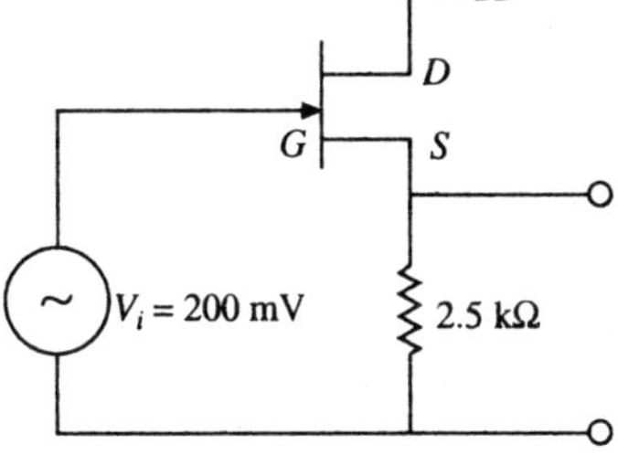

Fig. 8.22

Short Questions and Answers and Fill in the Blanks

Q.1. What is amplification?
Ans. The process of converting a weak signal into a strong signal is called amplification.

Q.2. Why do we wish to replace actual devices with ficticious models?
Ans. It is easy to analyse the amplifier and oscillator circuits using circuit theory if actual devices are replaced with ficticious circuit model.

Q. 3. Why must the base be narrow for the *BJT* action?
Ans. Beta is the ratio of collector current to base current. Base current becomes less if the width of the base is narrow. Higher value of beta can be obtained with lower value of base current.

Q. 4. Describe how amplification and switching are achieved by a *BJT*.
Ans. For amplification, *BJT* must operate in the active region. For switching, cut-off and saturation regions of output characteristics are used. When it is saturated, it acts as a closed switch. When it is in cut-off state, it acts as an open switch. We know that $V_{CE} = V_{CC} - I_C R_C$. When a pulse is given at the base, the transistor conducts and the voltage V_{CC} drops across R_C. V_{CE} is zero since $V_{CE} = I_C R_C$. *BJT* acts as a closed switch.

Q. 5. What is a load line?
Ans. Line drawn on the output characteristics with a slope of tan inverse of $(-1/R_c)$ is called *D.C.* load line. Line drawn on the output characteristics with a slope of $\tan^{-1}(-1/R_{AC})$ is called A.C. load line. R_{AC} is the parallel combination of R_C and R_L.

Q. 6. What is the locus of operating points of collector dissipation curve?
Ans. Hyperbola. $P = V_{CE} I_C$ and $I_C = \dfrac{P}{V_{CE}}$. Therefore $y \, \alpha \, \dfrac{1}{x}$. The graph is a rectangular hyperbola. Operating point must lie below the dissipation curve.

Q. 7. What is the effect of cascading number of stages on the frequency response?
Ans. The product of gain and bandwidth is constant. The gain increases with the increase in the number of stages. The bandwidth decreases with cascading.

Q. 8. Which amplifier has lowest input impedance?
Ans. Common base amplifier has a low input impedance.

Q. 9. What is Miller effect?
Ans. In an inverting amplifier with a voltage gain *A*, common impedance *Z* can be replaced with input impedance $Z/(A - 1)$ and output impedance $ZA/(A - 1)$.

Q. 10. What is the overall current gain for a Darlington pair?
Ans. It is equal to square of beta.

Q. 11. What factors affect *BJT* biasing?
Ans. Change in temperature, V_{BE} and beta affect *BJT* biasing.

Q. 12. How does one define small signals?
Ans. Small signal produces small fluctuations in collector current compared to its quiescent value.

Q. 13. Why models are necessary?
Ans. By replacing device with the model, we can calculate input impedance, output impedance, current gain and voltage gain of an amplifier.

Q. 14. How does one define cut-off frequency (or frequencies) for an amplifier?
Ans. The frequencies at which voltage gain reduces to 70.7% of the maximum voltage gain are called cut-off frequencies. At the critical or cut-off frequencies, the power gain is 3*db* less than the maximum power gain.

Q. 15. Define three stability factors.

$$S = \partial I_C / \partial I_{CO}$$

$$S^i = \partial I_C / \partial V_{BE}$$

$$S^{ii} = \partial I_C / \partial \beta$$

Q. 16. What is the value of cut-in voltage for a silicon *BJT*?
Ans. 0.7 volts.

Q. 17. What operating modes are possible for a *BJT*?

Ans. Three operating modes are active mode, saturation mode and cut-off mode.

Q. 18. List the disadvantages of fixed bias method.

Ans. Thermal runaway problem is not eliminated.
Stability factor is too high.

Q. 19. What are the advantages of using transformers to cascade amplifier?

Ans. Less power loss
Less battery voltage
D.C. isolation

Q. 20. What is harmonic distortion?

Ans. If the harmonics in the output are not exactly similar to those of input, there is a harmonic distortion.

Q. 21. Why is a capacitive coupling used to connect a signal source to an amplifier? To block D.C. and to couple A.C. signal.

Q. 22. State (a) stabilization (b) compensation techniques

Ans. (a) Variation in collector current is stabilized using voltage divider bias. Stable *Q*-point can be maintained using voltage divider bias.

(b) Various compensation methods are
  (i) Thermistor compensation
  (ii) Sensistor compensation
  (iii) Diode compensation

Q. 23. Define thermal resistance.

Ans. The resistance offered by the *BJT* to the flow of heat is called thermal resistance. To reduce the thermal resistance heat sink is provided over the casing.

Q. 24. Why common collector configuration finds wide application as a buffer?

Ans. It has high input impedance. It prevents loading of the system. The voltage gain is approximately equal to one.

Q. 25. Define the terms bandwidth and gain bandwidth product.

Ans. Bandwidth is the difference between upper and lower cut-off frequencies. It gives the range of frequencies in which voltage gain is above 70.7% of maximum voltage gain. It indicates the range of frequencies in which the power gain is not less than 3 db below the maximum power gain. The product of gain and bandwidth is called figure of merit. It is always constant.

Q. 26. List the disadvantages of *R–C* coupled amplifier.

Ans. 1. Impedance matching is not possible
2. Gain reduces in low and high frequency ranges
3. Less conversion efficiency

Q. 27. Why is it considered necessary to stabilize the operating point of a transistor amplifier?

Ans. If operating point is not stabilized, collector current increases. If it increases beyond rated current, *BJT* may be damaged.

Q. 28. What are the merits and demerits of direct coupled amplifier?

Merits

1. No coupling capacitor

2. Flat low frequency response
3. Suitable for D.C. amplification

Demerits
1. No D.C. isolation between the stages
2. Drift gets transmitted to the next stage.

Q. 29. Mention the uses of emitter follower.
It can be used as a buffer.
It can be used as an impedance matching circuit.

Q. 30. Define the *h*-parameters of a *CE* transistor.

$$h_{ie} = \frac{V_b}{I_b};\ h_{fe} = \frac{I_C}{I_b};\ h_{re} = \frac{V_b}{V_C};\ h_{oe} = \frac{I_C}{V_C}$$

Q. 31. Why common emitter configuration needs biasing?
Ans. It has a stability factor of $(1 + \beta)$. To reduce the stability factor, voltage divider bias is used.

Q. 32. For a cascade amplifier having three non-inverting stages of identical higher and lower 3*db* frequency f_H and f_L what is the 3*db* frequency for the complete cascade amplifier?

$$f_H = f_H\sqrt{2^{1/n} - 1};\ f_L = \frac{f_L}{\sqrt{2^{1/n} - 1}}$$

Ans. Substitute $n = 3$ to get f_H and f_L for 3-stage amplifier.

Q. 33. Define 3 db bandwidth.
Ans. It is the difference between upper and lower 3*db* frequencies.

Q. 34. What is the maximum efficiency of class A power amplifier?
Ans. *R–C* coupled power amplifier has maximum efficiency of 25%. Transformer coupled power amplifier has maximum efficiency of 50%.

Q. 35. What is the maximum conversion efficiency of class B power amplifier?
Ans. 78.5%

Q. 36. What is the maximum conversion efficiency of class B push pull amplifier?
Ans. 78.5%.

Q. 37. In class A power amplifier, maximum power in the device occurs under *Zero Signal Condition*.

Q. 38. The slope of the static load line in a transformer coupled amplifier is *90°*.

Q. 39. The collector of transistor is normally connected to the case, electrically to ensure *easy heat dissipation*.

Q. 40. Class *C* amplifiers are normally used as *Tuned amplifiers*.

Q. 41. Common emitter amplifier has both *voltage gain* and *current gain* greater than unity.

Q. 42. Class *AB* amplifier corresponds to a conduction interval slightly *greater than 180°*.

Q. 43. At cut off frequencies, the gain is down by 3 db.

Q. 44. In a transistor common base amplifier, the current gain is approximately *equal to unity*.

Q. 45. The main purpose of coupling capacitor in an amplifier is to *block D.C.*

Q. 46. The ratio of output power to input power is 1000. What is the gain in db?

Ans. Gain in db = 20 log 1000 = 60 db.

Q. 47. Collector junction in a transistor amplifier circuit has *reverse bias* at all times.

Q. 48. The maximum theoretical efficiency of class *B* push pull amplifier is *78.5%*.

Q. 49. A source follower using FET usually has a voltage gain which is *slightly less than unity*.

Q. 50. The most preferred type of amplifier for A.F. amplifier is *class B push-pull type*

Q. 51. The circuit shown in fig is a source follower having $A_v = V_o / V_i$. The input resistance of the circuit is

By Miller's theorem

$$R_i = \frac{Rg}{1 - A}$$

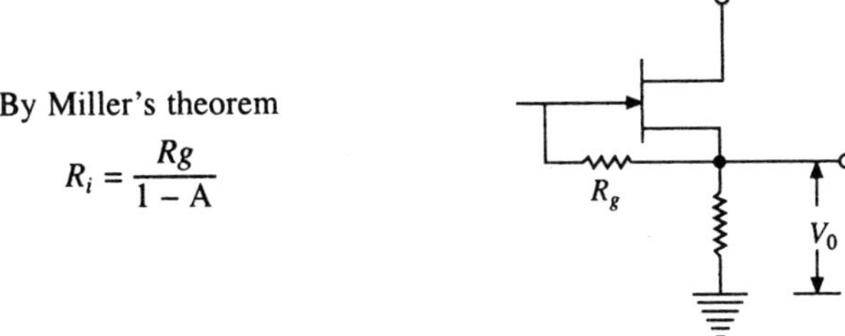

Fig. 8.23 Source Follower

Q. 52. Transformer coupling is preferred to *R-C* coupling because *it can match impedance*.

Q. 53. A signal may have frequency which lie in the range 0.001 Hz to 10 Hz. Which one of the following types of coupling is used

Ans. Direct coupling.

Q. 54. Class *AB* operation is often used in power amplifier to *overcome cross-over distortion*.

Q. 55. The overall bandwidth of two identical voltage amplifiers connected in cascade *is larger than that of single stage*.

Q. 56. A class-B push pull amplifier uses $V_{CC} = 25$ V, $R_L = 8\ \Omega$. If peak output voltage is 16 V, what would the power drawn from the source be?

Ans.

$$V_{RMS} = \frac{16}{\sqrt{2}} = 11.3$$

$$P_O = \frac{11.3^2}{8} = 16 \text{ W}$$

$$P_i = 16 \times \frac{4}{\pi} = 20.3 \text{ W}$$

Q. 57. If properly biased, JFET will act as *voltage controlled current source*.

Q. 58. FET source follower circuit has g_m of 2 $m\Omega$ and $r_d = 50\ k\Omega$. If the resistance is 1 $k\Omega$, what is the output resistance of the amplifier?

Ans.

$$\mu = g_m \times r_d = 2 \times 50 = 100$$

$$R_O = \frac{r_d}{\mu + 1} = \frac{50}{101} \simeq 0.5\ k\Omega$$

9

Power and Negative Feedback Amplifiers

9.1 Power Amplifier

The power amplifier circuit is shown in Fig. 9.1. Final stage of multistage amplifier system is always a power amplifier. Stage 1 and stage 2 amplifiers are called signal amplifiers or voltage amplifiers. They use low power transistors with high beta value. The power amplifier circuit is similar to voltage amplifier circuit except that it uses low beta high power transistor and an impedance matching transformer in the output stage. Power amplifier converts the D.C. power in the battery into useful A.C. power. Sum of A.C. output power and losses are equal to D.C. input power since A.C. power input to the amplifier is negligible. Conversion efficiency is defined as the ratio of A.C. power output to the D.C. power input.

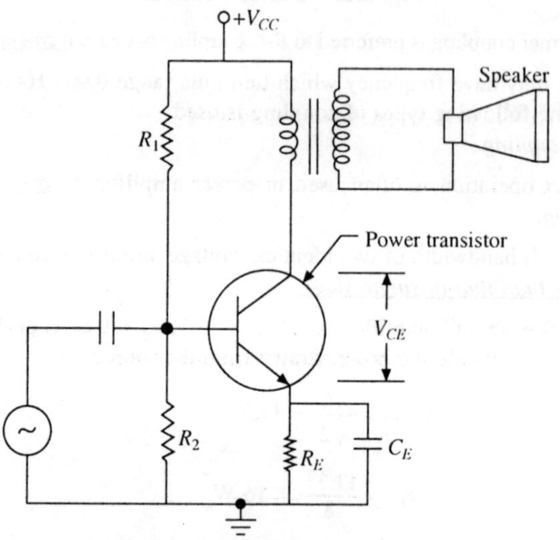

Fig. 9.1 Power amplifier circuit.

9.2 Decibel Notation

Let input and output resistances be R.

$$\underset{(\text{Ap})}{\text{Power gain}} = \frac{\text{Output power}}{\text{Input power}} = \frac{V_0^2 / R}{V_1^2 / R} = \left(\frac{V_0}{V_i}\right)^2$$

Hence

$$Ap = \left[\frac{V_0}{V_i}\right]^2$$

V_i — A — V_o — R_o

Take log on both sides

$$\log Ap = 2 \log [V_0/V_i]$$

one Bel $$= \log_{10} \left(\frac{V_0}{V_i}\right)^2$$

1 Bel is log of power gain

∴ No of decibels = 10 × No of Bels.

$$\text{Power gain in } dB = 10 \log_{10} \left(\frac{V_0}{V_i}\right)^2$$

$$\text{Power gain in } dB = 20 \log_{10} [V_0/V_i]$$

$$= 20 \log_{10} A_v$$

where A_v is voltage gain

$$\boxed{\text{Power gain in } dB = 20 \log_{10} A_v}$$

Advantages of *dB* Notation

(i) Multiplication reduces to addition.
(ii) Small and large numbers can be easily represented.
(iii) Sound is ultimately measured in *dB*. Gain gives an indication of sound level.

9.3 Maximum Efficiency of R–C Coupled Class-A Amplifier

The waveforms of class-A R–C coupled amplifier are shown in Fig. 9.2. The efficiency is defined as the ratio of A.C. output power to the D.C. input power.

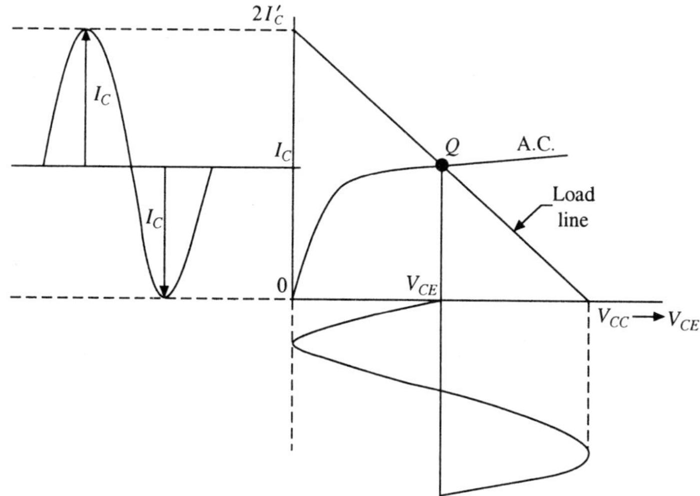

Fig. 9.2 R–C coupled class-A amplifier.

The total power input to the amplifier is the sum of D.C. power and the A.C power. The A.C. power is negligible, since the input voltage is of the order of millivolts. The power amplifier converts the D.C. power of the battery into useful A.C. power.

$$\text{Efficiency} = \frac{\text{Output A.C. power}}{\text{Input D.C. power}} = \frac{V_{RMS} I_{RMS}}{V_{CC} I_C}$$

From the Fig. 9.2 $\quad I_{RMS} = \frac{I_C}{\sqrt{2}}$ and $V_{RMS} = \frac{V_{CC}/2}{\sqrt{2}}$

$$\eta = \frac{\frac{V_{CC}}{2\sqrt{2}} \times \frac{I_C}{\sqrt{2}}}{V_{CC} I_C} = \frac{1}{4} \tag{9.1}$$

Maximum % efficiency = 25%.

The disadvantage is that the efficiency is less. This is because of the losses in resistance R_C and *BJT*.

9.4 Maximum Efficiency of Transformer Coupled Class-A Amplifier

The waveforms of transformer coupled class A power amplifier are shown in Fig. 9.3. We know that the slope of D.C. load line is $\tan^{-1}(-1/R_C)$. The resistance of transformer is zero. Therefore θ is 90°. The value of V_{CE} is equal to V_{CC} since the voltage drop in the primary is negligible.

$$\text{Efficiency} = \frac{\text{Output A.C power}}{\text{Input D.C power}} = \frac{V_{RMS}\, I_{RMS}}{V_{CC} I_C}$$

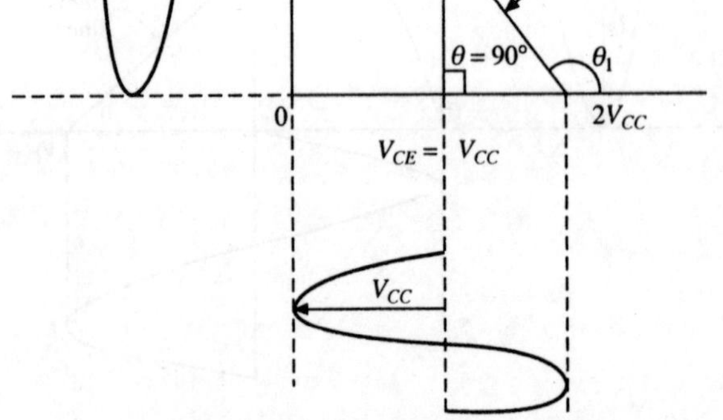

Fig. 9.3 Transformer coupled class-A amplifier.

From Fig. 9.3, $V_{RMS} = V_{CC}/\sqrt{2}$ and $I_{RMS} = I_C/\sqrt{2}$

$$\text{Efficiency} = \frac{\dfrac{V_{CC}}{\sqrt{2}} \times \dfrac{I_C}{\sqrt{2}}}{V_{CC} I_C} = \frac{1}{2} \tag{9.2}$$

Theoretically maximum efficiency = 50%.

Practical value of peak to peak current is less than $2I_C$. Therefore the maximum conversion efficiency is slightly less than 50% in practice.

9.5 Maximum Efficiency of General Class-B Power Amplifier

The waveforms of class-B power amplifier are shown in Fig. 9.4.

$$\text{Efficiency} = \frac{\text{A.C. Power output}}{\text{D.C. Power input}}$$

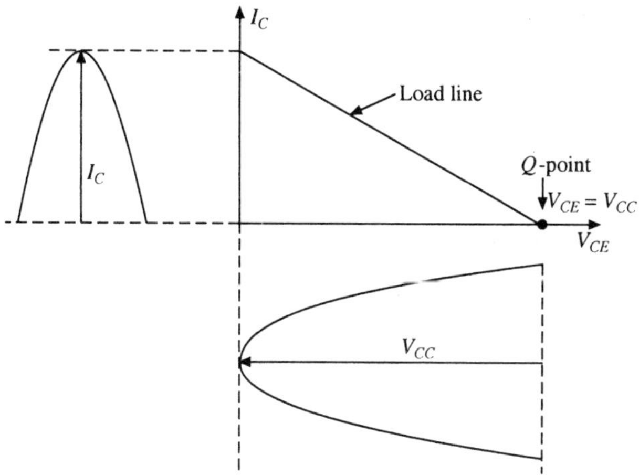

Fig. 9.4 Waveforms of class-B power amplifier.

The output A.C. power is available for half cycle only, since the device conducts only for half cycle.

In class B amplifier, Q-point coincides with cut off point.

From Fig. 9.4, A.C. power output $= \dfrac{1}{2}\dfrac{V_{CC}}{\sqrt{2}}\dfrac{I_C}{\sqrt{2}} = \dfrac{V_{CC} I_C}{4}$.

In class A amplifier, the Q-point is selected at the middle of the load line. The D.C. power at the zero signal condition is dissipated as heat by the power transistor. In class-B amplifier, the D.C. power at the zero signal condition is zero since the collector current is zero.

The average power or the D.C. power = D.C. voltage × D.C. current.

D.C. current is nothing but the average current. The average value of current can be obtained by integrating the current waveform. The current waveform is half wave rectified wave in a class-B amplifier.

$$I_{ave} = \frac{1}{2\pi}\int_0^{\pi} i d\theta = \frac{1}{2\pi}\int_0^{\pi} I_m \sin\theta \, d\theta$$

$$= \frac{I_m}{2\pi}[-\cos\theta]_0^{\pi} = \frac{I_m}{2\pi}|\cos\theta|_{\pi}^{0} = \frac{I_m}{2\pi}[1-(-1)]$$

$$I_{ave} = \frac{I_m}{\pi}$$

D.C. power Input = $V_{CC} \times I_{av} = V_{CC} \times \frac{I_m}{\pi}$.

From Fig. 9.3, $I_C = I_m$.

Therefore, $\quad$ Input = $\frac{V_{CC} I_C}{\pi}$

$$\text{Efficiency} = \frac{V_{CC} I_C / 4}{V_{CC} I_C / \pi} = \frac{\pi}{4} = 78.5\% \qquad (9.3)$$

The efficiency of class-B amplifier is higher than that of class A amplifier. The efficiency is higher because the quiescent collector current is zero in class-B amplifier. The current supplied by the battery is half wave rectified sine wave. Therefore the average value is I_m/π.

9.6 Maximum Efficiency of Class-B Push Pull Power Amplifier

The output circuit of push pull power amplifier is shown in Fig. 9.5(a). From the circuit, it can be seen that V_{CE1} and V_{CE2} are in opposite directions. I_{C1} and I_{C2} are in opposite directions. The waveforms of class B push pull amplifier are shown in Fig. 9.5(b).

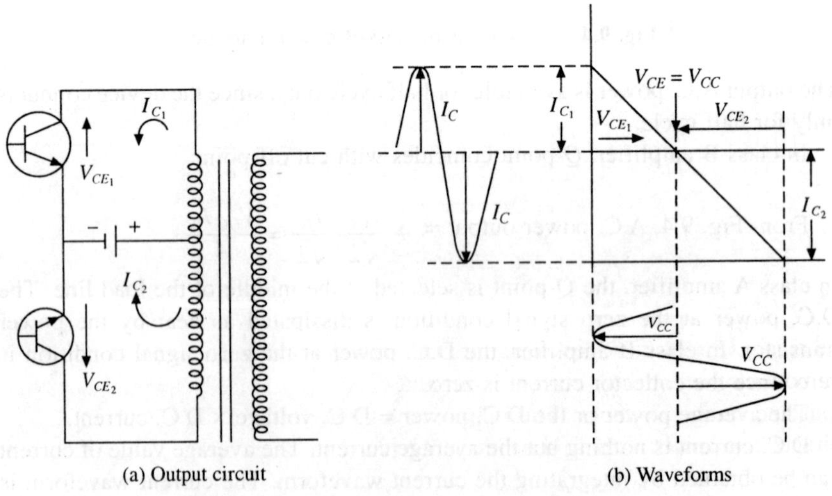

Fig. 9.5 Class-B push pull power amplifier.

$$\text{A.C. power output} = V_{rms} I_{rms} = \frac{V_{CC}}{\sqrt{2}} \cdot \frac{I_C}{\sqrt{2}} \tag{9.4}$$

$$= \frac{V_{CC} I_C}{2}$$

$$\text{D.C power input} = V_{CC} \cdot I_{av}$$

The current supplied by the battery is full wave rectified current. Hence the average value is $2I_m/\pi$.

$$= V_{CC} \frac{2I_m}{\pi} \tag{9.5}$$

From the waveforms, $I_m = I_c$

$$= \frac{\text{A.C. Power Output}}{\text{D.C. Power Input}} = \frac{V_{CC} I_C / 2}{2 V_{CC} I_C / \pi} = \frac{\pi}{4}$$

Efficiency = 78.5%.

9.7 Distortion

If the output waveform of an amplifier is not the exact replica of the input waveform, the amplifier is said to introduce distortion in the output. Various types of distortion are as follows:

(i) Frequency Distortion

An ideal amplifier amplifies the input voltage of various frequencies uniformly. From the frequency response curve of the R-C coupled amplifier, we observe that the output voltage decreases in the high frequency range. Thus the high frequency signals are not amplified uniformly. The amplifier is said to be introducing frequency distortion. The corresponding spectrum is shown in Fig. 9.6.

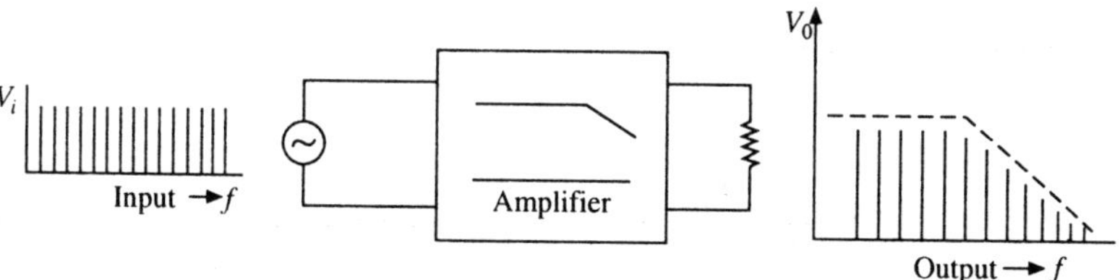

Fig. 9.6 Frequency distortion.

(ii) Phase Distortion

The reac lyve elements like capacitors and inductors introduce some delay. The components present between the input and output introduce delay between the input and output voltages. The output voltage is not in phase with the input voltage. The amplifier is said to introduce phase distortion.

(iii) Harmonic Distortion

The various frequency components of the input and output can be obtained by using the Fourier analysis. When some additional frequencies are present in the output which are not present in the input, the amplifier is said to introduce harmonic distortion.

9.8 Effect of Cascading of Amplifiers on Bandwidth

n number of amplifiers are cascaded as shown in Fig. 9.7(a).

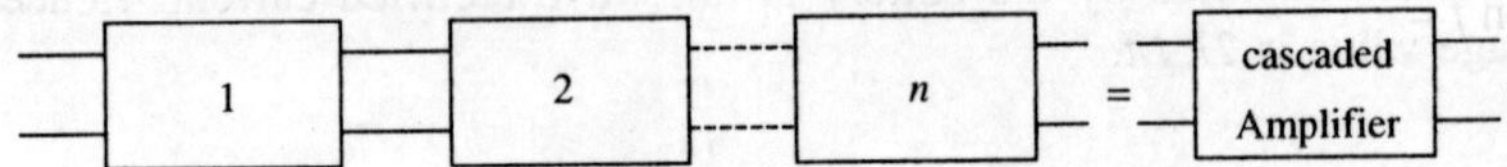

Fig. 9.7a Cascaded amplifier

n – No. of stages
f_L – Lower cut-off frequency of each stage
f_H – Higher cut-off frequency of each stage
f_L^* – Lower cut-off frequency of cascaded amplifier
f_H^* – Higher cut-off frequency of cascaded amplifier

We know that

$$A_{VH} = \frac{A_{Vm}}{1 + j\dfrac{f}{f_H}}$$

$$A_{VL} = \frac{A_{Vm}}{1 - j \cdot \dfrac{f_L}{f}}$$

$$|A_{VH}| = \frac{A_{Vm}}{\sqrt{1 + \left(\dfrac{f}{f_H}\right)^2}}$$

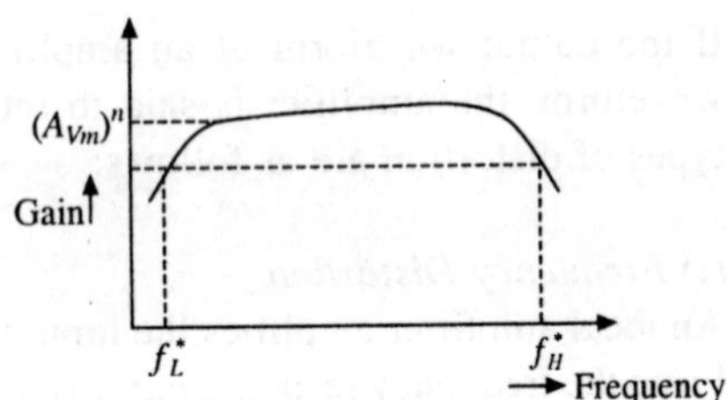

Fig. 9.7b Frequency response of cascaded amplifier

But for overall amplifier

$$\frac{A_{Vm}^n}{\sqrt{1 + \left(\dfrac{f}{f_H^*}\right)^2}} = \left[\frac{A_{Vm}}{\sqrt{1 + \left(\dfrac{f}{f_H}\right)^2}}\right]^n$$

$$\frac{A_{Vm}^n}{\sqrt{1 + \left(\dfrac{f}{f_H^*}\right)^2}} = \frac{A_{Vm}^n}{\left[\sqrt{1 + \left(\dfrac{f}{f_H}\right)^2}\right]^n}$$

Cancel numerator terms and equate the denominator terms.

$$\left(\sqrt{1+\left(\frac{f}{f_H^*}\right)^2}\right) = \left[\sqrt{1+\left(\frac{f}{f_H}\right)^2}\right]^n$$

$$\sqrt{1+\left(\frac{f}{f_H^*}\right)^2} = \left[1+\left(\frac{f}{f_H}\right)^2\right]^{n/2}$$

when $f = f_H^*$

$$\sqrt{1+1^2} = \left[1+\left(\frac{f_H^*}{f_H}\right)^2\right]^{n/2}$$

$$(\sqrt{2})^{2/n} = 1+\left(\frac{f_H^*}{f_H}\right)^2$$

$$2^{1/n} = 1+\left(\frac{f_H^*}{f_H}\right)^2$$

$$2^{1/n} - 1 = \left(\frac{f_H^*}{f_H}\right)^2$$

$$\sqrt{2^{1/n} - 1} = \frac{f_H^*}{f_H}$$

$$f_H^* = f_H\sqrt{2^{1/n} - 1} \tag{9.6}$$

when $n = 1$, $f_H^* = f_H\sqrt{2^{1/1} - 1}$

$$f_H^* = f_H$$

when $n = 2$, $f_H^* = 0.6\, f_H$.

It can be seen that upper cut-off frequency decreases with the increase in the number of stages.

The other relation

$$A_{VL} = \frac{A_{Vm}}{1 - j\left(\frac{f_L}{f}\right)}$$

$$|A_{VL}| = \frac{A_{Vm}}{\sqrt{1+\left(\frac{f_L}{f}\right)^2}}$$

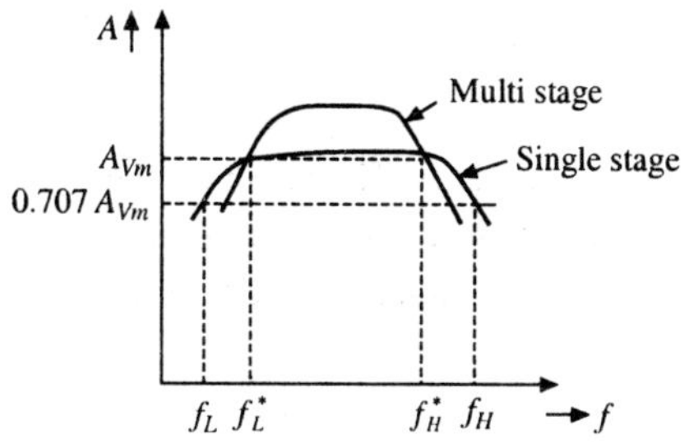

Fig. 9.7c Frequency response.

For cascaded amplifier

$$\frac{(A_{Vm})^n}{\sqrt{1+\left(\frac{f_L^*}{f}\right)^2}} = \frac{A_{Vm}}{\sqrt{1+\left(\frac{f_L}{f}\right)^2}} \cdots n \text{ times}$$

$$\frac{A_{Vm}^n}{\sqrt{1+\left(\frac{f_L^*}{f}\right)^2}} = \left(\frac{A_{Vm}}{\sqrt{1+\left(\frac{f_L}{f}\right)^2}}\right)^n$$

$$\sqrt{1+\left(\frac{f_L^*}{f}\right)^2} = \left[\sqrt{1+\left(\frac{f_L}{f}\right)^2}\right]^n = \left[1+\left(\frac{f_L}{f}\right)^2\right]^{n/2}$$

when $f = f_L^*$

$$\sqrt{1+(1)^2} = \left[1+\left(\frac{f_L}{f_L^*}\right)^2\right]^{n/2}$$

$$2^{1/n} = 1+\left(\frac{f_L}{f_L^*}\right)^2$$

$$2^{1/n} - 1 = \left(\frac{f_L}{f_L^*}\right)^2$$

$$\sqrt{2^{1/n} - 1} = \left(\frac{f_L}{f_L^*}\right)$$

$$\therefore \quad f_L^* = \frac{f_L}{\sqrt{2^{1/n} - 1}} \tag{9.7}$$

From the above equation it can be seen that f_L^* increases with the increase in the number of stages. Bandwidth of cascaded amplifier is

$$B_n = f_H^* - f_L^* \tag{9.8}$$

As number of stages increases, f_H^* decreases and f_L^* increases. Hence, band width decreases. The same can be observed from Fig. 9.7(c). Gain increases and bandwidth decreases with cascading. f_L^* is negligible compared to f_H^*. Therefore bandwidth is approximately equal to f_H^*.

$$B_n \simeq f_H \sqrt{2^{1/n} - 1} \tag{9.9}$$

Product of gain and bandwidth is constant. This constant is called figure of merit.

9.9 Two Stage R–C Coupled Amplifier

Two single stage R–C coupled amplifiers are cascaded to obtain two stage amplifier. Two stages are cascaded to increase the voltage gain. The signal from the microphone is given to the first stage through a coupling capacitor. This signal is amplified using transistor T_1. The output of stage-1 is given to the stage-2 through coupling capacitor C_{c2}. This is further amplified by using second stage. The output of second stage is given to the load through coupling capacitor C_{c3}. Each stage introduces a phase shift of 180 degrees. Total phase shift is 360 degrees (or) 0 degrees. Therefore, the output of amplifier is in phase with the input. The circuit and the arrangement to get frequency response are shown in Fig. 9.8.

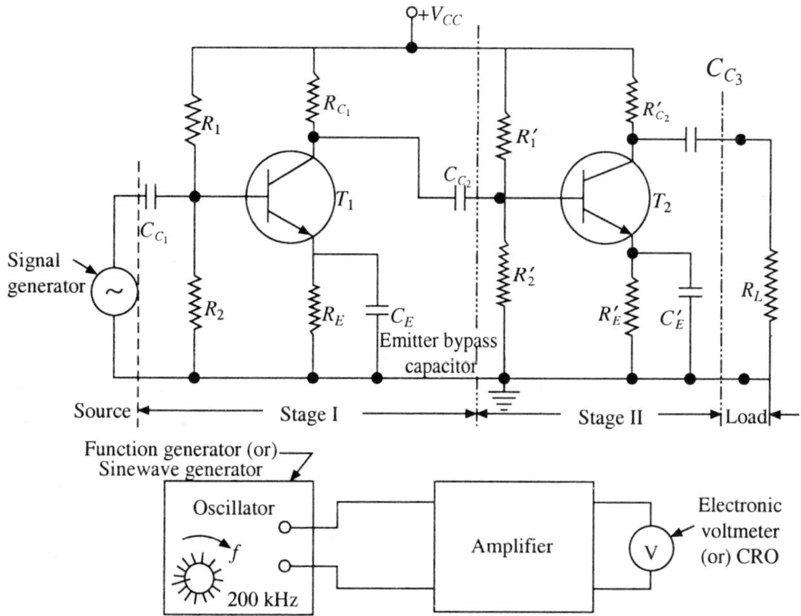

Fig. 9.8 Two stage R–C coupled amplifier.

9.10 Two Stage Transformer Coupled Amplifier

Two stage transformer coupled amplifier is shown in Fig. 9.9. A.C. input signal is applied to the base of through the transformer TR_1. The capacitor C_1 provides A.C. ground for the bottom terminal of the transformer. Without C_1, some A.C. signal will be dropped across R_2.

Secondary of TR_1 acts as a short for D.C. The potential at the top of R_2 is applied to the base through the secondary of TR_1. In the first stage amplifier, the voltage is applied at the input. This voltage is amplified and given to input of second stage through transformer TR_2 which provides D.C. isolation between stage 1 and stage 2. The voltage is further amplified by stage 2. The transformer in the output stage is called impedance matching transformer. The resistance R_L can be transferred to primary. Let the value referred to primary be R_L'. According to

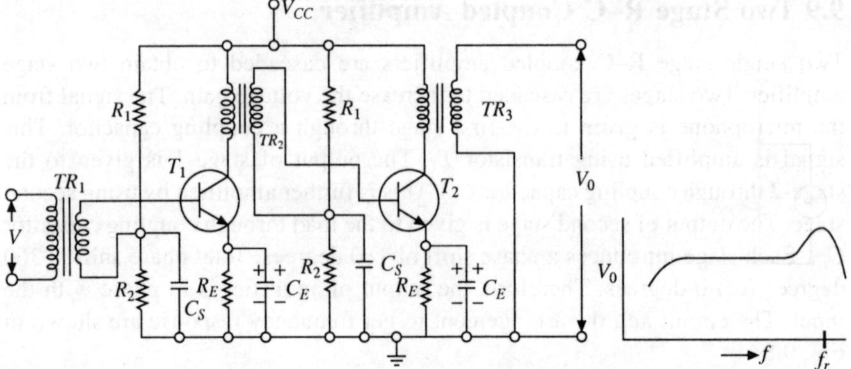

Fig. 9.9 Two stage transformer coupled amplifier.

Steinmetz, the power loss must be same whether the resistance is present in the primary or secondary $R'_L = R_L/K^2$ where $K = N_2/N_1$. For a step down transformer, K is always less than 1. By selecting proper value of turns ratio K (0.01 to 0.1), R'_L can be made almost equal to output resistance of BJT. According to maximum power transfer theorem, maximum power flows from source to the load if source resistance and load resistance are equal. Transformer is called impedance matching transformer since it helps to match reflected load impedance (R'_L) with the output impedance of the amplifier. Output voltage reduces in low frequency range since reactance of the transformer gets reduced in the low frequency range. In the high frequency range, the inductance of the transformer and the wiring capacitance form series resonant circuit. At f_r, the natural frequency of LC circuit is equal to the operating frequency of the amplifier. Resonance takes place and voltage increases. The bandwidth of transformer coupled amplifier is less than that of R–C coupled amplifier due to resonauce at high frequency.

9.11 Two Stage Direct Coupled Amplifier (or) DC Amplifier

If coupling capacitors of R–C coupled amplifier are shorted, we get two stage direct coupled amplifier. In R–C coupled amplifier, the gain in the low frequency range is reduced due to coupling capacitors and emitter bypass capacitor. In a direct coupled amplifier, these capacitors are removed and A.C. signal is directly given without any coupling capacitors. The D.C. from the first stage is directly given to the second stage since there is no coupling capacitor. This makes design of biasing resistors complicated. A drift in the operating point of first stage is carried to the second stage since there is no D.C. isolation. This configuration is used in the fabrication of IC amplifiers since this circuit does not contain capacitors. Generally it is difficult to fabricate large capacitors and inductors using IC technology.

From the frequency response curve, it can be seen that gain and output voltage are constant in the low frequency range since coupling capacitors are removed. The gain reduces in high frequency range due to wiring capacitance, BJT internal capacitance and reduction of beta at high frequencies. Thus this amplifier is best suited for D.C. and low frequency A.C. amplification.

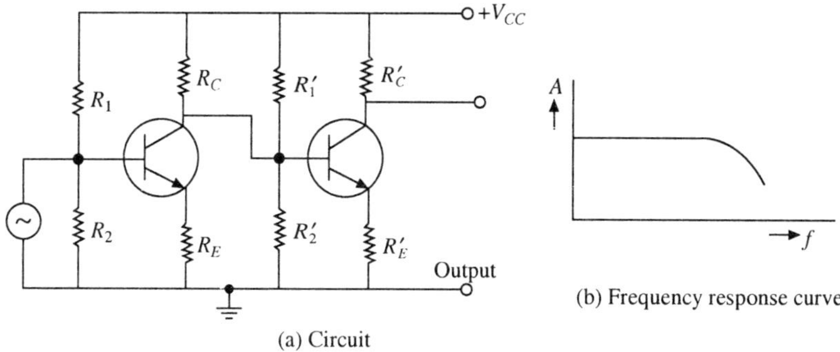

(a) Circuit

(b) Frequency response curve

Fig. 9.10 Two-stage direct coupled amplifier.

9.12 Analysis of Two-stage R–C Coupled Amplifier

Two stage R–C coupled amplifier is shown in Fig. 9.8(c). A.C. equivalent circuit is drawn by using the rules given in Section 8.9.3. Each transistor is represented by using approximate model. In Fig. 9.11a parallel combination of R_1, R_2 and h_{ie} is R_i. Parallel combination of R_C, R_1, R_2 and h_{ie} is R_{AC1}. Parallel combination of R_C and R_L is R_{AC2}. Simplified equivalent circuit with this notation is shown in Fig. 9.11b.

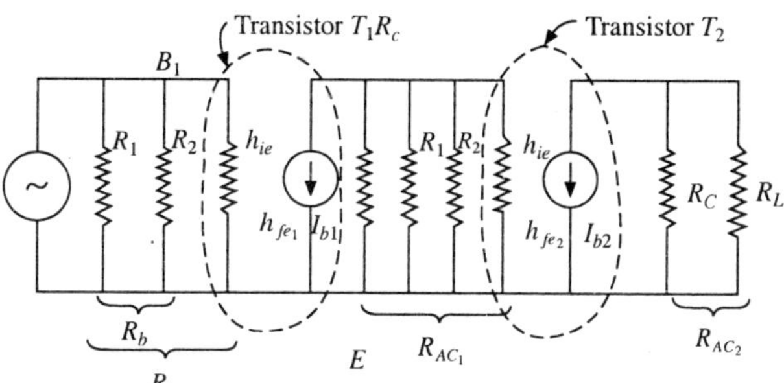

Fig. 9.11a AC equivalent circuit

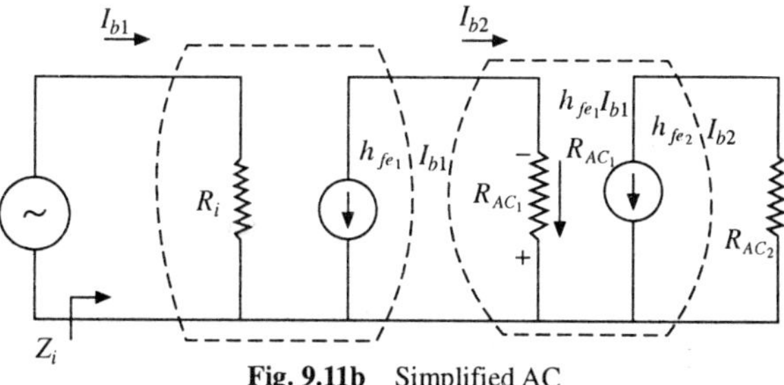

Fig. 9.11b Simplified AC

Gain of single stage R–C coupled amplifier is given by equation (8.4)

$$A_V = -\frac{h_{fe}\, R_L}{h_{ie}}$$

Overall gain of this stage amplifier is product of gains of each stage.

$$A_V = A_{V_1} \cdot A_{V_2} = \frac{-h_{fe_1} \cdot R_{AC1}}{R_i} \times \frac{-h_{fe2} \cdot R_{AC2}}{R_{AC1}}$$

Overall voltage gain = $A_V = \dfrac{h_{fe1} \cdot h_{fe2} \cdot R_{AC2}}{R_i}$ (9.10)

Current gain = $AI_1 \cdot AI_2$

$$A_I = (-h_{fe_1})(-h_{fe_2})$$

$$A_I = h_{fe_1} \cdot h_{fe_2} \quad (9.11)$$

if $h_{fe_1} = h_{fe_2}$

$$A_I = h_{fe2}^2$$

Input impedance $Z_i = \dfrac{V_S}{I_{b1}}$ $Z_i = R_i$ (9.12)

Output impedance $Z_0 = R_{AC_2}$ (9.13)

Output voltage is in phase with input voltage because of two inversions.

9.13 Push Pull Amplifiers

9.13.1 Reduction of Harmonics

In Push Pull amplifiers, two currents i_1 and i_2 with a phase shift of 180° are flowing.

$$i_1 = I_0 + I_1 \sin \omega t + I_2 \sin 2\omega t + I_3 \sin 3\omega t + \ldots \quad (9.14)$$

$$i_2 = I_0 + I_1 \sin (\omega t + \pi) + I_2 \sin (2\omega t + \pi) + I_3 \sin (3\omega t + \pi) + \ldots$$

Hence $i_2 = I_0 - I_1 \sin \omega t + I_2 \sin 2\omega t - I_3 \sin 3\omega t + \ldots$ (9.15)

Multiply with – 1 throughout

$$-i_2 = -I_0 + I_1 \sin \omega t - I_2 \sin \omega t + I_3 \sin 3\omega t \ \ldots \quad (9.16)$$

Add (9.14) and (9.16)

$$i_1 - i_2 = 2I_1 \sin \omega t + 2I_3 \sin 3\omega t + 2I_5 \sin 5\omega t + \ldots$$

$$i_1 - i_2 = 2\,(I_1 \sin \omega t + I_3 \sin 3\omega t + I_5 \sin 5\omega t + \ldots) \quad (9.17)$$

From this equation, it can be seen that D.C. and even harmonics are cancelled. Add (9.14) and (9.15)

$$i_1 + i_2 = 2\,(I_0 + I_2 \sin 2\omega t + I_4 \sin 4\omega t + \ldots) \quad (9.18)$$

Here fundamental ($I_1 \sin \omega t$) is getting cancelled. Since the fundamental component is not existing, the emitter bypass capacitor C_E is not required in a push pull amplifier circuit.

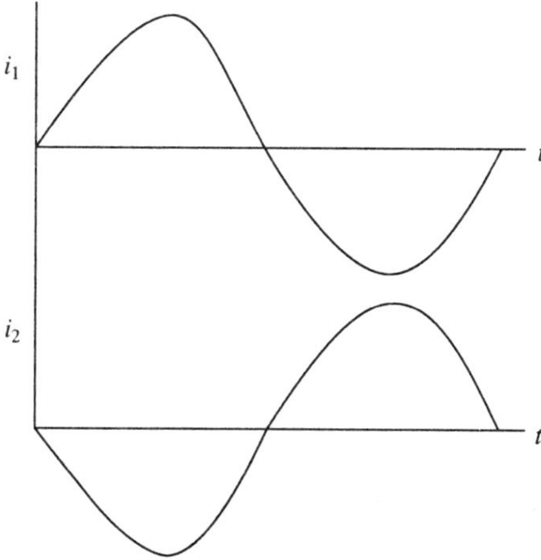

Fig. 9.12 Waveforms

9.13.2 Class-A Push-Pull Amplifier

Class-A push-pull amplifier circuit is shown in Fig. 9.13. The input to the transistors Q_1 and Q_2 is applied from the centre tap transformer T_1. The centre tap transformer produce two voltages with 180° phase shift. R_1 and R_2 provide the biasing for both the transistors. The dc voltage across R_2 is applied to the bases of Q_1 and Q_2 through the winding sections AB and BC respectively. The positive voltage V_{CC} is applied to the collectors of Q_1 and Q_2 through the upper and lower sections of the primary winding of the output transformer T_2. Q_1 and Q_2 are biased in Class A mode. The current through the primary of T_2 is $(i_1 - i_2)$. Therefore the D.C. components gets cancelled and the AC current gets doubled. When the current through Q_1 is pushed up, the current through Q_2 is pulled down. Hence this amplifier is named as push-pull amplifier. From Fig. 9.12, it can be seen that i_2 and i_2 are out of phase by 180°.

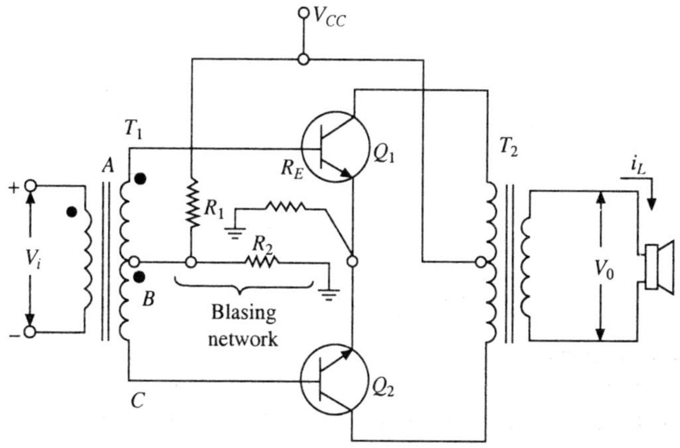

Fig. 9.13 Class-A Push-pull amplifier.

Advantages

1. The power amplified (transferred) is high.
2. All even harmonics are cancelled. Hence the distortion is reduced in the output.
3. The emitter bypass capacitor is not required.
4. The D.C. current in the upper half and lower half gets cancelled. So the size of the impedance matching transformer is reduced.
5. The humming is reduced as the ripple in the D.C. power supply in the two halves of the winding gets cancelled.

Disadvantages

1. The circuit requires two centre tap transformers.
2. The cost and the floor area of the transformers are high.

9.13.3 Class-B Push-Pull Amplifier

Class B push pull amplifier circuit is shown in Fig. 9.14(a). The input voltage to the transistors Q_1 and Q_2 is supplied from the output of a centre tap transformer. This is similar to class A push-pull amplifier. The transistors Q_1 and Q_2 are biased in Class-B mode. During the first half cycle, Q_1 conducts for 180° only. The output at the collector is inverted as Q_1 is connected in CE configuration. During the next half cycle, Q_2 conducts and Q_1 does not conduct. The voltage across the load will be a full sine wave. Class B operation is more efficient than class A. This advantages of class-B and the advantages of push-pull connection are combined to form class-B push-pull amplifier. Practically this type of amplifier is used as power amplifier.

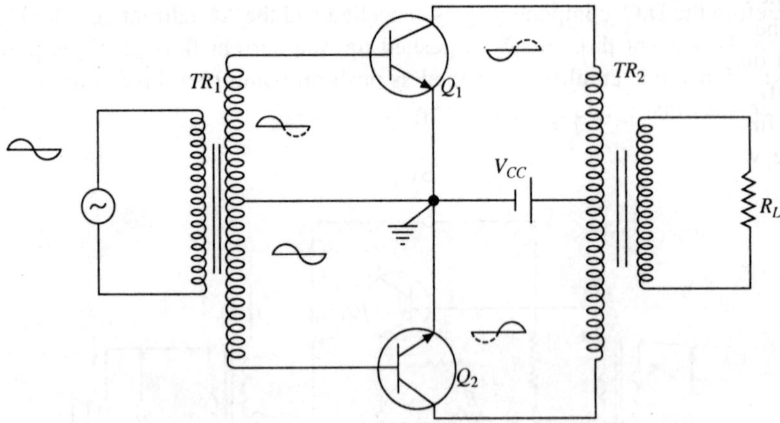

Fig. 9.14(a) Class-B push-pull amplifier.

Cross Over Distortion

Cross over distortion is shown in Fig. 9.14b(b). In class B push-pull amplifier, each transistor is supposed to conduct for 180°. The operating point is selected at the cut-off point. But in actual circuit, the transistor does not conduct for 180°. It conducts for less than 180° due to their requirement of 0.7 V between base and emitter. As shown in the waveform of the output, the zero crossing is not continuous. This type of distortion is known as cross over distortion. This can be eliminated by selecting the Q-point slightly before the cut-off point. In other words, the amplifier must be operated in class AB mode.

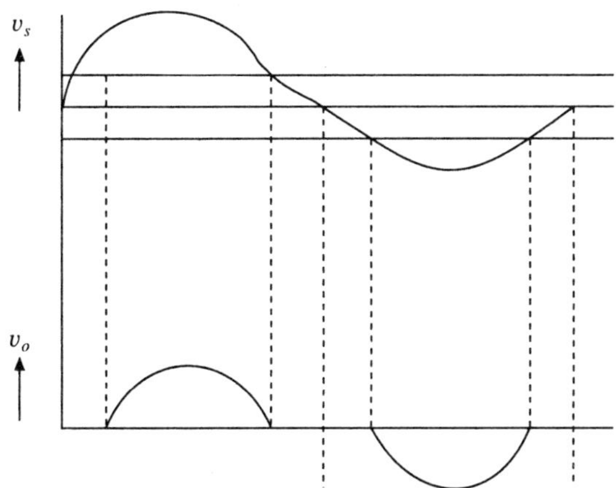

Fig. 9.14(b) Cross over distortion

9.13.4 Complementary Symmetry Push-pull Amplifier

Complementary symmetry push-pull amplifier is shown in Fig. 9.15. PNP and NPN transistors are complementary transistors. T_1 and T_2 arc cqually rated transistors. During the first half cycle, T_1 conducts and produces a negative half at the load. During the second half cycle, T_2 conducts and produces positive half at the load. This indicates a complementary push-pull amplifier without input and output transformers. T_1 and T_2 are connected in (CC) common collector configuration. V_{CC1} and V_{CC2} bias the transistors T_1 and T_2 respectively. During the first half cycle, T_1 conducts and during the next half cycle T_2 conducts. Full sine wave is available across the load resistance. The D.C. biasing currents

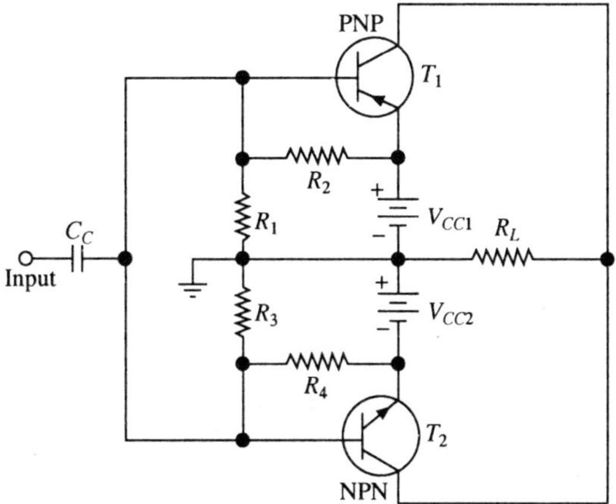

Fig. 9.15 Complementary symmetry push-pull amplifier.

through R_L get cancelled and only A.C. is available across it. Thus input and output transformers can be eliminated using this amplifier. Impedance matching is not possible since transformer is absent.

9.14 Negative Feedback Amplifiers

In this amplifier, the feedback signal gets subtracted from the source signal. They are classified as

(i) voltage series negative feedback
(ii) voltage shunt negative feedback
(iii) current series negative feedback
(iv) current shunt negative feedback

9.14.1 Voltage Series Negative Feedback Amplifier

The circuit of voltage series negative feedback amplifier is shown in Fig. 9.16. In this amplifier, part of the output voltage is made available across R_{E2}. The voltage across R_{E2} is in series opposition with the source voltage V_s. The amplifier is voltage series negative feedback amplifier since the output voltage is sensed and the feedback voltage is in series opposition with the source voltage. The output of the second stage drives a current through R_{E2} in such a way that the feedback voltage is in series opposition with the source voltage. Thus the negative feedback operation is possible with two stages.

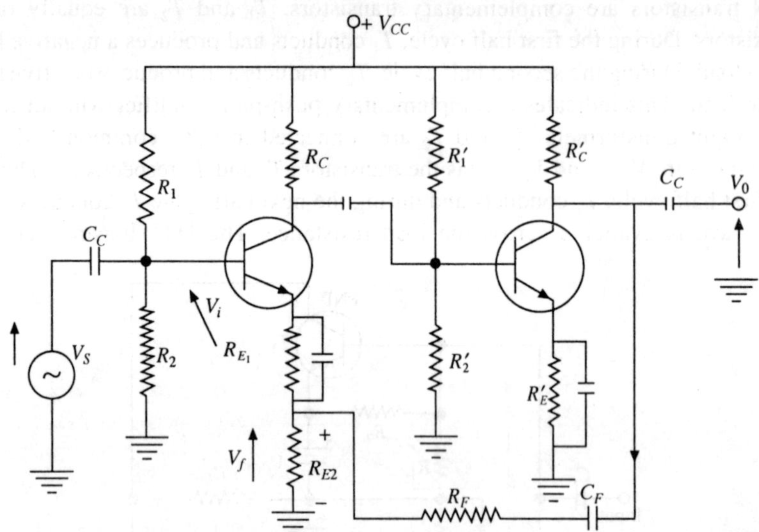

Fig. 9.16 Voltage series negative feedback amplifier.

9.14.2 Voltage Shunt Negative Feedback Amplifier

The circuit of voltage shunt negative feedback amplifier is shown in Fig. 9.17. The circuit of negative feedback amplifier is similar to the circuit of collector to the base bias.

$$I_s = I_i + I_f; \quad I_i = I_s - I_f$$

The amplifier is negative feedback amplifier since the feedback current is getting subtracted from the source current. KCL at input junction gives the output voltage is sensed. The voltage across R_c is applied in parallel with V_s through R_f. Hence this amplifier is called voltage shunt negative feedback amplifier.

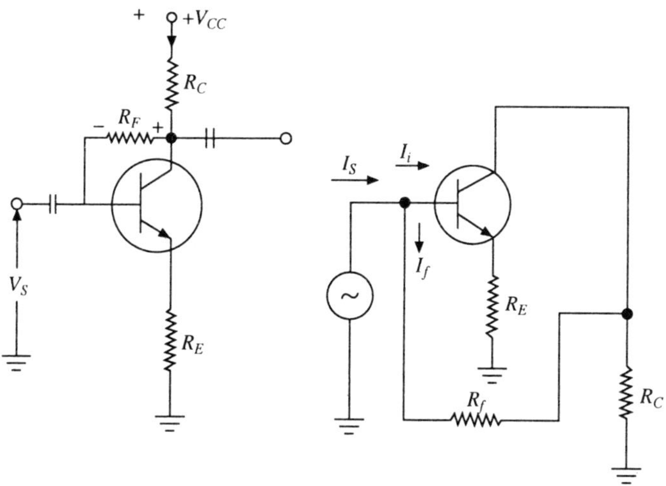

Fig. 9.17 Voltage shunt negative feedback amplifier.

9.14.3 Current Series Negative Feedback Amplifier

The circuit of current series negative feedback amplifier is shown in Fig. 9.18. This circuit is nothing but the voltage divider biasing circuit without the emitter bypass capacitor. A signal proportional to the output current is made available across R_E. The polarity of the feedback voltage is in series opposition with the source voltage. Hence this amplifier is a current series negative feedback amplifier.

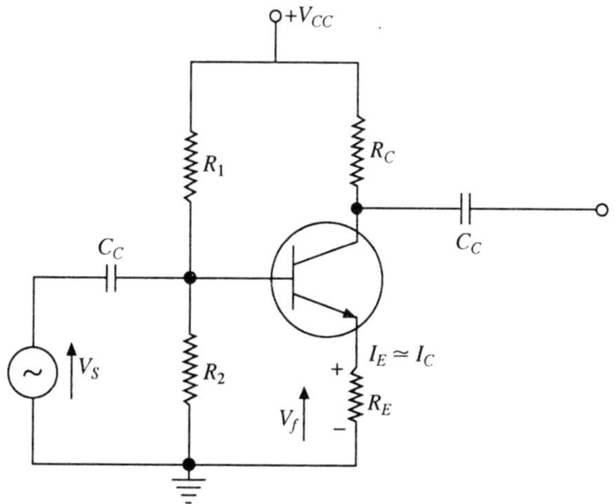

Fig. 9.18 Current series negative feedback amplifier.

9.15 Analysis of Voltage Series Negative Feedback Amplifier

Fig. 9.19 shows block diagram of a voltage series negative feedback amplifier. Output voltage of the second stage is sensed. When KVL is applied in the output loop, we get

$$V_i = V_s - V_f$$

It is a negative feedback connection due to the negative sign in the above equation. Apply KVL for the circuit in Fig. 4.17

$$V_s - V_i - V_f = 0 \tag{9.19}$$

Hence

$$V_i = V_s - V_f$$

$$A = \frac{V_o}{V_i} \tag{9.20}$$

$$A_f = \frac{V_o}{V_s}$$

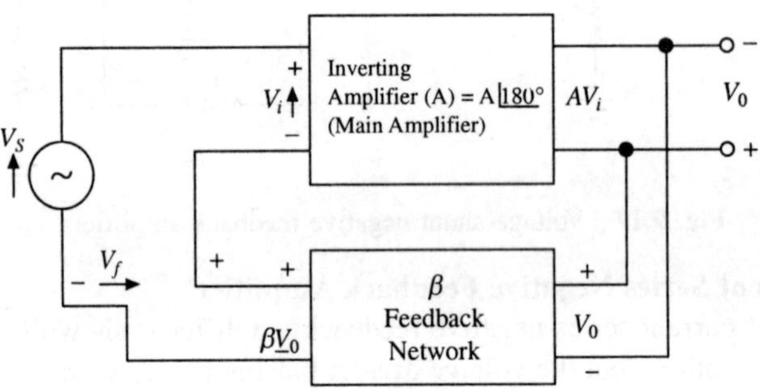

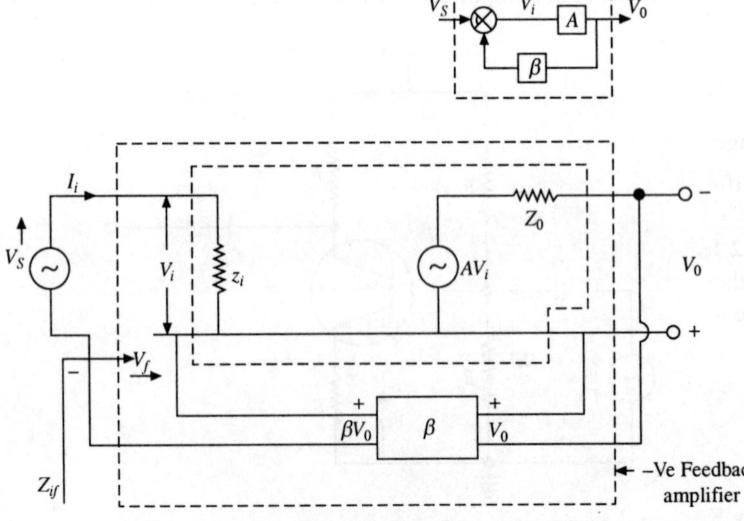

Fig. 9.19 Voltage series negative feedback amplifier.

$$V_o = AV_i$$

Substitute Eqn. (9.19) here

$$V_o = A(V_s - V_f) \qquad \text{substitute } V_f = \beta V_o$$

$$= A(V_s - \beta V_o); = AV_s - A\beta V_o$$

$$V_o + A\beta V_o = AV_s$$

Therefore, $V_o(1 + A\beta) = AV_s$

Hence $$\frac{V_o}{V_s} = \frac{A}{1 + A\beta}$$

Thus $$A_f = \frac{A}{1 + A\beta}. \qquad (9.21)$$

It can be seen that the gain reduces with negative feedback.

If $A\beta >> 1$, $$A_f = \frac{A}{A\beta} = \frac{1}{\beta}$$

It shows that gain is stable and independent of A if $A\beta >> 1$.

9.16 Advantages of Negative Feedback

9.16.1 Gain is Stable

Differentiate A_f in (9.21) with respect to A.

$$\frac{dA_f}{dA} = \frac{(1 + A\beta)1 - A\beta}{(1 + A\beta)^2}$$

Thus $$\frac{dA_f}{dA} = \frac{1}{(1 + A\beta)^2}$$

or $$dA_f = \frac{dA}{(1 + A\beta)^2}$$

Change in overall gain is $\frac{1}{(1 + A\beta)^2}$ times the change in the gain of the main amplifier.

9.16.2 Input Impedance Increases with Feedback

Z_i is the impedance of the amplifier without feedback. Z_{if} is the input impedance of the amplifier with feedback.

$$Z_i = \frac{V_i}{I_i} \qquad (9.22)$$

$$Z_{if} = \frac{V_s}{I_i} \qquad (9.23)$$

Apply KVL in the input side of Fig. 9.19.

$$V_s - V_i - V_f = 0; \quad V_i = V_s - V_f$$

$$V_i = V_s - \beta V_o \quad V_i = V_s - \beta A V_i$$

$$V_i(1 + \beta A) = V_s$$

Divide throughout by I_i

$$\frac{V_i}{I_i}(1 + A\beta) = \frac{V_s}{I_i}$$

Substitute (9.22) and (9.23) here

$$Z_{if} = Z_i(1 + A\beta)$$

If $Z_i = 1$ KΩ, and $A\beta = 1000 \times 0.1 = 100$

$$Z_{if} = 101 \text{ K}\Omega$$

9.16.3 Output Impedance Decreases with Voltage Series Negative Feedback

Z_o is the output impedance of the amplifier without feedback. Z_{of} is the output impedance with feedback. To find out the output impedance of the negative feedback amplifier, apply a known voltage V_o. Let the current be I_o and Z_{of} be the ratio of V_o to I_o when independent source alone is reduced to zero. This is the rule to be used to find out Thevenin impedance of a circuit having both dependent and independent sources.

Apply KVL in the output loop

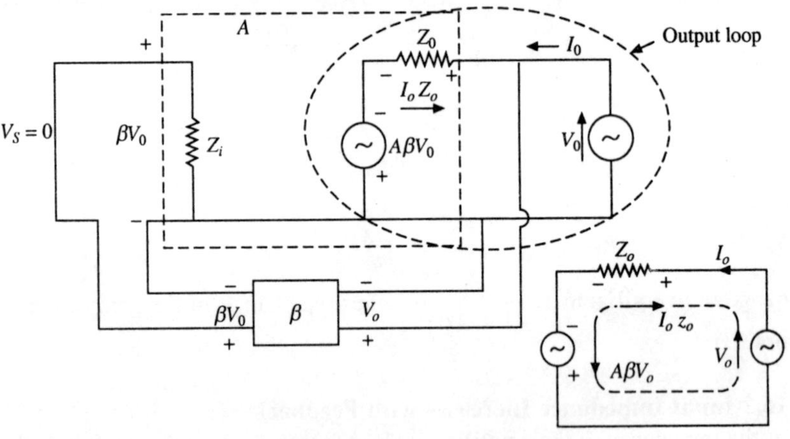

Fig. 9.20 Voltage series negative feedback amplifier.

$$V_o - I_o Z_o + A\beta V_o = 0$$

$$V_o(1 + A\beta) = I_o Z_o$$

Therefore,

$$Z_o = \frac{V_o}{I_o}(1 + A\beta)$$

$$Z_o = Z_{of}(1 + A\beta); \quad Z_{of} = \frac{Z_o}{1 + A\beta} \tag{9.24}$$

Thus the output impedance reduces with feedback.

9.16.4 Noise is Reduced

Let N be the noise without feedback and N_f is the noise with feedback. A derivation similar to the gain of negative feedback amplifier can be done by considering noise voltage alone. We get,

$$N_f = \frac{N}{1 + A\beta}$$

9.16.5 Distortion Reduces

D is the distortion without feedback and D_f is the distortion with feedback. Considering distortion alone, we get

$$D_f = \frac{D}{1 + A\beta}$$

9.16.6 Bandwidth Increases

Let A be the gain without feedback. A_f is the gain with feedback. Let BW be the bandwidth of the amplifier without feedback. BW_f be the bandwidth with the feedback. We know that the gain bandwidth product is always constant for an amplifier system.

Figure of merit without feedback = Figure of merit with feedback

$$A(BW) = A_f\, BW_f$$

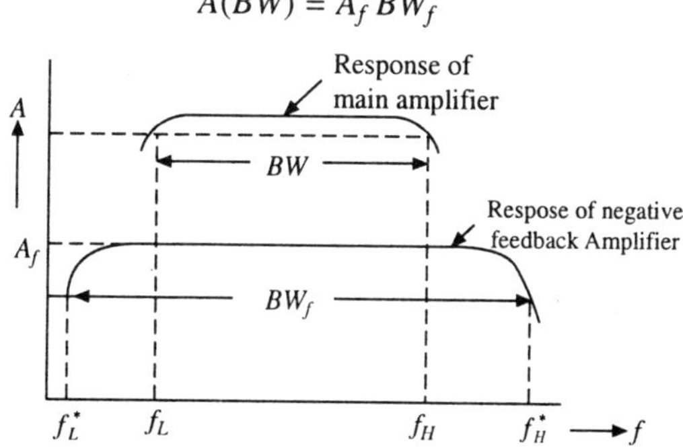

Fig. 9.21 Bandwidth increases.

Substitute (9.21) here

$$A(BW) = \frac{A}{1 + A\beta} BW_f$$

A on either sides get cancelled.

Hence

$$\boxed{BW_f = BW\,(1 + A\beta)} \tag{9.25}$$

Therefore band width increases with feedback.

9.17 Analysis of Current Series Feedback

9.17.1 Input Impedance with Feedback (Z_{in})

Transfer function A here is defined as the ratio of output current to input voltage V_i. The input impedance with feedback amplifier is shown in Fig. 9.22.

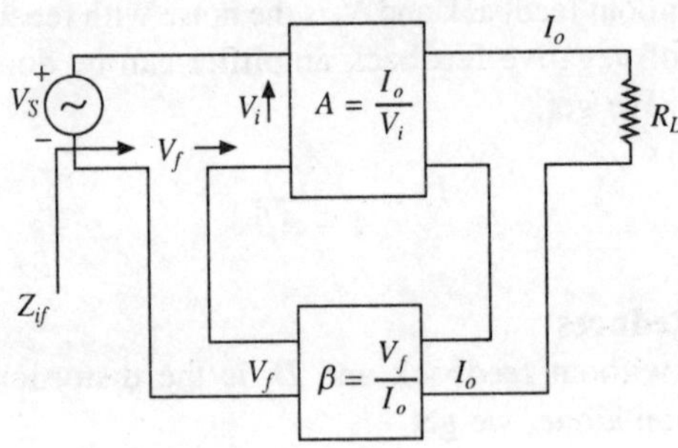

Fig. 9.22 Block diagram.

Apply KVL in input loop

$$V_s - V_i - V_f = 0$$

$$V_s = V_i + V_f$$

$$Z_{if} = \frac{V_s}{I_i} \quad \text{substitute } V_s = V_i + V_f$$

$$Z_{if} = \frac{V_i + V_f}{I_i} \quad \text{substitute } V_f = \beta I_o$$

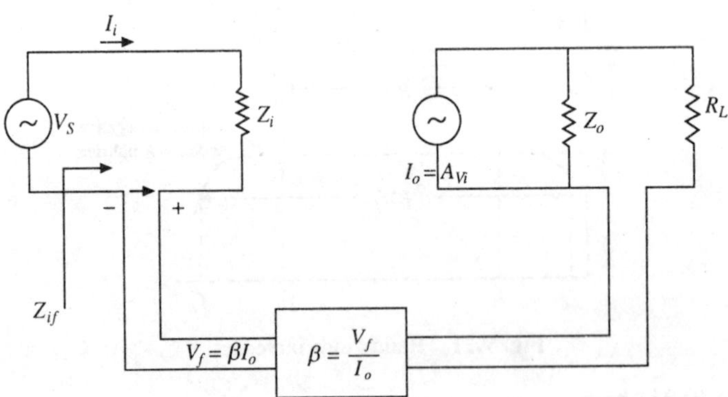

Fig. 9.23 Equivalent circuit

$$Z_{if} = \frac{V_i + \beta I_o}{I_i}$$

substitute $I_o = AV_i$

$$Z_{if} = \frac{V_i + \beta A V_i}{I_i}$$

$$Z_{if} = \frac{V_i(1 + A\beta)}{I_i}$$

substitute $\frac{V_i}{I_i} = Z_i$

$$Z_{if} = Z_i\,(1 + A\beta) \tag{9.26}$$

From the above equation, it can be seen that input impedance increases with feedback which is an advantage.

9.17.2 Output Impedance with Feedback (Z_{of})

Output impedance is defined as the ratio of voltage applied (V_0) to the current I. The output impedance with feedback circuit is shown in Fig. 9.24.

$$Z_{of} = \left.\frac{V_o}{I}\right|_{V_s = 0}$$

Apply KCL at P

$$I + I_o = \frac{V_o}{Z_o}$$

$$I = \frac{V_o}{Z_o} - I_o$$

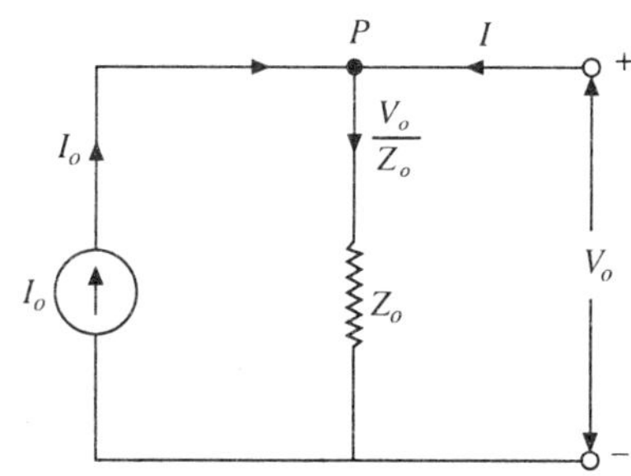

Fig. 9.24 Output circuit.

Substitute $I_o = A_{Vi}$

$$I_o = \frac{V_o}{Z_o} - AV_i$$

Therefore, $I = \frac{V_o}{Z_o} - A_{Vi}$ substitute $V_i = V_f$ since $V_s = 0$

$$I = \frac{V_o}{Z_o} - AV_f$$ substitute $V_f = \beta I_o$

$$I = \frac{V_o}{Z_o} - A\beta I_o$$

$$I = \frac{V_o}{Z_o} - A \cdot \beta I$$

$$I(1 + A \cdot \beta) = \frac{V_o}{Z_o}$$

$$\frac{V_o}{I} = Z_o\,(1 + A\beta)$$

$$Z_{of} = Z_o\,(1 + A\beta) \tag{9.27}$$

Output impedance increases with negative series current feedback. This is not a desirable feature.

9.18 Current Shunt Feedback Amplifier

The current shunt feedback amplifier is shown in Fig. 9.25. Current transfer function A is defined as ratio of output current to input current I_i.

Equivalent circuit of current shunt feedback amplifier is shown in Fig. 9.26.

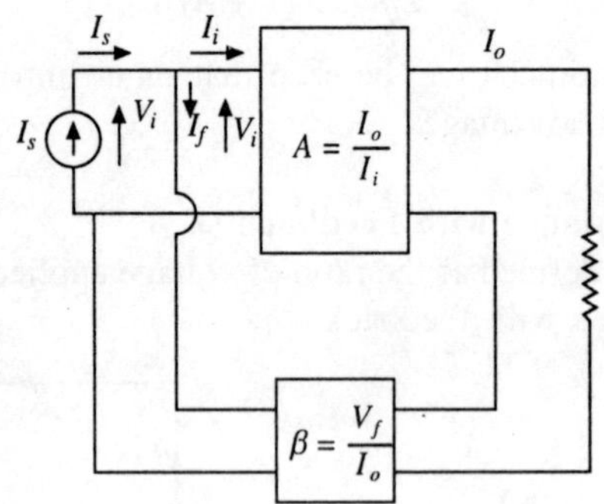

Fig. 9.25 Block diagram.

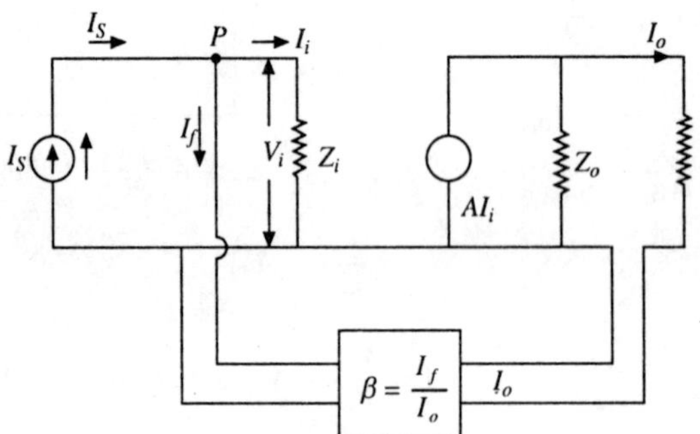

Fig. 9.26 Equivalent circuit.

9.18.1 Input Impedance with Feedback

$$Z_{if} = \frac{V_i}{I_s}$$

$$Z_{if} = \frac{Z_i I_i}{I_s}$$

KCL at P gives $I_i = I_s - I_f$.

Substitute this in the above equation

$$= \frac{Z_i I_s - Z_i I_f}{I_s}$$

substitute $I_f = \beta I_o$

$$Z_{if} = \frac{Z_i I_s - Z_i \beta I_o}{I_s}$$

substitute $I_o = AI_i$

$$Z_{if} = Z_i - \frac{Z_i \beta A I_i}{I_s}$$

substitute $Z_i I_i = V_i$

$$Z_{if} = Z_i - \frac{A\beta}{I_s} V_i$$

substitute $\frac{V_i}{I_s} = Z_{if}$

$$Z_{if} = Z_i - A\beta\, Z_{if}$$

$$Z_{if} + A\beta\, Z_{if} = Z_i$$

$$Z_{if}(1 + A\beta) = Z_i$$

$$Z_{if} = \frac{Z_i}{1 + A\beta} \tag{9.28}$$

Input impedance decreases with feedback. Therefore this method is not used in practice.

9.18.2 Output Impedance with Feedback

The output impedance with feedback circuit is shown in Fig. 9.27.

$$Z_{of} = \left.\frac{V_o}{I}\right|_{I_s = 0}$$

Output current is reversed since an external voltage source at the output drives a current in opposite direction.

When $I_s = 0$

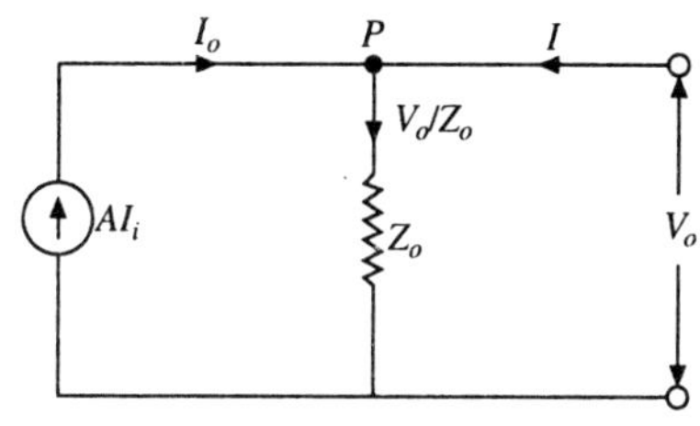

Fig. 9.27 Output circuit.

$I_i = I_f$

Apply KCL at node P.

$$I_o + I = \frac{V_o}{Z_o}$$

$$I = \frac{V_o}{Z_o} - I_o$$

$$I = \frac{V_o}{Z_o} - AI_i \qquad \text{substitute } I_i = I_f$$

$$I = \frac{V_o}{Z_o} - AI_f \qquad \text{substitute } I_f = \beta I_o$$

$$I = \frac{V_o}{z_o} - ABI_o; I(1 + A\beta) = \frac{V_o}{Z_o}$$

$$Z_o(1 + A\beta) = \frac{V_o}{I}$$

$$Z_o(1 + A\beta) = Z_{of}$$

$$Z_{of} = Z_o\,(1 + A\beta) \qquad (9.29)$$

Output impedance increases with feedback. Therefore this amplifier is inferior to the amplifier without feedback. Curent shunt feedback amplifier is the most unpopular amplifier.

Worked Problems

Ex. 9.1 A class-A power amplifier uses a transformer as a coupling device. The transformer has a turns ratio of 10:1 and secondary load resistance is 10 Ω. If the zero signal collector current is 100 mA, find the power output.

$$K = \frac{N_2}{N_1} = \frac{1}{10} = 0.1$$

Quiscent or zero signal collector current I_C = 100 mA.

Resistance referred to primary (R'_L) is R_L/K^2.

$$R'_L = \frac{10}{(0.1)^2} = \frac{10}{\left(\frac{1}{10}\right)^2} = 1000\ \Omega$$

Hence $$I_{rms} = \frac{I_C}{\sqrt{2}} = \frac{100}{\sqrt{2}}\ \text{mA}$$

Thus $$\text{output power} = I_{rms}^2\, R'_L = \left(\frac{100 \times 10^{-3}}{\sqrt{2}}\right)^2 1000 = 5\ \text{Watts}.$$

Ex. 9.2 Power transistor working in a class *A* amplifier has a supply voltage of 10 Volts. If the maximum change in collector current is 50 mA, calculate the power transferred to 4Ω speaker when (*a*) the speaker is directly connected. (b) the speaker is connected through a transformer to the transistor circuit. The turns ratio is 0.1.

$$k = 0.1;\quad \text{Max. change means } 2I_C$$

$$V_{CC} = 10\ \text{V}$$

(a) $2I_C$ = 50 mA

Thus $$I_C = 25\ \text{mA (peak value)}$$

$$I_{rms} = \frac{25}{\sqrt{2}};\quad P_o = I_{rms}^2\, R_L$$

$$P_o = \left(\frac{25 \times 10^{-3}}{\sqrt{2}}\right)^2 \times 4 = \frac{625 \times 10^{-6}}{2} \times 4$$

$$= 1250 \times 10^{-6} = 1.25\ \text{mW}.$$

(b) with transformer

$$R'_L = \frac{R_L}{K^2} = \frac{4}{(0.1)^2} = \frac{4}{\left(\frac{1}{10}\right)^2} = 400\ \Omega$$

Hence $P_o = I_{rms} R'_L = \left(\frac{25 \times 10^{-3}}{\sqrt{2}}\right)^2 = 400 = 25 \times 10^{-6} = 200 = 125$ mW

It can be seen that more power can be transferred to the load through an impedance matching transformer.

Ex. 9.3 Determine the approximate frequency f_T for which the magnitude of the current gain is unity for a BJT whose hybrid parametres are: $g_n = 0.05$, $r_\pi = 2k\Omega$, $C_\pi = 19.5$ PF and $C_\mu = 0.5$ PF.

$$f_T = h_{fe} f_b$$

$$h_{fe} = g_m r_\pi = 0.05 \times 2 \times 10^3 = 100$$

$$f_b = \frac{1}{2\pi r_\pi (C_\pi + C_\mu)} = \frac{1}{2\pi 2 \times 10^3 (19.5 + 0.5) 10^{-12}}$$

$$= 3.97 \text{ MH}_z$$

$$f_T = h_{fe} f_b = 100 \times 3.97 = 397 \text{ MHz}$$

Ex. 9.4 In a BJT amplifier system there is a total harmonic distortion of 6.75%, a current gain of 38 and a voltage gain of 241. Determine the amount of feedback required to have 1% harmonic distortion.

$$A_I = 38 \quad A_V = 241$$

$$D = 6.75\% = 0.0675 \quad D_f = 1\% = 0.01$$

$$D_f = \frac{D}{1 + A\beta}; \; 0.01 = \frac{0.0675}{1 + 241\beta}$$

$$\beta = 2.4\%$$

Ex. 9.5 Find the voltage gain and the feedback factor for a feedback amplifier that produces 90 watts of output with 0.05% distortion. The load resistance is 8Ω. When output supplies 90 watts to 8 Ω load in a non feedback case 4% harmonic distortion is produced. The input signal for 90 watts of output is to be 1 Volt.

$$P_o = 90 \text{ watts} \quad R_L = 8\Omega$$

$$D_f = 0.05\% \qquad D = 4\%$$

$$P_o = \frac{V_o^2}{R}; \qquad 90 = \frac{V_o^2}{8}$$

$$V_o = 26.83 \text{ V.}$$

Hence Voltage gain $A_V = \frac{V_o}{V_i} = \frac{26.83}{1} = 26.83$

$$D_f = \frac{D}{1 + A\beta}$$

$$\frac{0.05}{100} = \frac{4/100}{1 + A\beta}$$

$$0.05 \times 10^{-2} = \frac{4 \times 10^{-2}}{1 + 26.83\,\beta}$$

Therefore, $\beta = 2.94$

Ex. 9.6 An amplifier has a mid frequency gain of 100 and a bandwidth of 200 kHz. (a) What will be the new bandwidth and gain if 5% negative feedback is introduced? (b) What should be the amount of negative feedback if the bandwidth required is 1 MHz?

$A = 100$
$B.W = 200$ kHz
(a) $\beta = 5\% = 0.05$

$$\text{New bandwidth } (B.W') = (1 + A\beta)\, BW$$
$$= (1 + 100 \times 0.05)\, 200$$
$$= 1200 \text{ kHz} = 1.2 \text{ MHz}$$

$$A_f = \frac{A}{1 + A\beta} = \frac{100}{1 + 100 \times 0.05} = \frac{100}{6} = 16.67$$

(b) $B.W'' = 10^3$ kHz.

$$\text{Actual B.W} = 200 \text{ kHz.}$$
$$\text{B.W}'' = (1 + A\beta)BW$$
$$10^3 = (1 + 100\,\beta)\, 200$$
$$5 - 1 = 100\,\beta;\ \beta = 0.04 = 4\%$$

Ex. 9.7 A non-feedback amplifier has a gain of 100. Design a negative feedback amplifier with a feedback factor of 10%. Find the number of stages required to get a gain of 100 with feedback.

$$A = 100;\quad A_f = 100$$
$$\beta = 0.1;\quad A_f = \frac{A}{1 + A\beta}$$
$$100 = \frac{A}{1 + 0.1A}$$
$$100\,(1 + 0.1A) = A$$
$$100 + 10A = A$$
$$9A = -100 \quad A = -11.1$$
$$A_1 = \frac{A}{1 + A\beta} = \frac{100}{1 + 100 \times 0.1}$$
$$A_1 = 9.09$$

With n number of stages $\quad A_f = A_1^n$

Take ln on both sides $\quad \ln A_f = n \ln A_1$

$$n = \frac{\ln A_f}{\ln A_1} = \frac{\ln 100}{\ln 9.09} = 2$$

Two stages are needed to get an overall gain of 100 with negative feedback.

Ex. 9.8 An A.C. amplifier has a mid frequency voltage gain of 100 and lower and upper 3-*db* frequencies of 10 kHz and 100 kHz (resly). If a 10% negative feedback is applied, what would be (i) the mid frequency gain (ii) lower and upper 3-*db* frequencies?

$$A = 100 \quad \beta = 0.1$$

(i) $A_f = \frac{A}{1 + A\beta} = \frac{100}{1 + 100 \times 0.1} = 9.09$

(ii) $f_L' = \frac{f_L}{1 + A\beta}$

$$f_L' = \frac{10}{11} = 0.9 \text{ kHz}$$

$$f_H' = f_H (1 + A\beta) = 100(11) = 1100 \text{ kHz}$$

Ex. 9.9 A negative feedback of $\beta = 0.002$ is applied to an amplifier of gain 1000. Calculate change in overall gain of the feedback amplifier if the internal amplifier is subjected to a gain reduction of 15%.

$$Af_1 = \frac{A}{1 + A\beta} = \frac{1000}{1 + 1000 \times 0.002} = \frac{1000}{3} = 333.3$$

$$Af_2 = \frac{A^1}{1 + A^1\beta} = \frac{850}{1 + 850 \times 0.002} = 314.8$$

$$\text{Change} = 333.3 - 314.8 = 18.51$$

Ex. 9.10 An amplifier with gain of 60 db uses $\frac{1}{20}$ of its output in negative feedback. Calculate gain with feedback in db.

$$\text{Gain in db} = 20 \log A$$

$$60 = 20 \times \log A$$

$$\log A = 3$$

$$\log A = 3 \times 2.3 = 6.9$$

$$A = e^{6.9} = 1000$$

$$A_f = \frac{A}{1 + A\beta} = \frac{1000}{1 + 1000 \times \frac{1}{20}} = 19.6$$

Feedback gain in dB = 20 × 1.29 = 25.8 db

Short Questions

1. In common emitter amplifier, the unbypassed emitter resistance provides *current series feedback.*
2. An amplifier with mid band gain $A = 500$ has $\beta - 0.01$. If upper cut off frequency without feedback his at 60 kHz, then with –ve feedback it becomes

$$f_H' = f_H (1 + A\beta) = 60 (1 + 5) = 360$$

3. In the current series feedback amplifier, the basic amplifier and the feedback

network are in parallel at the input and series connected at the output. The signal sampled and summing done will be *current and voltage.*

4. A transistor amplifier has A = 50, Z_i = 1 kΩ of Z_o = 40 kΩ. The amplifier is now provided with 10% –ve feedback. A_f, Z_{if} and Z_{of} will be . . . . . .

$$A_f = \frac{A}{1 + A\beta} = \frac{50}{1 + 50 \times 0.1} = 8.3$$

$$Z_{if} = 6 \times 1 = 6 \text{ k}\Omega$$

$$Z_{of} = \frac{40}{6} = 6.66 \text{ k}\Omega$$

10

Oscillators

Oscillators convert D.C. power into A.C. power. It generates A.C. output without A.C. input signal. Alternator can be used to produce low frequency A.C. power. it is not possible to generate high frequency A.C. power using an alternator since large number of poles are required.

10.1 Positive Feedback Amplifier

The basics of positive feedback amplifier are required to understand oscillator circuits since oscillator is a positive feedback amplifier. The block diagram of positive feedback amplifier is shown in Fig. 10.1. The +ve sign in the summer indicates that feedback voltage gets added with the input signal V_s.

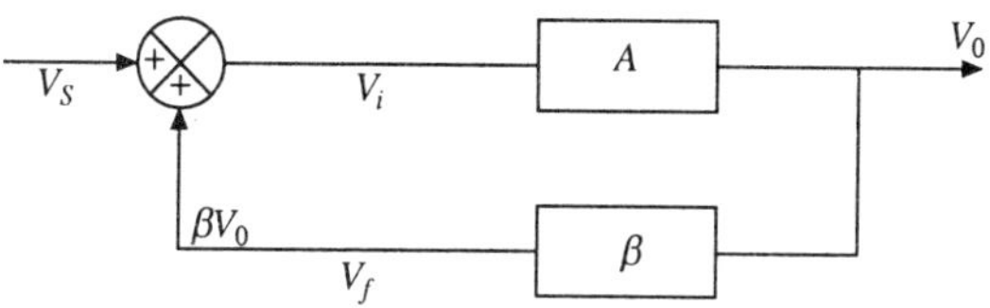

Fig. 10.1 Positive feedback amplifier.

Gain of main amplifier $A = \frac{V_o}{V_i}$

Gain of feedback amplifier $A_f = \frac{V_o}{V_s}$

$$A_f = \frac{V_o}{V_s} = \frac{AV_i}{V_s}$$

Substitute $V_i = V_s + V_f = V_s + \beta V_o$

$$A_f = \frac{A(V_s + \beta V_o)}{V_s}$$

Divide term by term

$$A_f = A + A\beta\frac{V_o}{V_s}$$

Substitute $\frac{V_o}{V_s} = A_f$

$$A_f = A + A\,\beta A_f; \qquad A_f - A\beta A_f = A$$

$$A_f(1 - A\beta) = A$$

$$A_f = \frac{A}{1 - A\beta} \tag{10.1}$$

From the above equation, it can be seen that the overall gain increases with positive feedback.

10.2 Bharkhausen's Criteria

For a positive feedback amplifier

$$A_f = \frac{A}{1 - A\beta}$$

When $A\beta = 1$, A_f tends to infinity.

This implies that output voltage can be obtained without input A.C. voltage. A small noise voltage is available in the circuit. Initially $A\beta$ must be greater than one to increase the level of oscillations to the required value. This can be seen from the output voltage waveform shown in Fig. 10.2. Then magnitude of $A\beta$ must reduce to unity to produce sustained oscillations. Magnitude and phase angle conditions are given below:

$$A\beta = 1 + j\,0$$

$$|A\beta| \angle\theta = 1 \angle 0$$

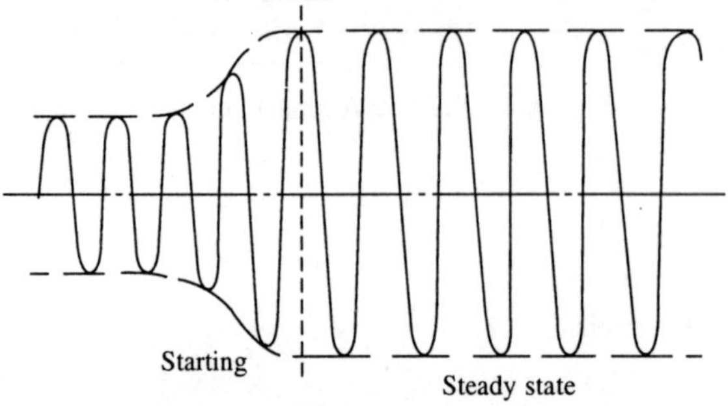

Starting

Steady state

Fig. 10.2 Output voltage of oscillator.

Zero angle implies that the feedback voltage must be added in phase with the noise voltage. The total loop phase shift must be zero degrees or integral multiples of 2π. Thus a positive feedback amplifier is used to add the feedback voltage in phase with the noise voltage.

According to Bharkhausen's criteria, sustained oscillations can be maintained if closed loop gain $A\beta$ is unity and loop phase shift is zero.

The following types of oscillators are discussed in this chapter.

1. Hartley oscillator
2. Colpitt's oscillator
3. Crystal oscillator

4. *R-C* phase shift oscillator
5. Wein Bridge oscillator

10.3 Hartley Oscillator

From circuit theory, we know an that ideal *L-C* circuit produces sustained oscillations. Ideal *L-C* parallel circuit is similar to ideal pendulum which can oscillate for ever. Practical *L-C* circuit is similar to practical pendulum. It oscillates for sometime and stops. If a charged capacitor is connected in parallel with an inductor, it can not maintain sustained oscillations due to the losses. Amplifier in the *L-C* oscillator circuit supplies the losses in the *L-C* circuit since it converts D.C. power in the battery into useful A.C. power.

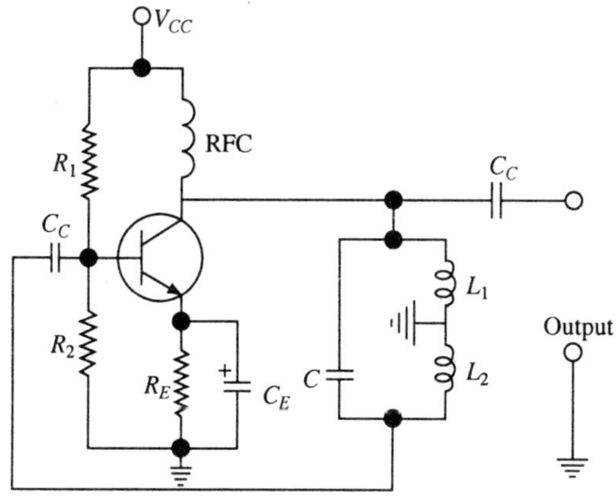

Fig. 10.3 Hartley oscillator.

The output of the transistor amplifier is fed back to the input through the *LC* network. The *LC* network consists of a centre tapped inductor and a capacitor connected in parallel. The *CE* amplifier introduces a phase shift of 180°. The centre tapped inductor introduces another 180° phase shift. The total loop phase shift is 360° or 0°. Thus the feedback voltage is in phase with the noise voltage.

According to Bharkhausen's criterion, oscillations are produced if $A\beta$ is greater than one initially and the total loop phase shift is zero. Capacitors C_{C1} and C_{C2} block the D.C. current towards the tank circuit. The RFC acts as short for D.C. and hence full D.C. voltage is available at the collector. This acts as almost open for A.C. Thus ac cannot pass through the battery and it has to pass through the tank circuit. The tank circuit operates at resonant frequency.

10.4 Colpitt's Oscillator

Colpitt's oscillator circuit is shown in Fig. 10.4.

In this oscillator, a centre tapped capacitor is used in parallel with an inductor. A small noise voltage is available in R_2. This voltage is amplified by the transistor amplifier. The output voltage has 180° phase shift with respect to the

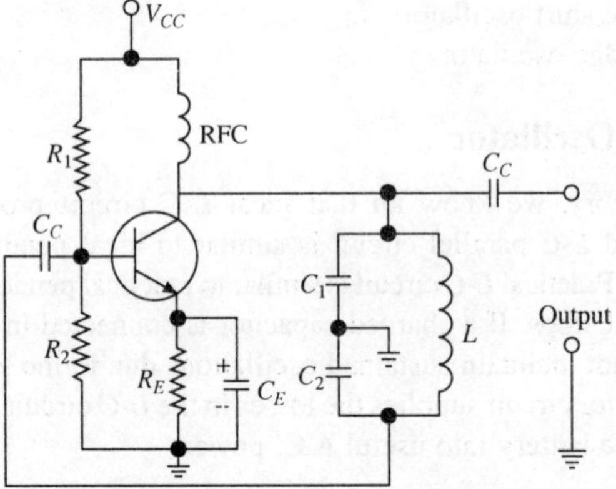

Fig. 10.4 Colpitt's oscillator.

noise voltage. The centre tapped capacitor introduces a total phase shift of 180° and hence the feedback voltage is in phase with the noise voltage. This satisfies the Bharkhausen's criteria since the feedback is a positive feedback. Wide variation of frequency is possible since two variable capacitors are used in the circuit.

The *L-C* oscillators operate at resonant frequency. At resonance, net susceptance is zero.

$$b_L - b_C = 0$$

$$\frac{1}{\omega L} = \omega C$$

$$\omega^2 = \frac{1}{LC}$$

$$2\pi f = \omega = \frac{1}{\sqrt{LC}}$$

$$f = \frac{1}{2\pi\sqrt{LC}}$$

In Hartley oscillator,

$$L = L_1 + L_2 + 2M$$

In colpitts oscillator, $$C = \frac{C_1 C_2}{C_1 + C_2}$$

10.5 Crystal Oscillator

When mechanical force is applied on a crystal in one direction, the electrical oscillations are produced in the perpendicular direction and vice versa. This effect is known as Piezo electric effect. The advantage of the crystal is that it produces stable oscillations over a long period. The circuit configuration of this oscillator is almost similar to the Colpitt's oscillator except that the inductor is replaced by the crystal. It is shown in Fig. 10.5.

The output energy is fed to the input through the crystal. The common emitter amplifier produces 180° phase shift and the centre tapped capacitor produces another 180° phase shift. The total loop phase shift is zero. If $A\beta > 1$, the oscillations are produced in the circuit.

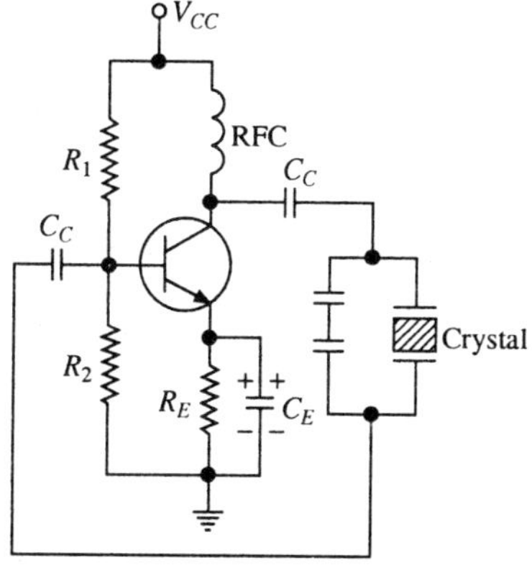

Fig. 10.5 Crystal oscillator.

10.6 R-C Phase Shift Oscillator

R-C phase shift oscillator circuit is shown in Fig. 10.6. The *L-C* oscillators are not suitable for low frequencies since the capacitor value required in the circuit is very large. *R-C* oscillators are best suited for low frequency operation. Part of the output voltage is fed to the input through the *R-C* phase shift network or ladder network. At a particular frequency, the three *R-C* sections together produce a phase shift of 180°. Thus the total loop phase shift is zero or feedback voltage is in phase with the noise voltage. Three sections produce 180° phase shift. The frequency can be varied by varying *R* or *C*. This oscillator cannot be used for high frequency operation since the capacitors act as short circuit at high frequencies

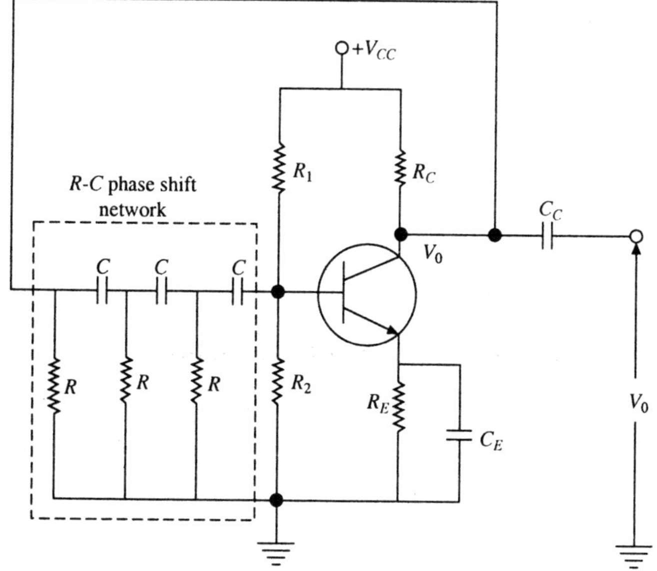

Fig. 10.6 *R-C* phase shift oscillator.

and load the amplifier heavily. The expression for frequency is derived in section 10.8. Phase shift produced by each *R-C* section is not equal since the currents are not equal. At a particular frequency, total phase shift produced by *R-C* ladder network is 180°.

10.7 Wein Bridge Oscillator

In this oscillator, a total loop phase shift of 360° is obtained by using two stage common emitter amplifier. A negative feed back is provided from the first stage and a positive feedback is provided from the output of second stage. Initially the lamp is cold and the resitance R_2 is less. The unbalance will be higher. Higher positive feed back will be given such that $A\beta$ is greater than unity. As the lamp gets heated, the resistance increases, the unbalance decreases and the $A\beta$ value drops to unity. Thus a combination of positive and negative feedbacks are used for initiating the oscillations and maintaining sustained oscillations. The frequency of oscillations is given by

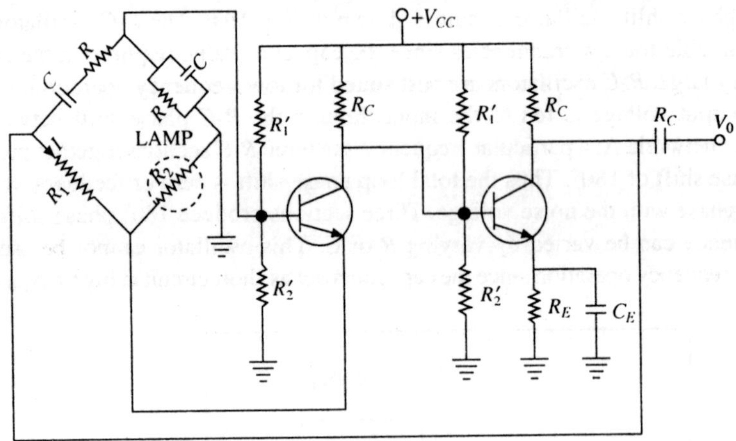

Fig. 10.7 Wein Bridge oscillator.

$$f = \frac{1}{2\pi RC}$$

This expression is derived in Section 10.9.

10.8 Analysis of *R-C* Phase Shift Oscillator

KVL equations in matrix form by inspection of the network shown in Fig. 10.8 are as follows: Let $x = -jx_c$

$$(Z)\ (I) = (V)$$

$$\begin{bmatrix} (X+R) & -R & 0 \\ -R & (X+2R) & -R \\ 0 & -R & (X+2R) \end{bmatrix} \begin{bmatrix} I_1 \\ I_2 \\ I_3 \end{bmatrix} = \begin{bmatrix} V_o \\ 0 \\ 0 \end{bmatrix}$$

$$I_3 = \frac{\Delta_3}{\Delta} = \frac{\text{Det of} \begin{bmatrix} (X+R) & -R & V_o \\ -R & (X+2R) & 0 \\ 0 & -R & 0 \end{bmatrix}}{\text{Det of} \begin{bmatrix} (X+R) & -R & 0 \\ -R & (X+2R) & -R \\ 0 & -R & (X+2R) \end{bmatrix}}$$

$$I_3 = \frac{V_o R^2}{|Z|} \tag{10.1a}$$

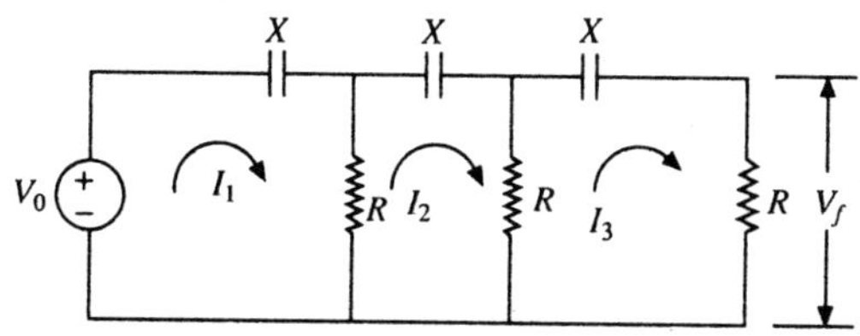

Fig. 10.8 R-C phase shift network.

$$|Z| = (X + R)[(X + 2R)^2 - R^2] + R[-R(X + 2R) - 0] + 0$$

$$= (X + R)[X^2 + 4R^2 + 4XR - R^2] + R[-RX - 2R^2]$$

$$= (X + R)[X^2 + 3R^2 + 4XR] - R^2X - 2R^3$$

$$= X^3 + 3R^2X + 4X^2R + RX^2 + 3R^3 + 4XR^2 - R^2X - 2R^3$$

$$|Z| = X^3 + 6R^2X + 5X^2R + R^3$$

$$V_f = I_3R$$

Substitute $I_3 = \dfrac{V_o R^2}{|Z|}$ in the above equaton.

$$V_f = \frac{V_o R^2}{|Z|} R; \quad \frac{V_f}{V_o} = \frac{R^3}{|Z|}$$

Hence $\beta = \dfrac{R^3}{X^3 + 6R^2X + 5X^2R + R^3}$

Divide numerator and denominator by R^3.

Thus $$\beta = \frac{1}{\dfrac{X^3}{R^3} + 6\dfrac{X}{R} + 5\dfrac{X^2}{R^2} + 1}$$

where $$X = \frac{1}{j\omega C}; \quad \frac{X}{R} = \frac{1}{j\omega CR}$$

$$X = -jX_C = -\frac{j}{\omega C} = \frac{1}{j\omega C}$$

Substiute $X = \frac{1}{j\omega C}$

$$\beta = \frac{1}{\left(\frac{1}{j\omega CR}\right)^3 + 6\frac{1}{j\omega CR} + 5\left(\frac{1}{j\omega CR}\right)^2 + 1}$$

Let, $$\frac{1}{\omega CR} = \alpha \tag{10.2}$$

$$\beta = \frac{1}{\frac{\alpha^3}{j^3} + 6\frac{\alpha}{j} + 5\frac{\alpha^2}{j^2} + 1}$$

$$\frac{1}{j^3} \times \frac{j}{j} = \frac{j}{j^4} = j$$

since $j^4 = (-1 \times -1) = 1$

$$\beta = \frac{1}{j\alpha^3 - j6\alpha - 5\alpha^2 + 1}$$

Multiply on both sides with A

$$A\beta = \frac{A}{j\alpha^3 - j6\alpha - 5\alpha^2 + 1} = \frac{A}{(1 - 5\alpha^2) + j(\alpha^3 - 6\alpha)}$$

Let $a = 1 - 5\alpha^2$, $b = \alpha^3 - 6\alpha$

Let $$A\beta = \frac{A}{a + jb}$$

$$A\beta = \frac{A}{a + jb} \times \frac{a - jb}{a - jb}$$

Therefore, $$A\beta = \frac{A(a - jb)}{a^2 + b^2} = \frac{Aa}{a^2 + b^2} - \frac{jAb}{a^2 + b^2}$$

condition for sustained oscillations. Bharkhausen's criteria is $|A\beta| = 1$.

Hence $$A\beta = 1 + j0 = \frac{Aa}{a^2 + b^2} - \frac{jAb}{a^2 + b^2} = 1 + j0$$

Equate real terms and imaginary terms of LHS with those of RHS.

Therefore, $$\frac{Aa}{a^2 + b^2} = 1, \frac{jAb}{a^2 + b^2} = 0$$

From this $b = 0$

$$\alpha^3 - 6\alpha = 0$$

$$\alpha^3 = 6\alpha$$

$$\alpha = \sqrt{6} \tag{10.3}$$

$$\frac{1}{\omega CR} = \sqrt{6}$$

$$\omega = \frac{1}{RC\sqrt{6}}$$

$$2\pi f = \frac{1}{RC\sqrt{6}}$$

$$f = \frac{1}{2\pi RC\sqrt{6}}$$

This is the expression for frequency of oscillation for R-C phase shift oscillator. Latest value of loop gain is

$$A\beta = \frac{A}{1 - 5\alpha^2}.$$ Since imaginary part is zero.

$$\beta = \frac{1}{1 - 5\alpha^2}$$

Substitute $\alpha = \sqrt{6}$

$$\beta = \frac{1}{1 - 5(\sqrt{6})^2} = \frac{-1}{29}$$

Hence $$|\beta| = \frac{1}{29}$$

Condition for sustained oscillation

$$A\beta = 1$$

$$A \times \frac{1}{29} = 1$$

Therefore, $$A = 29 \qquad (10.4)$$

The amplifier should have a gain of 29 to maintain oscillations.

10.9 Frequency of Wein Bridge Oscillator

Wein Bridge circuit is shown in Fig. 10.9. From Fig. 10.9

$$Z_1 = R - jX_C$$

$$= R - \frac{j}{\omega C} = R + \frac{1}{j\omega C} = \frac{j\omega CR + 1}{j\omega C}$$

$$Z_3 = \frac{R\dfrac{1}{j\omega C}}{R + \dfrac{1}{j\omega C}} = \frac{\dfrac{R}{j\omega C}}{\dfrac{j\omega CR + 1}{j\omega C}} = \frac{R}{1 + j\omega CR}$$

In a balanced bridge product of impedance of opposite arms must be equal.

$$Z_1Z_4 = Z_2Z_3$$

$$\left(\frac{1 + j\omega CR}{j\omega C}\right) R_2 = \frac{RR_1}{1 + j\omega CR}$$

$$(1 - \omega^2 C^2 R^2 + 2 j\omega C R) = \frac{j\omega CRR_1}{R_2}$$

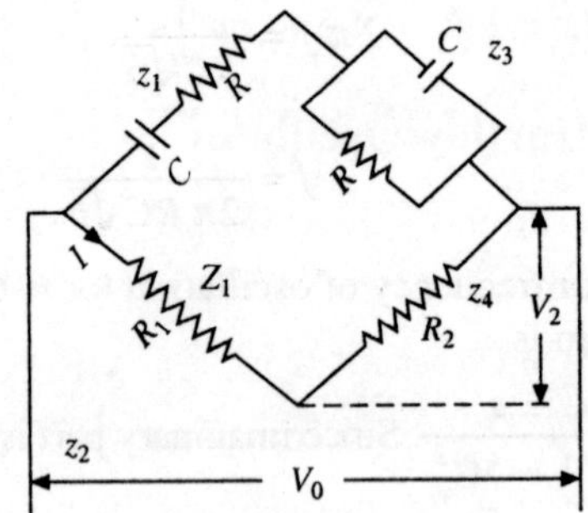

Fig. 10.9 Wein Bridge circuit.

Compare real terms of LHS and RHS

$$1 - \omega^2 C^2 R^2 = 0 \quad \text{Therefore,} \quad \omega^2 C^2 R^2 = 1$$

Hence $$\omega^2 = \frac{1}{C^2 R^2}; \quad \omega = \frac{1}{CR}; \quad 2\pi f = \frac{1}{RC}$$

Thus $$f = \frac{1}{2\pi RC} \tag{10.5}$$

This equation is the expression for frequency of oscillations.

Compare imaginary terms,

$$2\omega CR = \frac{\omega CRR_1}{R_2}$$

Hence $$\frac{R_1}{R_2} = 2$$

$$\beta = \frac{V_2}{V_o}$$

$$\beta = \frac{IR_2}{I(R_1 + R_2)} = \frac{R_2}{R_1 + R_2}$$

Divide numerator and denominator by R_2

$$= \frac{1}{\frac{R_1}{R_2} + 1} = \frac{1}{1 + \frac{R_1}{R_2}}$$

Substitute $$\frac{R_1}{R_2} = 2$$

$$\beta = \frac{1}{3}$$

Condition for oscillations is

$$A\beta = 1$$

$$A \times \frac{1}{3} = 1$$

Hence $$A = 3 \quad (10.6)$$

The amplifier should have a gain of 3 to maintain sustained oscillations.

10.10 Analysis of Hartley Oscillator

In Fig. 10.10, both the currents are leaving the dot. Therefore, sign of mutual inductance is positive.

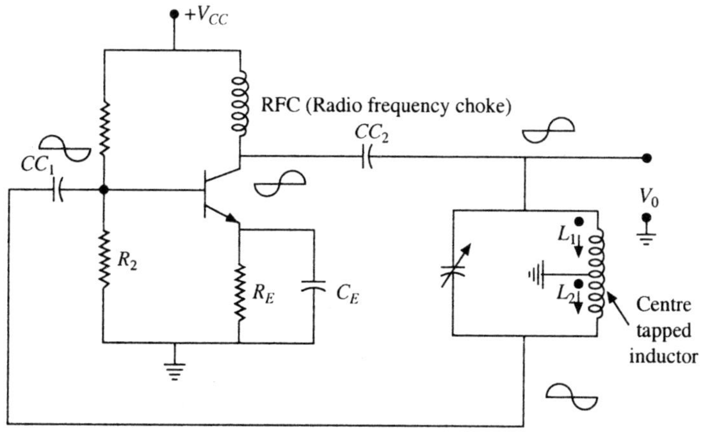

Fig. 10.10 Hartley oscillator.

$$L = L_1 + L_2 + 2M$$

At resonance $X_L = X_C$

$$\omega_o L = 1/\omega_o C; \;\; \omega_o^2 = \frac{1}{LC}$$

$$\omega_o = 1/\sqrt{LC} \Rightarrow 2\pi f_o = \frac{1}{\sqrt{LC}} \Rightarrow f_o = \frac{1}{2\pi\sqrt{LC}}$$

Substitute $L = L_1 + L_2 + 2M$

Therefore,

$$\boxed{f_o = \frac{1}{2\pi\sqrt{(L_1 + L_2 + 2M)C}}} \quad (10.7)$$

where f_o is the output frequency of Hartley oscillator.

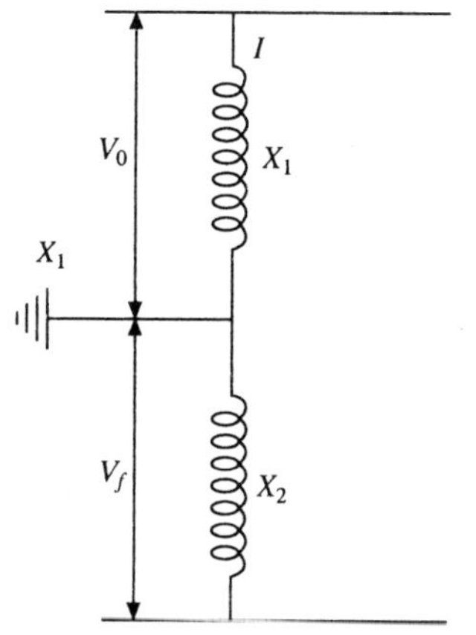

Fig. 10.10b Centre tapped reactor

Conditions for Oscillation

From the oscillator circuit, it can be seen that output voltage is available across centre tapped inductor.

Feedback voltage is taken across L_2.

$$\beta = V_f / V_o = \frac{Ix_2}{Ix_1} = \frac{\omega L_2}{\omega L_1}$$

Therefore,
$$\boxed{\beta = \frac{L_2}{L_1}} \tag{10.8}$$

Assume the value of L_2. The value of L_1 can be obtained by using this equation.

10.11 Gain of Transistor Amplifier

$$A_V = \frac{h_{fe} R_L}{h_{ie}}$$

R_L value is selected such that $R_L = h_{ie}$.

$$\boxed{\therefore A_V = h_{fe}}$$

Hence, condition for sustained oscillation, $A\beta = 1$.

$$h_{fe} \frac{L_2}{L_1} = 1 \Rightarrow \boxed{h_{fe} = L_1 / L_2} \tag{10.9}$$

At starting condition

$$\boxed{h_{fe} > \frac{L_1}{L_2}}$$

For producing oscillations the relation given in (10.9) must be satisfied.

10.12 Analysis of Colpitts Oscillator

Here the centre tapped capacitor and one inductor are in parallel. RFC prevents A.C. flowing into the battery. Centre tapped capacitor produces 180° phase shift and transistor amplifier produces 180°. Total phase shift = 360°. By varying the capacitance we get the wide range of frequencies.

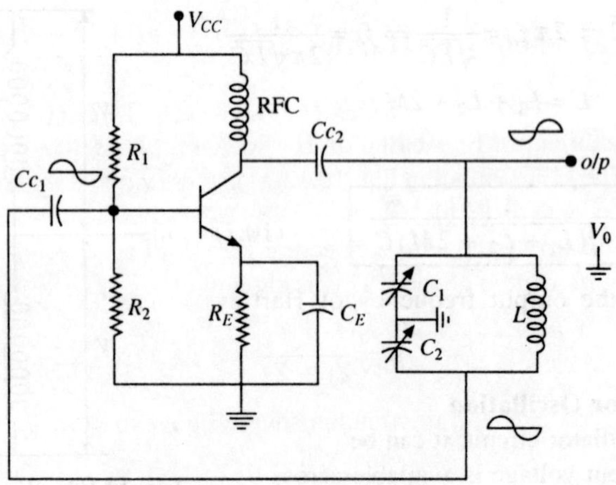

Fig. 10.11(a) corpitts oscillator.

Parallel resonant circuit offers high impedance at resonance. The current supplied by the amplifier to the tank circuit is negligible. Current through C_1 and C_2 is same. Therefore, C_1 and C_2 are in series.

$$\omega_o^2 = 1/LC \Rightarrow \omega_o = 1/\sqrt{LC};\ f_o = 1/2\pi\sqrt{LC}$$

$$\boxed{\therefore f_o = \frac{1}{2\pi\sqrt{L\dfrac{C_1C_2}{C_1 + C_2}}}} \quad \because\ C = \frac{C_1C_2}{C_1 + C_2} \qquad (10.10)$$

where f_o is the frequency of colpitts oscillator. Voltage across C_1 is output voltage. voltage across C_2 is feedback voltage.

$$V_o = Ix_1,\ V_f = Ix_2$$

$$\beta = \frac{V_f}{V_o} = \frac{Ix_2}{Ix_1} \Rightarrow x_2/x_1 \Rightarrow \frac{1/\omega C_2}{1/\omega C_1}$$

$$\boxed{\therefore \beta = C_1/C_2} = \frac{\text{Output capacitance}}{\text{Feedback capacitance}} \qquad (10.11)$$

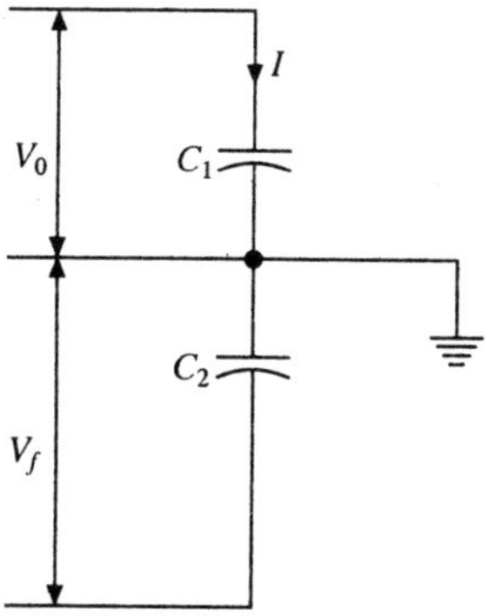

Condition for oscillation

$$A = \frac{h_{fe}R_L}{h_{ie}}$$

if $R_L = h_{ie}$, $A = h_{fe}$

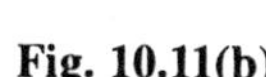

Fig. 10.11(b)

Condition for oscillations

$$A\beta = 1$$

$$h_{fe}\frac{C_1}{C_2} = 1;\ \boxed{h_{fe} = \frac{C_2}{C_1}} \qquad (10.12)$$

10.13 General Theory of Hartley and Colpitt's FET Oscillator

A.C. equivalent circuit of FET based on Hartley/colpitt's oscillator is shown in Fig. 10.12(a). The FET symbol in Fig. 10.12(a) is replaced by the model of FET. This is shown in Fig. 10.12(b).

The current drawn by the wire connecting gate terminal and node 3 is zero. Therefore this wire can be opened. This is shown in Fig. 10.12(c).

$$\beta = \frac{Z_1}{Z_1 + Z_3} \qquad (10.13)$$

We know the voltage gain of common source FET amplifier.

$$A = -\frac{\mu Z_L}{r_D + Z_L}$$

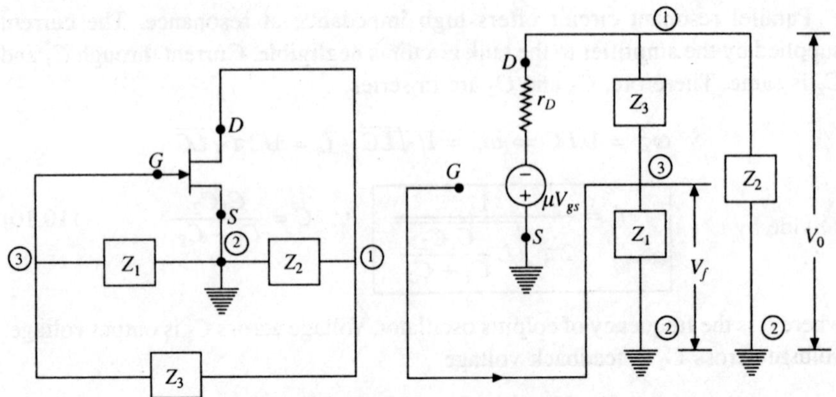

Fig. 10.12(a) AC Equivalent circuit **Fig. 10.12(b)** Equivalent circuit

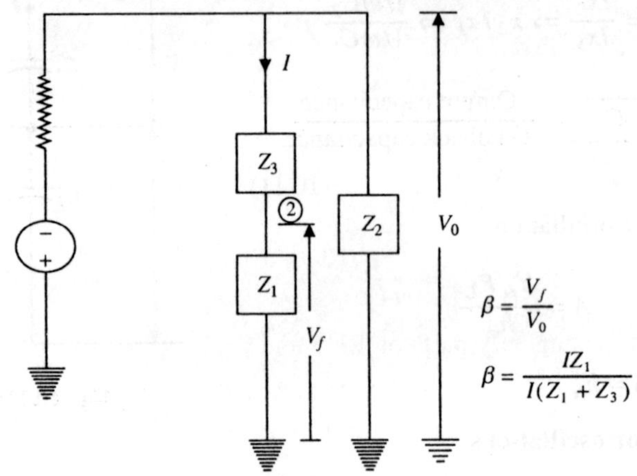

Fig. 10.12(c) Circuit ofter removing gate wire

where Z_L is the resultant impedance connected between node (1) and ground.

$$Z_L = Z_1 \parallel (Z_1 + Z_3)$$

$$\frac{1}{Z_L} = \frac{1}{Z_2} + \frac{1}{Z_1 + Z_3}$$

$$\frac{1}{Z_L} = \frac{Z_2 + (Z_1 + Z_3)}{Z_2(Z_1 + Z_3)}$$

$$\frac{1}{Z_L} = \frac{Z_1 + Z_3 + Z_2}{Z_2(Z_1 + Z_3)} \tag{10.13b}$$

Condition for sustained oscillations

$$A\beta = 1; \qquad A = \frac{1}{\beta}$$

$$A = \frac{Z_1 + Z_3}{Z_1}; \qquad A = 1 + \frac{Z_3}{Z_1}$$

$$1 + \frac{Z_3}{Z_1} - A = 0; \qquad 1 + \frac{Z_3}{Z_1} + \frac{\mu Z_L}{r_d + Z_L} = 0$$

$$\frac{Z_1(r_d + Z_L) + Z_3(r_d + Z_L) + \mu Z_1 Z_L}{Z_1(r_d + Z_L)} = 0$$

$$(r_D + Z_L)[Z_1 + Z_3] + \mu Z_1 Z_L = 0$$

Divide by $(Z_1 + Z_3)$

$$r_D + Z_L + \frac{\mu Z_1 Z_L}{Z_1 + Z_3} = 0$$

Substitute (10.13b) here

$$r_D + \frac{Z_2(Z_1 + Z_3)}{Z_1 + Z_2 + Z_3} + \frac{\mu Z_1}{(Z_1 + Z_3)} \frac{(Z_1 + Z_3) Z_2}{Z_1 + Z_2 + Z_3} = 0$$

$$r_D(Z_1 + Z_2 + Z_3) + Z_2 Z_1 + Z_2 Z_3 + \mu Z_1 Z_2 = 0$$

$$r_D(Z_1 + Z_2 + Z_3) + Z_1 Z_2(1 + \mu) + Z_2 Z_3 = 0 \qquad (10.14)$$

10.13.1 Hartley Oscillator

$$Z_1 = +j\,\omega L_1; \qquad Z_2 = j\,\omega L_2$$

$$Z_3 = \frac{-j}{\omega C_1}$$

Comparing the imaginary parts on RHS and LHS of (10.14)

$$r_D(Z_1 + Z_2 + Z_3) = 0 \qquad (10.14a)$$

$$r_D\left(j\omega L_1 + j\omega L_2 - \frac{j}{\omega C_1}\right) = 0$$

r_D cannot be zero

Hence $$j\omega L_1 + j\omega L_2 - \frac{j}{\omega C_1} = 0$$

$$\omega L_1 + \omega L_2 = \frac{1}{\omega C_1}$$

$$\omega(L_1 + L_2) = \frac{1}{\omega C_1}$$

Let $C = C_1$ since Hartley oscillator has only one capacitor in the tuned circuit.

$$\frac{1}{\omega^2 C} = L_1 + L_2$$

$$\omega^2 C = \frac{1}{L_1 + L_2} \qquad (10.14b)$$

$$\omega^2 C = \frac{1}{L_1 + L_2}$$

$$\omega^2 = \frac{1}{C(L_1 + L_2)}$$

$$\omega = \frac{1}{\sqrt{C(L_1 + L_1)}}$$

$$2\pi f = \frac{1}{\sqrt{C(L_1 + L_2)}}$$

Frequency of oscillation $$f = \frac{1}{2\pi\sqrt{(L_1 + L_2)C}} \tag{10.15}$$

By equating the real part of (10.14) to zero, we get the condition for oscillations

$$Z_1Z_2(1 + \mu) + Z_2Z_3 = 0$$

$$Z_2[Z_1(1 + \mu) + Z_3] = 0$$

Hence, $$Z_1(1 + \mu) + Z_3 = 0$$

$$j\omega L_1 (1 + \mu) - \frac{j}{\omega C} = 0$$

$$\omega L_1 (1 + \mu) - \frac{1}{\omega C} = 0$$

$$\omega^2 L_1 C (1 + \mu) - 1 = 0$$

$$\omega^2 L_1 C (1 + \mu) = 1$$

$$L_1 (1 + \mu) = \frac{1}{\omega^2 C}$$

From equation (10.14b), substitute $\frac{1}{\omega^2 C} = L_1 + L_2$

$$L_1 + L_1\mu = L_1 + L_2$$

We get condition for sustained oscillations as

$$\mu = \frac{L_2}{L_1} \tag{10.16}$$

This is the condition corresponding to sustained oscillations. At the time of start $A\beta$ must be greater than 1.

Thus $$\mu > \frac{L_2}{L_1}$$

10.13.2 Colpitts Oscillator

Let Z_1 and Z_2 be capacitive reactances, Z_3 be inductive reactance.

$$Z_1 = \frac{-j}{\omega C_1};\ Z_2 = \frac{-j}{\omega C_2}$$

$$Z_3 = j\omega L$$

From (10.14a)

$$r_D(Z_1 + Z_2 + Z_3) = 0$$

$$r_D\left(-\frac{j}{\omega C_1} - \frac{j}{\omega C_2} + j\omega_L\right) = 0$$

Hence $$j\left[-\frac{1}{\omega C_1} - \frac{1}{\omega C_2} + \omega_L\right] = 0$$

Thus $$\omega L = \frac{1}{\omega C_1} + \frac{1}{\omega C_2}$$

$$\omega^2 L = \frac{1}{C_1} + \frac{1}{C_2} = \frac{1}{C} \tag{10.17}$$

$$\omega^2 = \frac{1}{LC}$$

$$\omega = \frac{1}{\sqrt{LC}}; \quad 2\pi f = \frac{1}{\sqrt{LC}}$$

Hence $$f = \frac{1}{2\pi\sqrt{LC}} \tag{10.18}$$

By equating the real part

$$Z_1Z_2(1 + \mu) + Z_2Z_3 = 0$$

$$Z_1(1 + \mu) + Z_3 = 0$$

$$\frac{-j}{\omega C_1}(1 + \mu) + j\omega_L = 0$$

$$+\frac{1 + \mu}{C_1} - \omega^2 L = 0$$

$$\frac{1 + \mu}{C_1} = \omega^2 L$$

From equation (10.17), substitute $\omega^2 L = \frac{1}{C_1} + \frac{1}{C_2}$

$$\frac{1}{C_1} + \frac{\mu}{C_1} = \frac{1}{C_1} + \frac{1}{C_2}$$

Hence $$\mu = \frac{C_1}{C_2} \tag{10.19}$$

This is the condition for maintaining oscillations.

Worked Problems

Ex. 10.1 The tuned circuit of L-C oscillator has a fixed inductance of 50 μH and has to be tunable over a frequency band of 200 kHz to 1000 kHz. Find the range of variable capacitor required.

$$f_1\ 200\ \text{kHz};\ f_2 = 1000\ \text{kHz};\ L = 50 \times 10^{-6}\ \text{H}.$$

$$f_1 = \frac{1}{2\pi\sqrt{LC_1}};\ f_1^2 = \frac{1}{4\pi^2 LC_1};\ C_1 = \frac{1}{4\pi^2 \times 50 \times 10^{-6} \times (200 \times 10^3)^2}$$

$$C_1 = 12665.1\ \text{PF}$$

$$C_2 = \frac{1}{4\pi^2 L f_2^2} = \frac{1}{4\pi^2 \times 50 \times 10^{-6} \times (1000 \times 10^3)^2}$$

$$= 506.6\ \text{PF}$$

Ex. 10.2 The resonance circuit of a tuned oscillator has a resonant frequency of 4 MHz. If the capacitance is increased by 60%, calculate the new frequency.

$$4\ \text{MHz} = f_1 = \frac{1}{2\pi\sqrt{LC}}$$

For increased frequency

$$f_2 = \frac{1}{2\pi\sqrt{L(1.6)C}}$$

The ratio of f_1 to f_2 is

$$\frac{4 \times 10^6}{f_2} = \sqrt{1.6}$$

$$f_2 = \frac{4 \times 10^6}{\sqrt{1.6}} = 3.16\ \text{MHz}.$$

Ex. 10.3 Design a phase shift oscillator to operate at a frequency of 2 kHz using *JFET*. Assume $\mu = 50$; $r_D = 5\ \text{k}\Omega$. Find

(i) the minimum value of R_D to be used in the drain circuit.
(ii) *R-C* product.
(iii) Choose reasonable values of *R* and *C*.

$$f = 2000\ \text{Hz};\quad \mu = 50;\quad r_D = 5\ \text{k}\Omega;\quad R_D = ?\ R_C = ?$$

We know that

$$A = \frac{\mu R_D}{r_D + R_D}$$

For a *R-C* phase shift oscillator the minimum gain is 29 to produce oscillations

$$29 = \frac{50 \times R_D}{5 \times 10^3 + R_D}$$

$$145 \times 10^3 + 29\ R_D = 50\ R_D$$

$$145 \times 10^3 = 21\ R_D.$$

$$R_D = 6.9\ \text{k}\Omega.$$

(ii) $$f = \frac{1}{2\pi RC\sqrt{6}}$$

$$2000 = \frac{1}{2\pi RC\sqrt{6}}$$

Thus $$RC = 3.25 \times 10^{-5} = 32.5\ \mu\text{sec}$$

(iii) Assume standard value of capacitance and calculate the value of resistance.

Assume $C = 0.1\ \mu F$

$$R = \frac{RC}{C} = \frac{32.5 \times 10^{-6}}{0.1 \times 10^{-6}} = 325\ \Omega.$$

Ex 10.4 It is desired to design a phase shift oscillator using FET having $g_m = 5000$ micro Siemens. $r_D = 40$ K and $R = 10$ K. Select the value of capacitance for oscillator operation at 1 kHz and the value of R_D for a gain of 40 to ensure oscillator action.

$$\mu = g_m r_D = 5000 \times 10^{-6} \times 40 \times 10^3 = 200$$

Given $$A = 40$$

We know that $$A = \frac{\mu R_D}{r_D + R_D}$$

Thus $$40 = \frac{200 \times R_D}{40 \times 10^3 + R_D}$$

$$1600 \times 10^3 + 40\ R_D = 200\ R_D$$

$$1600 \times 10^3 = 160\ R_D$$

Therefore $$R_D = 10\ \text{k}\Omega$$

$$f = \frac{1}{2\pi RC\sqrt{6}}$$

Substitute $f = 1$ kHz and $R = 10$ K

$$10^3 = \frac{1}{2\pi \times 10 \times 10^3 \times C\sqrt{6}}$$

$$C = \frac{1}{2\pi \times 10 \times 10^6 \times \sqrt{6}} = 6.497 \times 10^{-9} = 6.49\ \text{nF}$$

Ex. 10.5 Find L_1 and L_2 of Hartley oscillator to produce 1 MHz sinusoidal output using BJT. Given $C = 1\ \mu F$. $h_{fe} = 100$ and M is negligible.

$$f = \frac{1}{2\pi\sqrt{LC}}$$

$$10^6 = \frac{1}{2\pi\sqrt{L \cdot 10^{-6}}}$$

$$10^{-6} = 2\pi\sqrt{L \cdot 10^{-6}}$$

$$10^{-12} = 4\pi^2 \times L \times 10^{-6}$$

$$L_1 + L_2 = L = 2.53 \times 10^{-8}$$

$$h_{fe} > \frac{L_1}{L_2}$$

$$h_{fe} = 1.1\frac{L_1}{L_2}$$

$$1.1\ L_1 = L_2\ h_{fe}$$

$$L_1 = L_2 \cdot \frac{h_{fe}}{1.1} = \frac{L_2\ 100}{1.1}$$

$$L_1 = 90.9\ L_2$$

$$90.9\ L_2 + L_2 = 2.53 \times 10^{-8}$$

$$L_2 = 2.5 \times 10^{-10}\ \text{H}$$

$$L_1 = 2.34 \times 10^{-8}\ \text{H}$$

Short Questions and Answers

Q. 1. Crystal oscillators are used where *stability* is necessary.

Q. 2. *Positive* feed back is used in oscillator circuits.

Q. 3. Wein Bridge oscillator uses *positive* and *negative* feedback.

Q. 4. In colpitt's oscillator, the frequency determining network has *two* capacitors and *one* inductor.

Q. 5. To make an amplifier oscillate, the conditions to be satisfied are that the loop gain be *unity* and loop phase shift be *zero*.

Q. 6. In a Hartley oscillator the frequency determining network has *two* inductors and *one* capacitors.

Q. 7. What is the condition to initiate oscillations?

Ans. $A\beta > 1$.

Q. 8. Will the oscillations take place in a Wein's Bridge oscillator if the bridge is balanced?

Ans. Yes.

Q. 9. What are the factors which affect the stability of an oscillator?
Change in values of *R*, *L*, *C* and beta.

Q. 10. What is the condition for oscillations of FET based Hartley and colpitt's oscillators?
Amplification factor $\mu > L_2/L_1$, $\mu > C_1/C_2$.

Q. 11. Why *L-C* oscillators are not used at audio frequencies?
L and *C* will be very bulky.

Q. 12. Why *L-C* circuit alone cannot produce sustained oscillations?
Due to losses in *L* and *C*.

Q. 13. Why is an *R-C* phase shift oscillator so called?
R-C network produces 180° phase shift. Hence it is called *R-C* phase shift oscillator.

Q. 14. What is the maximum phase shift realizable in a single stage *R-C* network?
Around 90°.

Q. 15. Do you get exactly 60° phase shift from each stage of *R-C* network?
No. The three *R-C* sections together produce a phase shift of 180°.

Q. 16. Match the following:

(A) R-C phase shift oscillator	(1) R-F oscillator
(B) UJT oscillator	(2) High frequency oscillator
(C) L-C oscillator	(3) Audio frequency oscillator
(D) Crystal oscillator	(4) Relaxation oscillator

Ans. (A) — (3)
(B) — (4)
(C) — (2)
(D) — (1)

Q. 17. For a *R-C* phase shift oscillator amplifier gain is *negative* and total phase shift introduced by the feedback network is 180°.

11

Tuned Amplifiers and Multivibrators

Tuned amplifiers use a tuned circuit in series with the collector. They are called narrow band amplifiers. Tuned amplifiers are classified as (i) single, tuned (ii) double tuned and (iii) stagger tuned.

11.1 Single Tuned Amplifier

In radio and television receivers, it is required to select a particular frequency and amplify the same. The tuned amplifier selects and amplifies the required signal. The impedance frequency response curve of a parallel resonant circuit is an inverted *V* curve. Therefore the frequency response curve of tuned amplifier is also an inverted *V* curve.

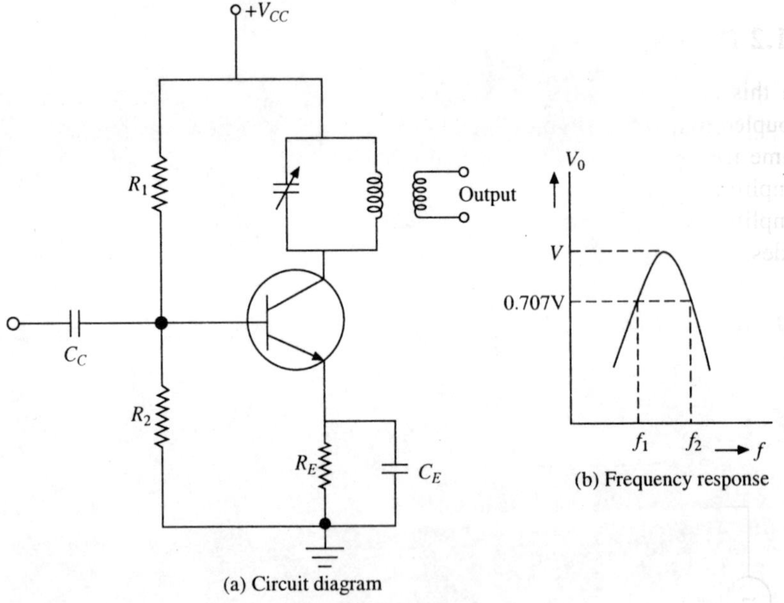

Fig. 11.1 Single tuned amplifier.

In this amplifier, *L-C* resonant circuit is connected in the place of resistance R_C. Output is taken from the secondary of the transformer in the tank circuit. The frequency can be varied by varying the value of capacitance.

We know the voltage gain of common emitter amplifier.

$$A_V = \frac{-h_{fe} R_L}{h_{ie}}$$

R_L has to be replaced with the impedance of parallel resonant circuit.

$$A_V = \frac{-h_{fe} Z_p}{h_{ie}}$$

The dynamic impedance of parallel resonant circuit at resonance is L/CR.

Substitute $Z_p = L/CR$.

$$A_V = \frac{-h_{fe} L}{h_{ie} CR}$$

Voltage gain of this amplifier is very high since the value of C is very less.

Tuned amplifier has train of narrow current pulses. Parallel *L-C* circuit converts narrow current pulses into sinusoidal voltage. The power lost in the *L-C* circuit is supplied by the amplifier in the form of current pulses. Hence, the output of tuned amplifier consists of sustained oscillations.

The frequency response curve is shown in Fig. 11.1(b). Let the output voltage at resonance be *V*. Draw a horizontal line at 0.707 V. The corresponding frequencies are f_1 and f_2. Bandwidth is $f_2 - f_1$. From the frequency response curve, it can be seen that the bandwidth of tuned amplifier is very narrow. Hence they are called narrow band amplifiers.

11.2 Double Tuned Amplifier

In this amplifier, voltage developed across the tuned circuit is magnetically coupled to another circuit (L_2 and C_2). Both the tuned circuits are tuned to the same frequency equal to the signal frequency. The bandwidth of double tuned amplifier is higher than that of single tuned amplifier. The figure of merit of this amplifier is also higher. It provides gain versus frequency curve having steeper sides.

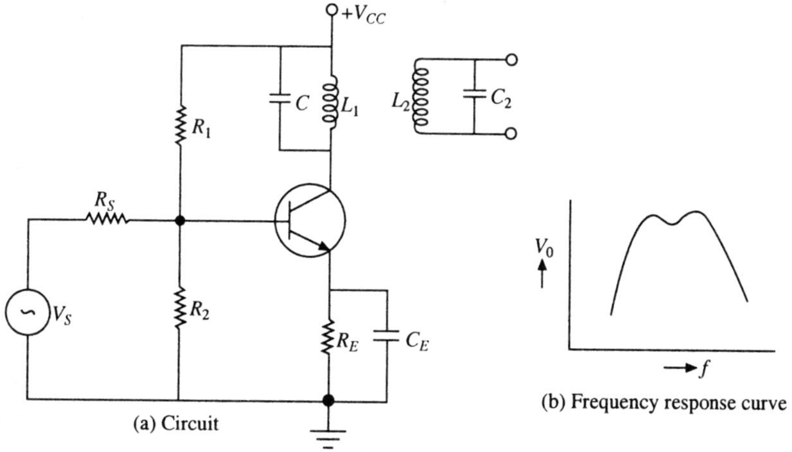

(a) Circuit

(b) Frequency response curve

Fig. 11.2 Double tuned amplifier.

When two resonant circuits are critically coupled, the response becomes maximally flat i.e. the response is flat over a broad range of frequencies than is possible with single tuned amplifier. Hence the bandwidth increases with double tuning.

11.3 Stagger Tuned Amplifier

In this amplifier, two single tuned amplifiers are cascaded. The resonant frequencies of the two tuned circuits are such that they are seperated by an amount equal to the bandwidth. It is called stagger tuned amplifier since the resonant frequencies are staggered. The bandwidth of stagger tuned amplifier is √2 times the bandwidth of single tuned amplifier.

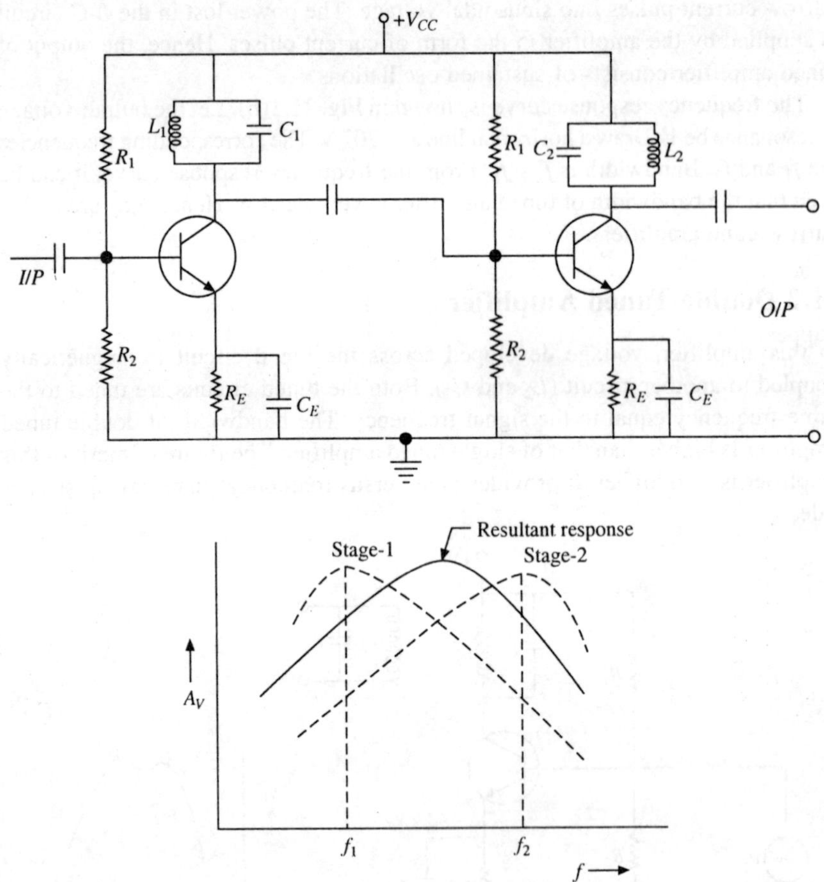

Fig. 11.3 Stagger tuned amplifier.

Stage 1 is operating at a frequency of f_1 and stage 2 is operating at a frequency of f_2. The frequency response curve of overall amplifier is obtained by super position principle. From this curve, it can be seen that the bandwidth is increased.

11.4 Multivibrators

To understand multivibrator circuits, it is necessary to know how the transistor acts as a switch.

For a transistor, we know that $V_{CE} = V_{CC} - I_C R_C$. When a pulse is not given at the base, the transistor does not conduct. $I_C = 0$; $V_{CE} = V_{CC}$ since the drop in R_C is zero. The device operates in cut off region. Thus a transistor acts as an open switch when base drive is not given.

When base drive is given, the transistor conducts and almost the entire supply voltage drops in $R_C \cdot V_{CE} = 0$ since $V_{CC} = I_C R_C$. The device operates in saturation region. It acts as a closed switch since voltage across it is zero. Thus transistor acts as a closed switch when base drive is given. The following circuits are discussed here.

(i) Astable multivibrator
(ii) Monostable multivibrator
(iii) Bistable multivibrator
(iv) Schmitt trigger

11.4.1 Astable Multivibrator

When the D.C. power supply (V_{CC}) is switched on, one of the transistor will start conducting earlier than the other due to some imbalance in the circuit. Let us suppose that transistor Q_1 starts conducting more than that of transistor Q_2. Then because of positive feedback, the transistor Q_1 will be driven into saturation and transistor Q_2 to cut-off. Thus at $t > 0$, the transistor Q_1 is ON and Q_2 is OFF. Thus at $t > 0$, $V_{B1} = V_{BE}$ (sat) (i.e. 0.7 V for silicon transistor), $V_{C1} = V_{CE}$ (sat) (i.e., 0.3 V for silicon transistor), V_{B2} is negative and $V_{C2} = V_{CC}$.

During the time $t > 0$ (i.e. when Q_1 is ON and Q_2 is OFF), the capacitor C_1 is charging towards the voltage V_{CC} through R_1. The charging takes place exponentially with time constant $T_1 = R_1 C_1$, since the base of transistor Q_2 is

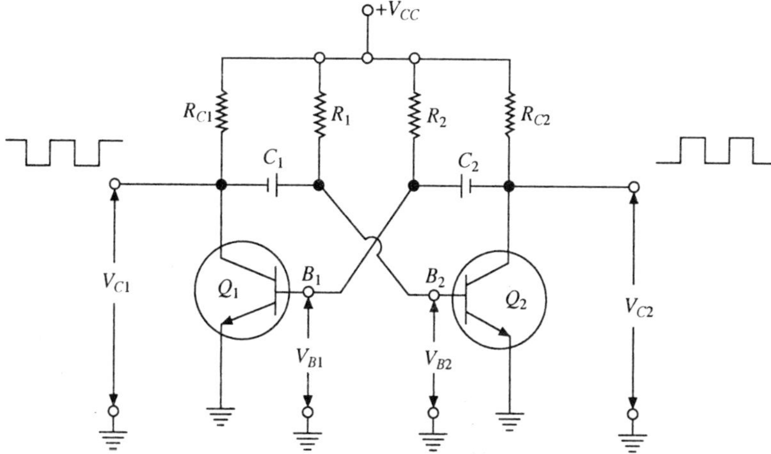

Fig. 11.4a Astable multivibrator.

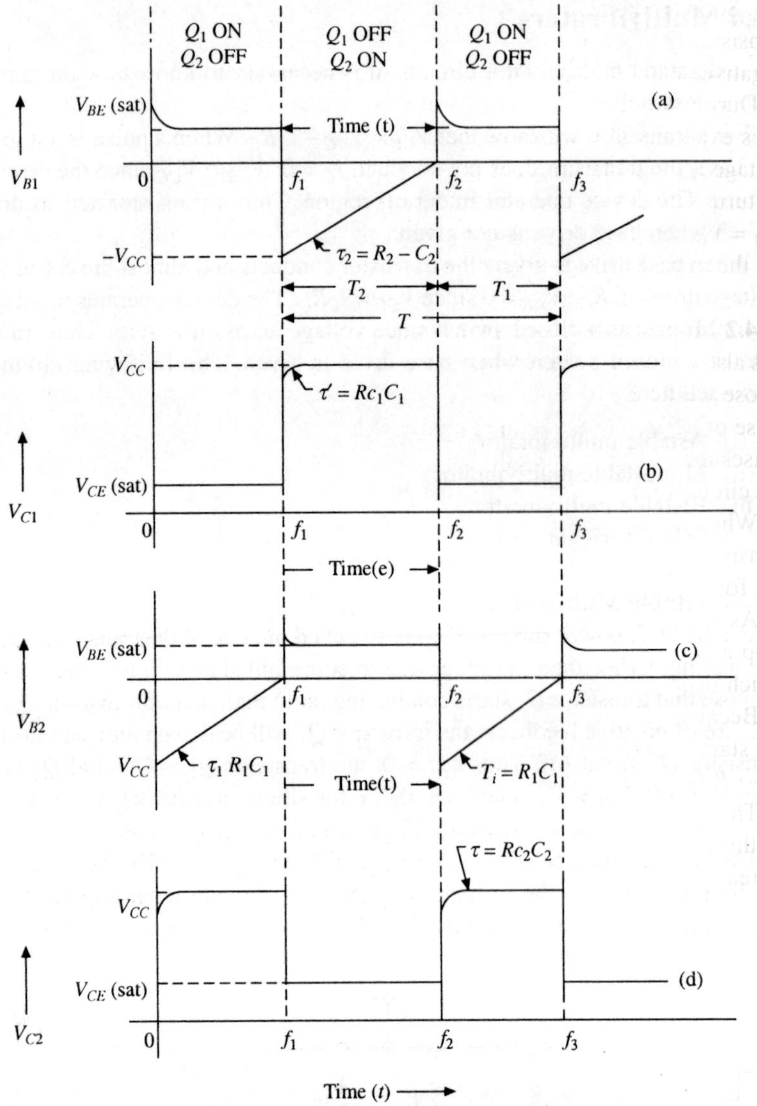

Fig. 11.4b Astable multivibrator.

directly connected to capacitor C_1, as shown in Fig. 11.4. Therefore, the voltage V_{B2} also increases exponentially towards V_{CC}.

As soon as the voltage V_{B2} increases above the cut in voltage (i.e. 0.7 V for silicon transistor), the transistor Q_2 starts conducting. It occurs at $t = t_1$. As the transistor Q_2 goes into saturation, its collector voltage (V_{C2}) falls to V_{CE} (sat). The fall in voltage V_{C2} causes an equal fall, (i.e., $V_{CC} - V_{CE}$ (sat) $= V_{CC}$) in voltage V_{B1} because the two are capacitively coupled. The fall in voltage V_{B1} cuts off the transistor Q_1 and its collector voltage V_{C1} starts rising towards V_{CC} with a time constant $\lambda = R_{CC1}$. The rise in voltage V_{C1} is coupled through capacitor C_1 to the base of the transistor Q_2, causing a small overshoot in voltage

V_{B2}. Soon the voltage V_{B2} settles at V_{BE} (sat) (i.e., 0.7 V) level. Thus at $t > t_1$, the transistor Q_1 is OFF and Q_2 is ON. The voltage levels at this instant are, V_{B1} is negative, V_{C1}, $V_{B2} = V_{BE} = V_{BE}$ (sat) and $V_{C2} = V_{CE}$ (sat).

During the time $t > t_1$ (i.e., when Q_1 is OFF and Q_2 is ON) the voltage V_{B1} rises exponentially with time constant $\lambda_2 = R_2 \cdot C_2$ towards V_{CC}. At $t = T_2$, the voltage V_{B1} reaches the cut in level and a reverse transition takes place (i.e., Q_1 turns ON and Q_2 turns OFF). The voltage levels for $t > t_2$ are $V_{B1} = V_{BE}$ (sat), $V_{C1} = V_{CE}$ (sat); V_{B2} is negative and $V_{C2} = V_{CC}$. Thus the voltage levels for $t > t_2$ are the same as for $t > 0$.

11.4.2 Monostable Multivibrator

It is also called one shot multivibrator and can be used to generate a gating pulse, whose width can be controlled. The monostable multivibrator provides a single pulse of desired duration in response to an external trigger. The external trigger causes the circuit to go to the quasi-stable state. After a certain interval of time, the circuit returns to its original stable state.

When a positive trigger pulse of sufficient amplitude is applied to the base of transistor Q_1, it overrides the reverse bias provided by the V_{BB} supply and gives it a forward bias. Because of this, the transistor Q_1 starts conducting.

As the transistor Q_1 conducts, its collector voltage falls due to the voltage drop across resistor R_{C1}. This fall in voltage is coupled through capacitor C_2, which decreases the forward bias of transistor Q_2.

Because of the reduced forward bias, the collector current of transistor Q_2 starts decreasing and its collector voltage rises exponentially towards $V_{CC} \cdot R_1(R_1 + R_{C2})$ with a time constant $\lambda_2 = C_1 \cdot (R_1 \cdot R_{C2})$.

The rising collector voltage of transistor Q_2 is coupled to the base of transistor Q_1 through resistor R_1, where it further increases its forward bias. Because of the increased forward bias, the transistor Q_1 conducts more. This action is cumulative

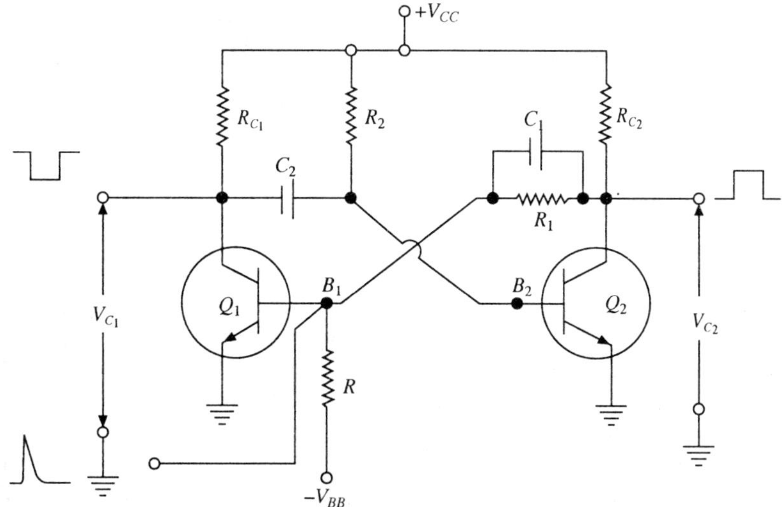

Fig. 11.5a Monostable multivibrator.

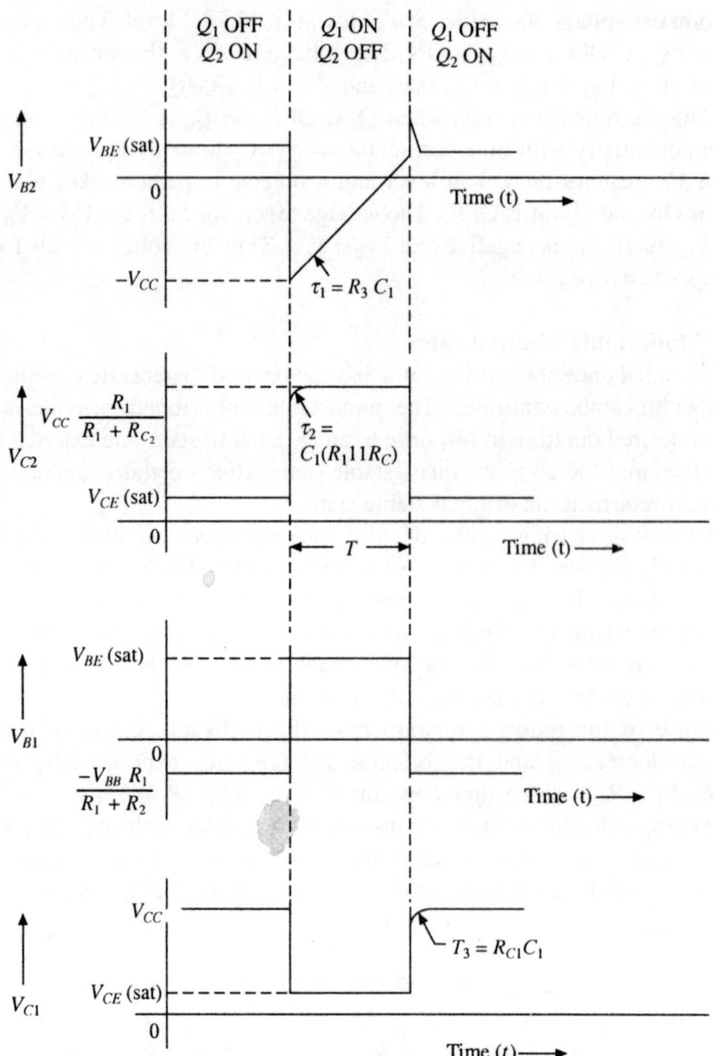

Fig. 11.5b Monostable multivibrator.

because of the positive feedback. The collector voltage of transistor Q_1 falls to V_{CE} (sat).

Immediately, the capacitor C_2 starts charging exponentially towards V_{CC} with a time constant $\lambda_1 = R_3\, C_2$. As C_1 charges, the voltage at the base of the transistor Q_2 decreases. As C_1 charges further, the transistor Q_2 is pulled out from the cut off as the reverse transition takes place i.e., Q_2 turns ON and Q_1 turns OFF.

When the transistor Q_2 starts conducting, its collector voltage falls because of the drop across resistor R_{C2}. This drop is coupled to the base of the transistor Q_1, whose collector voltage rises towards V_{CC} with time constant $\lambda_3 = R_{C1} \cdot C_2$. Finally, the transistor Q_1 turns fully ON and the transistor Q_1 goes OFF. The circuit remains in this stable state till another pulse is applied.

11.4.3 Bistable Multivibrator

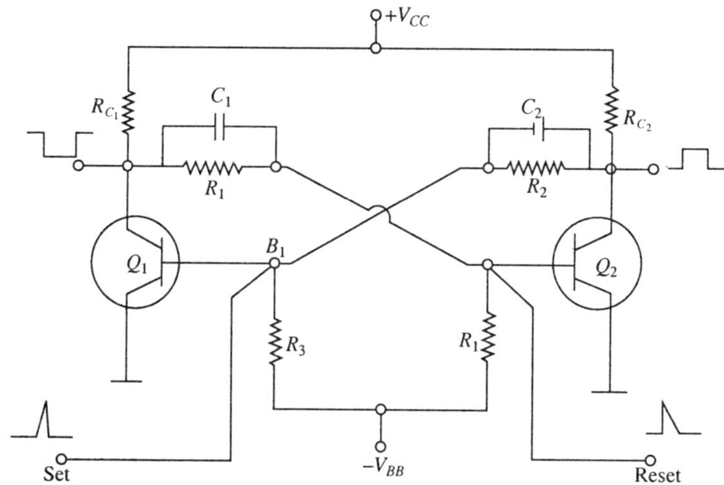

Fig. 11.6(a) Bistable multivibrator.

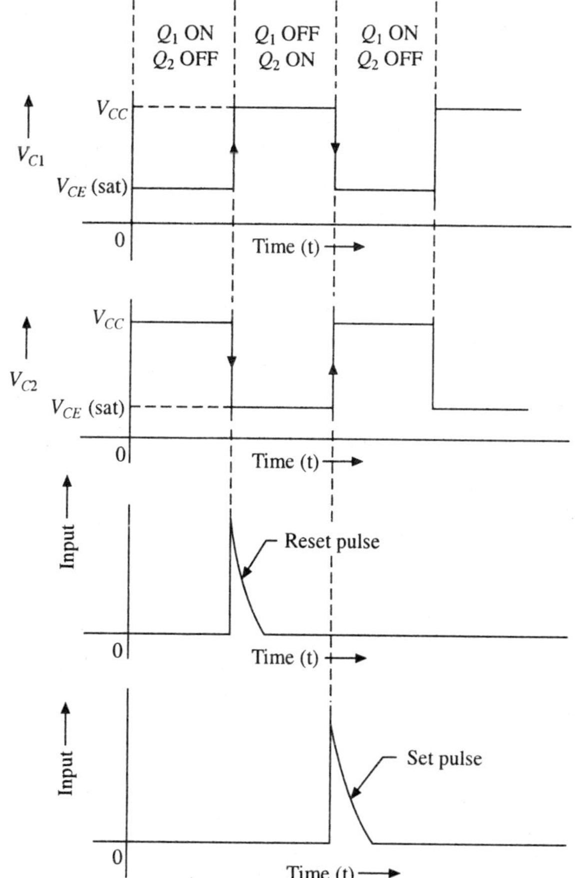

Fig. 11.6(b) Bistable multivibrator.

It is also called a flip-flop. The bistable multivibrator is used for counting and storing of binary infomation in computer circuits. It also finds extensive application in the generation and processing of pulse type waveforms. The bistable multivibrator has two stable states and can stay in any one of these two states, indefinitely, as long as the power is supplied. It changes to other state only when it receives a trigger (a short duration pulse) from outside.

The operation of the circuit is as given below:

When V_{CC} supply is switched on, one of the transistors will start conducting more than the other. Then because of the feedback action, this transistor will be driven into saturation and the other to cut-off. Let us assume that the transistor Q_1 is in saturation (i.e., ON) and Q_2 is cut-off (i.e., OFF). It is a stable state of the circuit and will remain in this state, till a trigger pulse is applied from outside. A negative pulse applied to the set input will turn the transistor Q_1 to OFF and Q_2 to ON. A similar action can also be achieved by applying a positive pulse at reset input.

If a positive pulse is applied at the reset input, it will cause the transistor Q_2 to conduct. As the collector voltage of Q_2 falls, it cuts-off the transistor Q_1. Thus the circuit switches to the other stable state i.e., a state where Q_1 is OFF and Q_2 is ON.

Now if a positive pulse is applied at the set input, it will switch the circuit back to its original stable state i.e., Q_1 ON and Q_2 OFF.

11.4.4 Schmitt Trigger

The circuit is, somewhat, like multivibrator circuits. The Schmitt trigger is used for wave shaping circuits. It can be used for generation of a square wave from a sine wave input. Basically, the trigger signal is a slowly varying A.C. voltage. The Schmitt trigger is level sensitive and switches the output state at two distinct

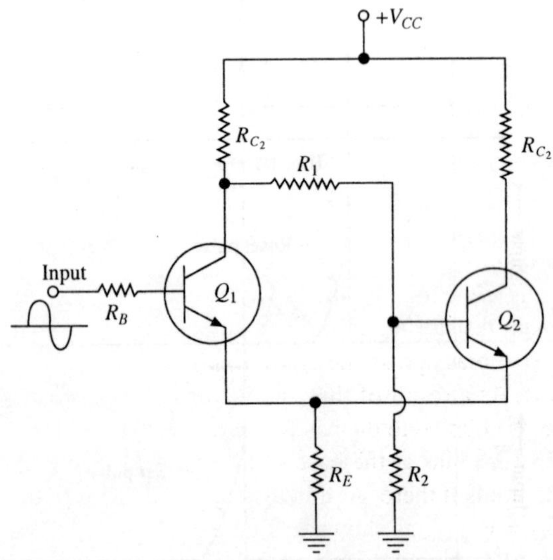

Fig. 11.7(a) Schmitt trigger.

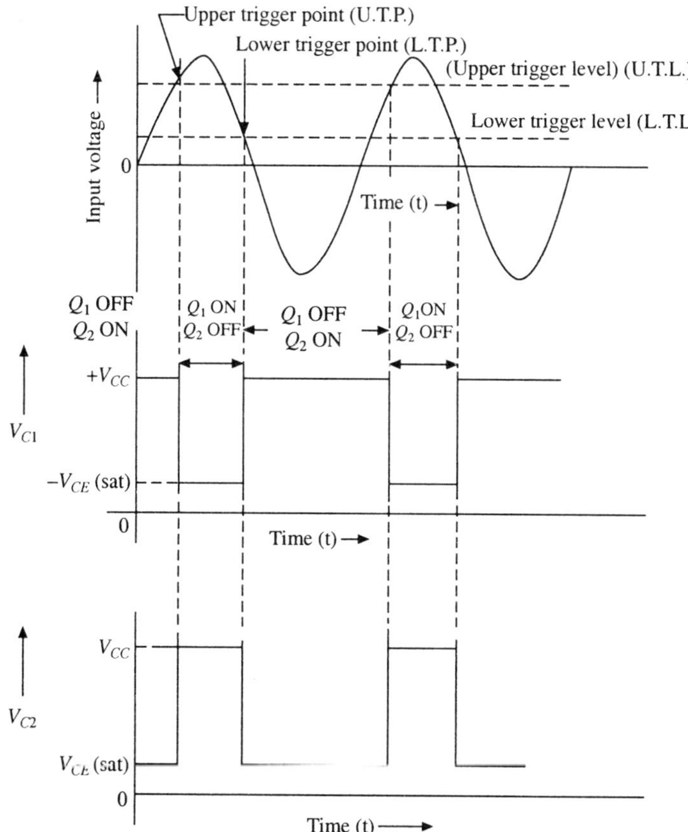

Fig. 11.7(b) Schmitt trigger.

trigger levels. One of the triggering levels is called a lower trigger level (abbreviated as L.T.L) and the other as upper trigger level (abbreviated as U.T.L)

The circuit of a Schmitt trigger consists of two identical transistors Q_1 and Q_2 coupled through an emitter resistance R_E. The resistors R_1 and R_2 form a voltage divider across the V_{CC} supply and ground. These resistors provide a small forward bias on the base of transistor Q_2. The operation of the circuit may be explained as follows:

Let us suppose that initially there is no signal at the input. Then as soon as the power supply V_{CC} is switched on, the transistor Q_2 starts conducting. The flow of its current through resistor R_E produces a voltage drop across it. This voltage drop acts as a reverse bias across the emitter base junction of transistor Q_1 due to which it cuts-off. As a result of this, the voltage at its collector rises to V_{CC}. This rising voltage is coupled to the base of transistor Q_2 through the resistor R_1. It increases the forward bias at the base of transistor Q_2 and therefore drives it into saturation and holds it there. At this instant, the collector voltage levels are $V_{C1} = V_{CC}$ and $V_{C2} = V_{CE(sat)}$ as shown in figure.

Now suppose an A.C. signal is applied at the input of the Schmitt trigger (i.e. at the base of the transistor Q_1). As the input voltage increases above zero nothing will happen till it crosses the upper trigger level (U.T.L). As the input voltage increases, above the upper trigger level, the transistor Q_1 conducts. The point at which it starts conducting, is known as upper trigger point (U.T.P). As the transistor Q_1 conducts, its collector voltage falls below V_{CC}. This fall is coupled through resistor R_1 to the base of transistor Q_2 which reduces its forward bias. This in turn reduces the current of transistors Q_2 and hence the voltage drop across the resistor R_E. As a result of this, the reverse bias of transistor Q_1 is reduced, and it conducts more. As the transistor Q_1 conducts more heavily, its collector further reduces due to which the transistor Q_2 conducts near cut-off. This process continues till the transistor Q_1 is driven into saturation and Q_2 into cut-off. At this instant, the collector voltage levels are $V_{C1} = V_{CE(\text{sat})}$ and $V_{C2} = V_{CC}$.

The transistor Q_1 will continue to conduct till the input voltage falls below the lower trigger level (L.T.L). When the input voltage becomes equal to the lower trigger level, the emitter-base junction of transistor Q_1 becomes reverse biased. As a result of this, its collector voltage starts rising towards V_{CC}. This rising voltage increases the forward bias across transistor Q_2, due to which it conducts. The point, at which transistor Q_2 starts conducting, is called lower trigger point (L.T.P). Soon the transistor Q_2 is driven into saturation and Q_1 to cut-off. This completes one cycle. The collector voltage levels at this instant are $V_{C1} = V_{CC}$ and $V_{C2} = V_{CE(\text{sat})}$. No change in state will occur during the negative half cycle of the input voltage.

It is evident from the above discussion that the output of a Schmitt trigger is a positive going pulse, whose width depends upon the time during which transistor Q_1 is conducting. The conduction time is set by the upper and lower trigger levels.

Short Questions and Answers

1. Which amplifier should be used to amplify a radio frequency signal?
   Turned voltage amplifier.
2. An amplifier has a pass-band from 745 kHz to 755 kHz and a resonant frequency of 750 kHz. Its bandwidth is *10 kHz.*
3. If the Q-factor of a resonant circuit is doubled, its bandwidth is *halved.*
4. A turned circuit has resonant circuit frequency of 1 MHz. To get a bandwidth of 20 kHz, its Q-factor must be equal to *50.*
5. Stagger-tuning the tuned amplifiers is achieved by using *tuned circuits which are tuned to slightly different frequencies.*
6. Astable multivibrator can be used as a *small square wave to large square wave* generator.
7. Bistable multivibrator is used as *flip-flop* element in shift register.
8. Schmitt trigger is a type of *astable* vibrator sensitive to *level* triggering?
9. If the frequency of the input pulse train to a bistable multivibrator is *f*, what is the output frequency.
   Frequency is *f.*
10. Multivibrators employ amplifier with *positive* feedback and produce square waveforms.
11. The astable multivibrators circuit involves 0 stable states, while Schmitt trigger has *2* stable states.

12. Is Schmitt trigger same as emitter coupled bistable multivibrator?
    No.
13. Time delayed pulses and square wave pulses can be obtained from a pulse train by using *monostable* circuit and *astable* circuit, respectively.
14. An astable multivibrator is called free running relaxation oscillator because it *runs and relaxes alternately*.
15. The frequency of oscillation of an astable multivibrator depends on the *values of R and C*.
16. A monostable multivibrator gives *one* output pulse for one trigger pulse.
17. A monostable multivibrator has *one* stable state.
18. A bistable multivibrator has *two* stable states.
19. *Astable* multivibrator is a square wave oscillator.

12

Differential Amplifier

12.1 Differential Amplifier

Block diagram of a linear differential amplifier is shown in Fig. 12.1. V_1 and V_2 are input signals. V_1 is amplified by A_1. V_2 is amplified by A_2. V_0 is defined as follows: it is an algebraic sum of the terms A_1V_1 and A_2V_2.

$$V_0 = A_1V_1 + A_2V_2 \tag{12.1}$$

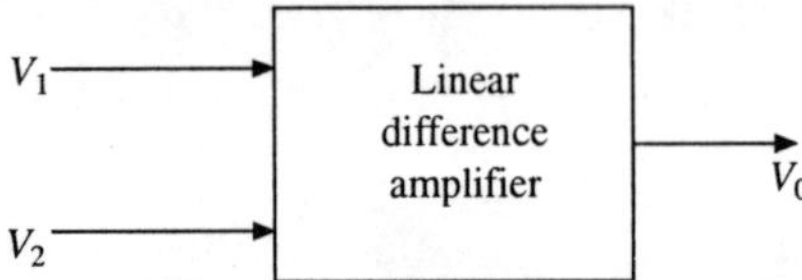

Fig. 12.1 Block diagram.

Common mode signal is defined as the average of the two input signals.

$$V_C = \frac{V_1 + V_2}{2} \tag{12.2}$$

Differential mode signal is defined as the difference between V_1 and V_2.

$$V_d = V_1 - V_2 \tag{12.3}$$

Solve for V_1 and V_2 from equations (12.2) and (12.3)

$$2V_C = V_1 + V_2; \quad V_d = V_1 - V_2 \text{ on adding}$$

$$2V_C + V_d = 2V_1$$

$$\therefore \qquad V_1 = \frac{2V_C + V_d}{2} \tag{12.4}$$

Equation (12.1) × –1 ⇒ $-2V_C = -V_1 - V_2$

$$V_d = V_1 - V_2$$

$$-2V_C + V_d = -2V_2$$

$$2V_C - V_d = 2V_2$$

Hence $$V_2 = \frac{2V_C - V_d}{2} \tag{12.5}$$

Substitute equations (12.4) and (12.5) in equation (12.1)

$$V_0 = A_1\left[\frac{2V_C + V_d}{2}\right] + A_2\left[\frac{2V_C - V_d}{2}\right]$$

$$V_0 = V_C[A_1 + A_2] + V_d\left[\frac{A_1 - A_2}{2}\right]$$

Let $$V_0 = V_C A_C + V_d \cdot A_d$$

$A_1 + A_2$ is defined as common mode gain (Ac).

Therefore, $$A_c = A_1 + A_2 \qquad (12.6)$$

$\frac{A_1 - A_2}{2}$ is defined as differential mode gain (Ad).

$$A_d = \frac{A_1 - A_2}{2} \qquad (12.7)$$

12.2 The Differential Amplifier

The differential amplifier (diff. amp.) is heavily used in linear ICs. A differential amplifier uses no coupling or bypass capacitors., all it requires are resistors and transistors, both easily integrated on a chip.

12.2.1 Tail Current

Figure 12.2(a) shows the basic form of a differential amplifier. There are two input signals and one output signal. Note that the output signal is the voltage between the collectors. Ideally, the circuit is symmetrical, each half is identical to the other half. Monolithic ICs can approach this symmetry because the integrated components are on the same chip and have virtually identical characteristics.

A differential amplifier is sometimes called a long tail pair because it consists of a pair of identical transistors connected to a common emitter resistor (the tail). The current through this common resistor is known as the tail current I_T.

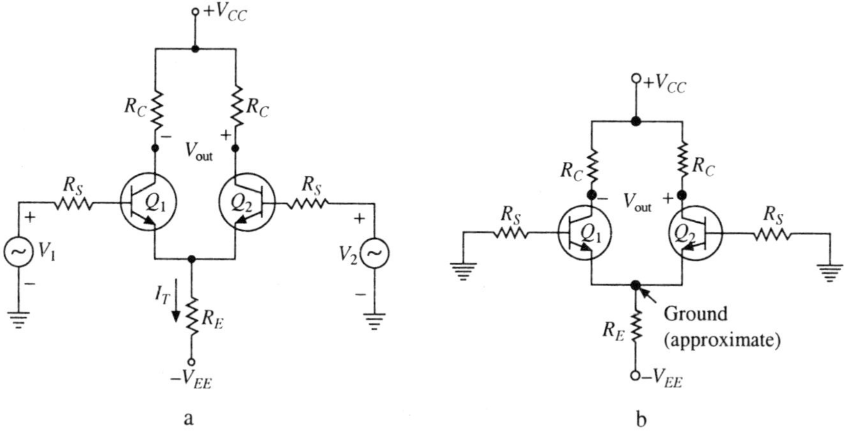

Fig. 12.2 DC analysis.

The differential amplifier of Fig. 12.2(a) uses emitter bias. The key to analyse emitter biased circuits is to remember that the top of the emitter resistor is an approximate ground point. The D.C. equivalent circuit of Fig. 12.2(b) emphasizes this idea. Because of the ground, almost all the V_{EE} supply voltage appears across R_E. Therefore, the D.C. tail current is

$$I_T = \frac{V_{EE}}{R_E} \tag{12.8}$$

When both transistors are identical, the tail current divides equally between them. That is, the D.C. emitter current in each transistor is half the tail current.

$$I_E = \frac{I_T}{2}$$

I_C equals I_E to a close approximation:

$$I_C = I_E$$

In Fig. 12.2(b) the D.C. voltage from the Q_1 collector to ground equals the V_{CC} supply voltage minus the drop across the collector resistor

$$V_{C1} = V_{CC} - I_C R_C$$

Similarly, the D.C. voltage from the Q_2 collector to ground is

$$V_{C2} = V_{CC} - I_C R_C$$

When the transistors and collector resistors are identical, V_{C1} equals V_{C2} so there is no voltage difference between the collectors. In other words, the D.C. output voltage is zero:

$$V_{\text{out}} = 0$$

12.3 AC Analysis of a Differential Amplifier

In Fig. 12.3(a), the output is taken between collectors. If the two halves of the differential amplifier are identical, the output voltage is zero when V_1 and V_2 are zero. If V_1 and V_2 change by exactly the same amount, the output voltage remains zero because of the symmetry. Only when there is a difference between V_1 and V_2, we get an output voltage. When V_1 is more positive than V_2, more collector current flows through the transistor on the left, because of this, the output voltage has the polarity shown in Fig. 12.3(a).

In a good differential amplifier design, V_{EE} and R_E set up an almost constant tail current and we can visualize the differential amplifier by the equivalent circuit of Fig. 12.3(b). If current source bias is used, Fig. 12.2(b) becomes a highly accurate equivalent circuit.

To get the A.C. equivalent circuit, visualize all D.C. sources of Fig. 12.2(b) reduced to zero. This is equivalent to A.C. grounding the V_{CC} supply point and opening the I_T current source. Since two A.C. sources drive the differential amplifier, we can use the superposition theorem to find the A.C. output voltage. This means taking one A.C. source at a time, finding its contribution to the A.C. output, and then algebraically adding the separate contributions.

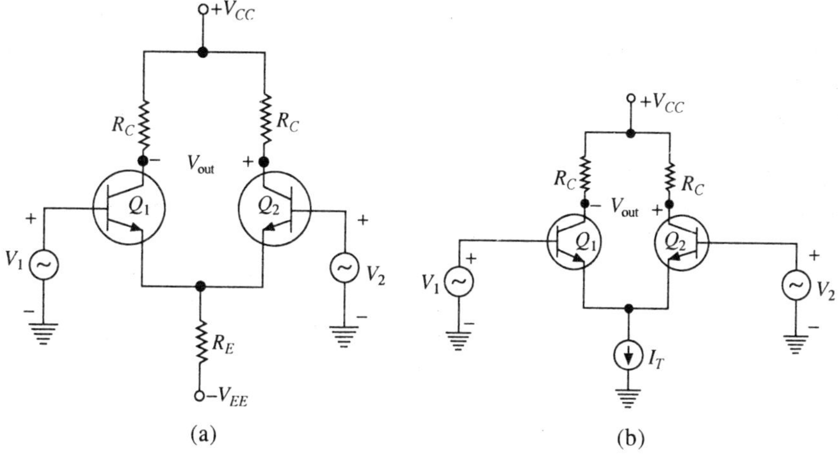

Fig. 12.3 A.C. analysis.

Start with V_1 active and V_2 off, (Fig. 12.4(a)). Q_1 acts like a *CE* amplifier, therefore, an inverted sinusoid appears at its collector. This A.C. voltage is between the Q_1 collector and ground. The A.C. emitter current from Q_1 has to flow into the emitter of Q_2 because the I_T current source is open to A.C. Therefore, Q_2 acts like a CB amplifier, and an in phase signal appears at its collector. The A.C. output voltage is taken between the collectors. Because it is the algebraic difference of two equal and opposite sine waves, thc output has twice the amplitude. So, the contribution of the first source acting alone is

$$V_{out(1)} = AV_1$$

Next, V_2 is made active and V_1 off (Fig. 12.4(c)). This time Q_2 acts like a CE amplifier and Q_1 like a CB amplifier. An in-phase signal appears at the Q_1 collector and an inverted signal at the Q_2 collector. The algebraic difference of these two collector voltages is the A.C. output (see Fig. 12.4(d)). This A.C. output voltage is the second contribution.

$$V_{out}(2) = -AV_2$$

This is the A.C. output produced by V_2 alone. The minus sign means the contribution has the opposite polarity from that of Fig. 12.4(b). The superposition theorem tells us that the A.C. output voltage when both sources act simultaneously is the algebraic sum of the two separate contributions.

$$V_{out} = V_{out(1)} + V_{out(2)}$$

or

$$V_{out} = AV_1 - AV_2 \qquad \text{(two inputs)}$$

Therefore when we want the A.C. voltage out of a differential amplifier, we multiply each input voltage by A and take the difference.

12.4 Inverting and Noninverting Inputs

In the differential amplifier of Fig. 12.4(a) the output voltage can drive another stage, possibly a differential amplifier. As we saw, a positive V_1 acting alone produces a positive output voltage. For this reason, the V_1 input is the noninverting input. On the other hand, a positive V_2 acting alone produces a negative output voltage. This is why the V_2 input is called the inverting input.

12.4.1 Differential Input

Besides not having coupling and bypass capacitors, the differential amplifier has other advantages. If we like, we can ground one input and drive the other input. This special case is called single ended input (similar to Fig. 12.4(a)).

Or, we can drive the differential amplifier with a signal between the bases as shown in Fig. 12.4(b). This is known as double ended or differential input. When we have a differential input, V_{in} is the difference of V_1 and V_2. In symbols,

$$V_{in} = V_1 - V_2$$

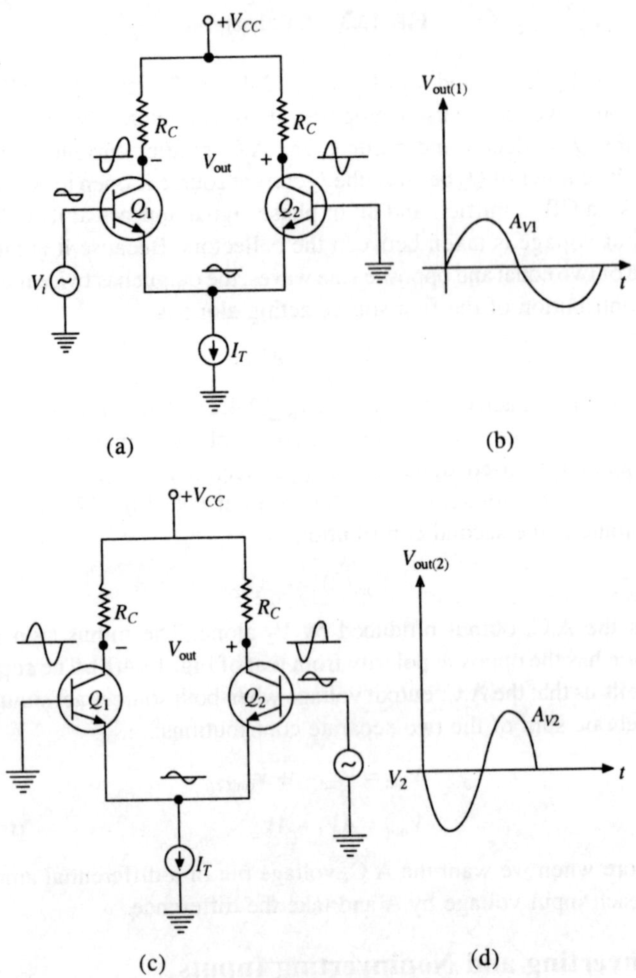

Fig. 12.4 Differential amplifier

With this view point, equation becomes

$$V_{out} = AV_1 - AV_2 = A(V_1 - V_2)$$

$$V_{out} = AV_{in} \text{ (differential input)}$$

If we have two separate signal sources, we think of the output voltage as $AV_1 - AV_2$. On the other hand, if we have only one signal source driving the differential amplifier in the differential mode, we think of the output as AV_{in}.

12.4.2 Common Mode Input

Figure 12.5(c) illustrates the common mode input, the same signal is applied to both inputs. If each half of the differential amplifier is identical, the A.C. output voltage will equal zero. But the only time we deliberately use a common mode input signal is when we are testing the differential amplifier to see how well balanced the two halves are.

The common mode rejection ratio (CMRR) is defined as $\frac{A_d}{A_c}$

$$CMRR = \frac{AV_{in(CM)}}{V_{out(CM)}} \quad (12.9)$$

where the numerator is calculated and the denominator is measured. As an example, suppose $V_{in(CM)}$ is 1 V in Fig. 12.5(c). Ideally, we should get no output, there may be a small output signal because of nonsymmetry.

If $A = 100$, $V_{out(CM)} = 0.01$

$$CMRR = \frac{100 \times 1}{0.01} = 10^4$$

CMRR tends to infinity means that common mode output tends to zero. The circuit rejects common mode input voltage. Hence it is named as CMRR.

12.5 Analysis of Differential Amplifier

A differential amplifier with CE amplifier is shown in the above Fig. 12.5(a). The output is taken across both the collectors which produces an output voltage proportional to the difference between the two input voltages.

Calculation as common mode gain

$$V_0 = A_C V_C + A_d V_d$$

$$A_C = \left.\frac{V_o}{V_C}\right|_{V_d = 0}$$

Let $V_1 = V_S$, $V_2 = V_S$ $\quad V_d = V_1 - V_2 = 0$

$$V_C = \frac{V_1 + V_2}{2}$$

$$\boxed{V_C = V_S}$$

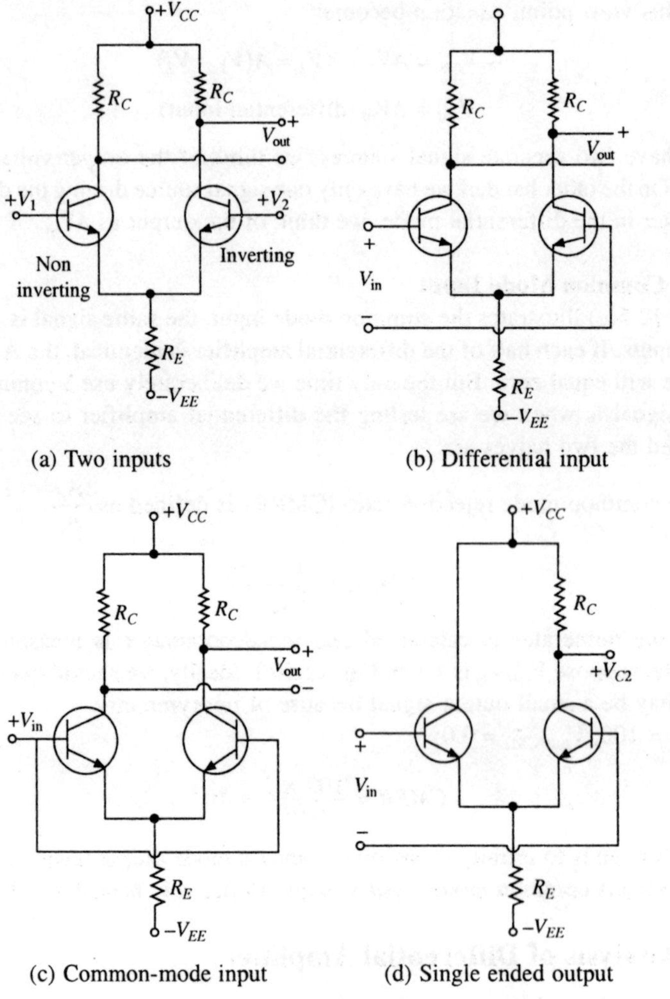

Fig. 12.5 Types of connections

The common mode gain can be calculated from the A.C. equivalent circuit and is shown below.

Replacing the transistor by its approximate hybrid model, we have the following equivalent circuit in Fig. 12.6(d)

$$V_o = -h_{fe} i_b R_C$$

$$V_S = (R_S + h_{ie})\, i_b + (1 + h_{fe}) i_b\, 2R_E$$

$$i_b = \frac{V_S}{R_S + h_{ie} + (1 + h_{fe})\, 2R_E}$$

$$V_o = \frac{-h_{fe}\, R_C V_S}{R_S + h_{ie} + 2(1 + h_{fe})\, R_E}$$

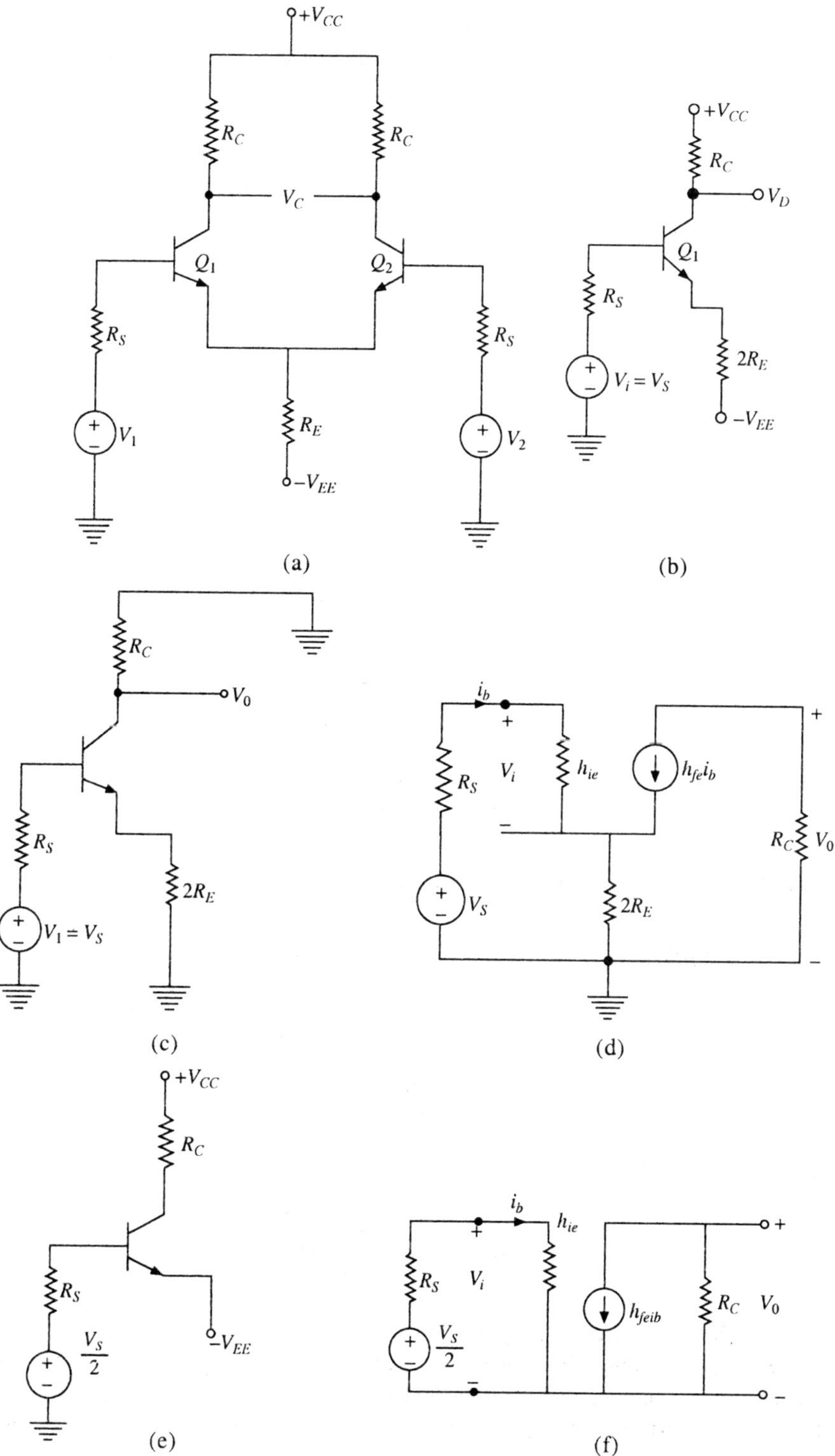

Fig. 12.6 Equivalent circuits

$$\frac{V_o}{V_S} = \frac{-h_{fe}R_C}{R_S + h_{ie} + 2Re(1 + h_{fe})}$$

$$\boxed{A_C = \frac{-h_{fe}R_C}{R_S + h_{ie} + 2Re(1 + h_{fe})}} \qquad (2.10)$$

Calculation of differential mode gain

$$V_1 = V_S/2,\ V_2 = -V_{S/2};\ V_C = \frac{V_1 + V_2}{2} = 0$$

$$V_d = V_1 - V_2 = \frac{V_S}{2} + \frac{V_S}{2}$$

$$\boxed{V_d = V_S}$$

Replacing the transistor by its hybrid model,

$$V_o = -h_{fe}\, i_b\, R_C$$

$$= -h_{fe}\, R_C\, \frac{V_S/2}{R_S + h_{ie}}$$

$$\frac{V_o}{V_S} = \frac{-h_{fe}\, R_C}{2(R_S + h_{ie})}$$

$$\boxed{A_d = \frac{-h_{fe}R_C}{2(R_S + h_{ie})}}$$

CMRR: (Common Mode Rejection Ratio):

$$\text{CMRR} = A_d/A_C = \frac{-h_{fe}R_C}{2(R_S + h_{ie})} = \frac{R_S + h_{ie} + 2R_e(1 + h_{fe})}{-h_{fe}R_C}$$

If $2Re(1 + h_{fe}) >> R_S + h_{ie}$, $h_{fe} >> 1$

$$\approx \frac{2\, Re\, h_{fe}}{2(R_S + h_{ie})}$$

if R_S is neglected

$$\text{CMRR} = \boxed{\rho = \frac{R_e\, h_{fe}}{h_{ie}}} \qquad (12.11)$$

The CMRR is directly proportional to R_e. To make the CMRR infinite, the value of R_e should be infinite which is impossible. The other option is instead of R_e, a current source can be used to improve the emitter resistance. Otherwise a Darlington configuration is placed across the emitter load to enhance the emitter resistance.

Worked Problems

Ex. 12.1 A differential amplifier has a differential mode gain of 100 and a common mode gain of 0.01. Calculate CMRR in decibels.

$$A_d = 100$$

$$A_c = 0.01$$

$$\text{CMRR} = A_d/A_c$$

$$\text{CMRR (db)} = 20 \log (A_d/A_c)$$

$$= 20 \log (100/0.01) = 20 \log (10^4)$$

$$= 80 \text{ db}$$

Ex. 12.2. A differential amplifier has input voltages V_1 = 10 mV, V_2 = 9 mV. It has a differential mode gain of 60 db and CMRR of 80 db. Find the output voltage.

$V_i = V_1 - V_2 = 10 - 9 = 1$ mV

$A_d = 60$ db

CMRR = 80 db

$20 \log (\rho) = 80; \rho = 10^4$

$20 \log (A_d) = 60; \log (A_d) = 3; A_d = 10^3$

$\rho = A_d/A_c$

$A_c = A_d/\rho = 10^3/10^4 = 0.1$

$V_o = A_d{*}\ V_d + A_c \cdot V_c$

$V_d = V_1 - V_2 = 1$ mV

$V_c = (V_1 - V_2)/2 = 5$ mV

$V_o = 10^3 \cdot 1 \cdot 10^{-3} + 10^{-1} \times 5 \cdot 10^{-3}$

$= 1.0005$ V

Ex. 12.3. When V_1 of 40 mV is applied to inverting input and V_2 of –40 mV is applied to non-inverting input, an output voltage of 100 mV is obtained. When $V_1 = V_2 = 40\ \mu$V, output voltage is 0.4 mV. Calculate A_d, A_c, and CMRR.

Differential Mode

$V_1 = 40\ \mu$v; $V_2 = -40\ \mu$v; $V_0 = 100$ mV; $V_d = V_1 - V_2 = 80\ \mu$V

$V_0 = A_d \cdot V_d$; $A_d = V_o/V_d$

$A_d = (100 \cdot 10^{-3})/(80 \cdot 10^{-6})$

$A_d = 1250$

Common Mode

$V_1 = V_2 = 40\ \mu$V;

$V_0 = (V_1 + V_2)/2 = 40\ \mu$V

$V_o = 0.4$ mV

$V_o = A_C \cdot V_c$

$A_c = V_o/V_c = (0.4 \cdot 10^{-3})/(40 \cdot 10^{-6}) = 10$

CMRR = $(A_d / A_c) = 1250/10 = 125$.

Short Questions and Answers

Q. 1. What is a differential amplifier? What are its advantages?

Ans. The output of a differential amplifier is proportional to difference of the two inputs. It has high gain and high CMRR.

Q. 2. What is an operational amplifier?

Ans. An operational amplifier is a difference amplifier having high gain, high CMRR, high bandwidth, high input impedance and low output impedance.

Q. 3. What is meant by CMRR of a differential amplifier?

Ans. In an ideal difference amplifier, output under common mode is zero since two sections are assumed to be identical. When output voltage tends to zero, CMRR tends to infinity. Ideal value of CMRR is infinity. CMRR tends to infinity implies that common mode output voltage tends to zero. The circuit rejects common mode input voltage. Hence it is named as CMRR. Smaller the value of output voltage (CM) larger the value of CMRR. CMRR = A_d/A_c.

$$= A \frac{V_{\text{in}}(CM)}{V_{\text{out}}(CM)}$$

Q. 4. What are the requirements of an operational amplifier?

Ans.
1. Input impedance has to be infinity.
2. Bandwidth must be infinity.
3. Output impedance has to be zero.
4. Open-loop voltage gain must be infinity.
5. Typical values of CMRR for good differential amplifier will be more than *10,000 or 80 db.*
6. CMRR for a differential amplifier circuit should be *as high as possible.*

13

Operational Amplifier

Theory and applications of operational amplifiers are discussed in this chapter.

13.1 Ideal Operational Amplifier

Requirements

1. Input impedance has to be infinity
2. Output impedance has to be zero
3. Open-loop voltage gain must be infinity
4. Bandwidth must be infinity

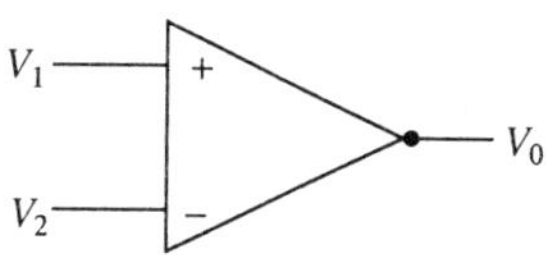

Fig. 13.1(a) Symbol.

V_1 is the input voltage with respect to ground
V_2 is another input voltage with respect to same ground
V_0 is output voltage taken between output terminal and ground.

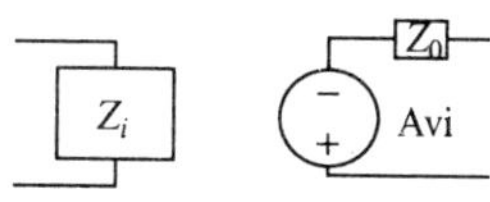

Fig. 13.1(b) Thevinin equivalent circuit.

In most of the circuits the ground is not shown. It is understood that the voltages are with respect to ground. In Thevenin equivalent circuit, z_i is the input impedance and it is of the order of mega ohms practically. Output voltage V_0 is in series with output impedance z_0, $V_0 = A(V_1 - V_2)$, where A is the open loop voltage gain. Practically the gain is of the order of 10^5.

13.2 Specifications of Operational-Amplifier

1. *Input biasing current:* Let I_{B1} and I_{B2} be the two currents through input lines. Input biasing current is defined as the average of two input currents.

2. *Input offset current drift:* It is defined as the ratio of change in input offset current to the change in temperature.

3. *Input offset voltage:* It is defined as the differential input voltage required to get quiescent null voltage at the output.

4. *Slew rate:* Rate of rise of voltage in a capacitor is given by the ratio i/c. Larger the current, faster the capacitor charges and smaller the capacitance, faster it gets charged. For an Operational-Amplifier, the maximum value of current is limited to I_m. Let the capacitance be C_c. Typical values are $I_m = 60$ μAmps, $C_c = 30$ pico. farads. Slew rate is 2×10^6 v/sec. The voltage should not change faster than this. Slew rate is defined as maximum rate of output voltage change. If dv/dt of input voltage is less than specified slew rate, output voltage is an amplified sine-wave. If the initial slope of sine wave is greater than the slew rate, output wave form gets distorted. Such distortion is called slew rate distortion.

5. *Power Bandwidth:* It is defined as the maximum possible frequency without slew rate distortion (typical value is 8 kHz).

6. *Output-offset voltage:* For common mode operation, both the input terminals are tied-up. For an ideal amplifier the output is zero. For a practical amplifier a small voltage exists in the output. This voltage is called output offset voltage.

7. *Output offset voltage drift:* It is defined as the ratio of change in output voltage to the change in temperature.

8. *Power supply rejection ratio (PSRR):* It is defined as the ratio of change in input offset voltage to the change in power supply voltage. PSRR should be as minimum as possible.

13.3 Virtual Ground

In electronic circuits conventional ground wire has a potential of zero volts. It can draw any amount of current since it acts as return path for battery current. For an operational-amplifier, the open-loop voltage gain is around 10^5. Output voltage is restricted to V_{cc}. Therefore the input voltage is of order of micro volts. This can be approximated to zero. At the input stage of operational-amplifier an FET is used. The input impedance of FET is around infinity. Therefore input current is negligible. Consider an operational-amplifier with its non-inverting terminal grounded as shown in figure 13.2. The inverting input point P_1 is called virtual ground. Terminal 2 is grounded. The drop across z_i is approximately zero since the current drawn is negligible. Therefore the potential of terminal one is same as the potential of terminal two. Hence, the point p_1 is at ground potential. Virtual ground is a ground for voltage but not for current. It cannot draw large current like conventional ground of electronic circuits. Its potential is zero like a conventional ground.

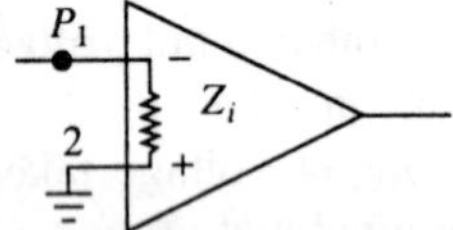

Fig. 13.2 Virtual ground.

Applications of Op-Amp

13.3.1 Non-Inverting Amplifier

The non-Inverting amplifier is shown in Fig. 13.3. In a non-inverting amplifier, the small input voltage is applied between non-inverting terminal and ground. Inverting terminal is grounded through a resistance R_1. Feedback resistance R_2 is connected between output and inverting input. The voltage across R_1 is approximately equal to V_{in} since the input voltage is negligible. Select the values of R_1 and R_2 to get the required gain. If $R_2/R_1 >> 1$, then $A = R_2/R_1$. Generally R_2 is chosen A times R_1. It is also called positive scalar circuit since the output voltage is $+k$ times input voltage.

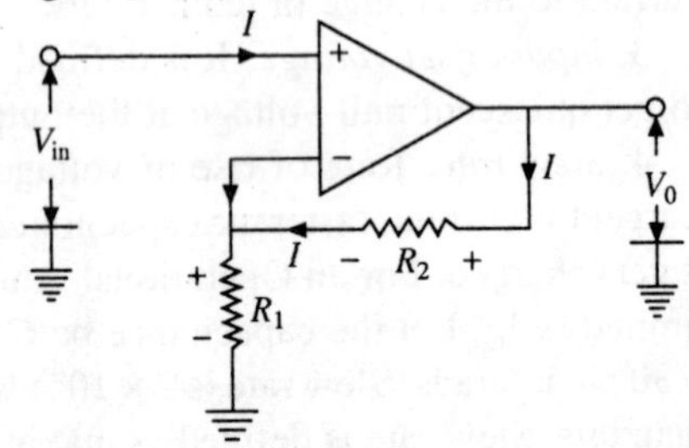

Fig. 13.3 Non-inverting amplifier.

By applying KVL in input loop, we get

$$V_{in} = IR_1$$

$$V_0 = I(R_1 + R_2)$$

$$V_0 = IR_1 + IR_2$$

$$A = \frac{V_0}{V_{in}} = \frac{I[R_1 + R_2]}{IR_1}$$

$$\boxed{A = 1 + \frac{R_2}{R_1}} \quad (13.1)$$

$$\text{Let } A = +k$$

$$V_0 = k\,V_{in}$$

13.4 Current Source

A current source circuit is shown in Fig. 13.4. The voltage across the branch R is V_{in} since the error voltage is negligible. The current through the load must be equal to current through R since the current drawn by operational-amplifier is negligible. Current $i = V_{in}/R$. Since V_{in} and R are maintained constant, i is constant. The value of current through the load is independent of load resistance, hence the system can be represented as a current source of value V_{in}/R. Thus a non-inverting amplifier can be used as a current source.

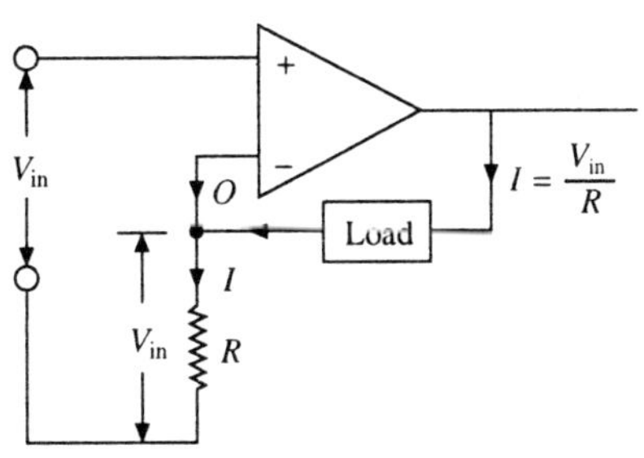

Fig. 13.4 Current source.

13.5 Inverting Amplifier

The inverting amplifier is shown in Fig. 13.5. In inverting amplifier, the input voltage is applied to inverting terminal through the resistance R_1. Non-inverting

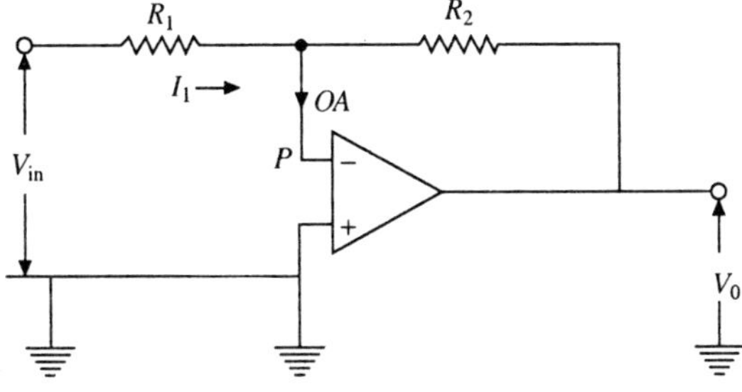

Fig. 13.5 Inverting amplifier.

input is grounded. Let the input current through R_1 be I_1. The current through R_2 has to be I_1 since current drawn by operational-amplifier is negligible. Voltage across R_1 is V_{in}, since error voltage can be neglected. It is called negative scalar since output voltage is product of $-K_1$ and input voltage.

Therefore, $$V_{in} = I_i R_1 \tag{13.1}$$

$$I_i = \frac{V_{in}}{R_1}$$

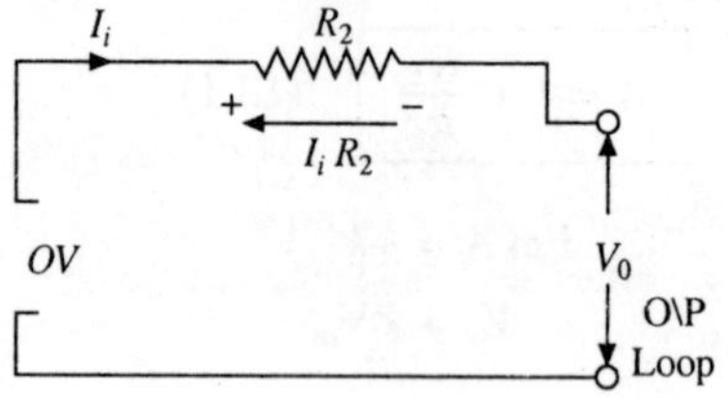

Apply KVL in output loop

$$V_0 + I_i R_2 = 0$$

$$V_0 = -I_i R_2 \tag{13.2}$$

Find the ratio of (13.2) to (13.1) to get gain

$$A = \frac{V_0}{V_{in}} = \frac{-I_i R_2}{I_i R_1} \qquad \boxed{A = \frac{-R_2}{R_1}} \tag{13.3}$$

Negative sign signifies 180° phase shift between input and output.

This amplifier is more popular, since the input impedance of this circuit can be defined by us.

$$Z_i = \frac{V_i}{I_i} = \frac{I_i R_1}{I_1} = R_1$$

$$\boxed{Z_i = R_1} \tag{13.4}$$

R_1 can be choosen to any value that is required.

$$A = \frac{-R_2}{R_1}; \; A = -k_1$$

$$\frac{V_o}{V_i} = -k_1; \quad V_o = -k_1 V_i \tag{13.5}$$

13.6 Summer or Adder

The summer or adder circuit is shown in Fig. 13.6. Many input signals can be given to the inverting input through the resistors R_1 and R_2 etc. The current through the feedback resistor must be sum of I_1 and I_2 since the current drawn by operational-amplifier is negligible. Current through R_1 is V_1/R_1 since error voltage

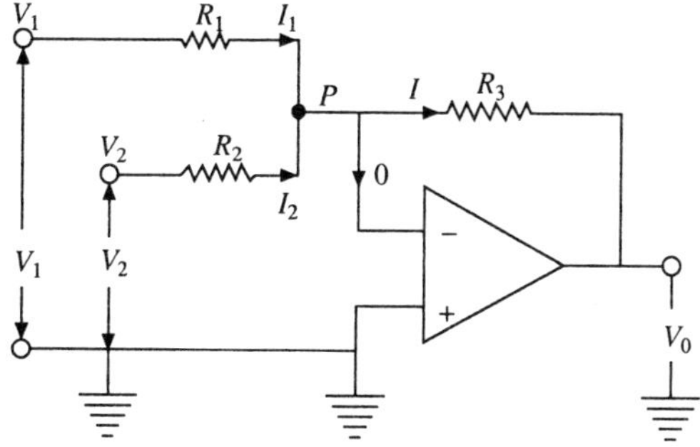

Fig. 13.6 Summer or adder.

is negligible. Voltage across R_2 is V_2, since input voltage is zero. Current through R_2 is V_2/R_2. If equal values of resistance are selected, magnitude of output voltage is sum of input voltages.

Apply KCL at p

$$I = I_1 + I_2$$

KVL in loop having R_3, V_o and input gives

$$V_0 = -IR_3$$

$$I = \frac{-V_0}{R_3}$$

$$\frac{-V_0}{R_3} = \frac{V_1}{R_1} + \frac{V_2}{R_2}$$

$$V_0 = -R_3\left[\frac{V_1}{R_1} + \frac{V_2}{R_2}\right] \tag{13.6}$$

if $$R_1, R_2, R_3 = R$$

$$\boxed{V_0 = -[V_1 + V_2]} \tag{13.7}$$

Output voltage is proportional to algebraic sum of input voltages.

13.7 Miller Integrator

The Miller integrator circuit is shown in Fig. 13.7. In an integrator circuit, output voltage is integral of the input signal. In the place of feedback resistor, a capacitor is used. The current through the capacitor is equal to input current since current drawn by operational-amplifier is negligible. This circuit is used for converting a square wave into triangular wave form. If a single positive step is applied a negative ramp(can be obtained at the output since integration of step voltage is ramp voltage. The slope of the ramp is negative due to the negative sign in the expression or due to the use of inverting amplifier.

$$i = C\frac{dV_c}{dt}; \quad \frac{i}{C} = \frac{dV_c}{dt}$$

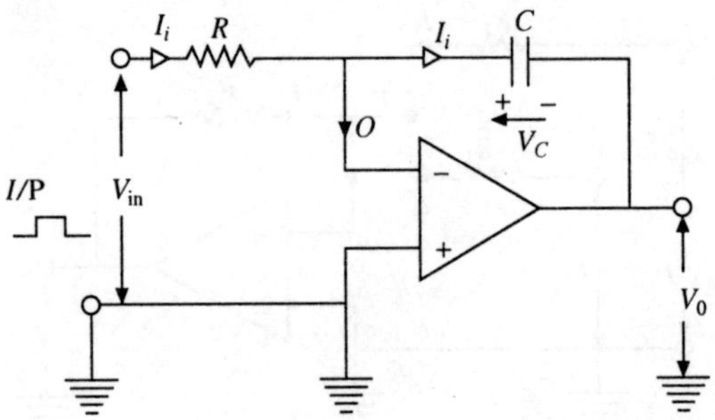

Fig. 13.7 Miller integrator.

$$dV_c = \frac{i}{C}dt; \quad V_c = \frac{1}{C}\int idt$$

KVL in output loop gives $V_c = -V_o$

$$-V_o = \frac{1}{C}\int idt; \quad V_o = -\frac{1}{C}\int I_i\,dt$$

$$V_o = \frac{-1}{C}\int \frac{V_{in}}{R}dt; \quad V_o = \frac{-1}{RC}\int V_{in}\,dt \tag{13.8}$$

From (13.8) it can be seen that output is proportional to integral of input.

13.8 Differentiator

The differentiator circuit is shown in Fig. 13.8. If the resistance and capacitance in the integrator circuit are interchanged we get differentiator circuit. Current through the resistance is equal to input current since the current through operational-amplifier is negligible. Voltage across capacitor is equal to V_{in} since the error voltage is negligible. Output voltage is proportional to derivative of input-voltage. This circuit converts triangular waveform into square wave.

$$i = C\frac{dv}{dt}; \quad I_i = C\frac{dV_{in}}{dt} \tag{13.9}$$

KVL in output loop

$$V_0 = -I_i R$$

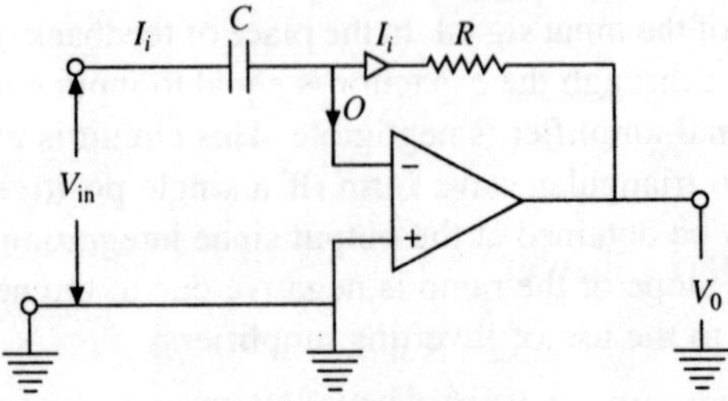

Fig. 13.8 Differentiator.

Substitute (13.9) here

$$V_0 = -RC\frac{dV_{in}}{dt} \qquad (13.10)$$

From (13.10), it can be seen that output voltage is proportional to derivative of input voltage. When a square wave is applied, spikes are produced in the output. The height of the spike can be reduced by increasing the value of R.

13.9 Comparator

The symbol, characteristics, comparator circuit and waveforms are shown in Fig. 13.9. Output of operational-amplifier goes to positive saturation voltage when V_1 is higher than V_2. Output of operational-amplifier goes to negative saturation, when V_2 is greater than V_1. A zero crossing detector circuit using operational-amplifier is shown below. ZCD converts sine wave into square wave. Output is high for 180° and low for 180°. Input signal V_1 is sine wave and V_2 is a small dc voltage. Comparator goes to positive saturation from t_1 to t_2 since $V_1 > V_2$. From t_2 to t_3, $v_2 > v_1$, comparator goes to negative saturation. This voltage forward biases the diode D. When diode conducts output voltage is approximately zero. The resistance limits the current through the diode. Thus the output of ZCD is a square wave with approximately 180° on and 180° off. If V_2 = 0, a square wave with exactly 180° on-and-off can be generated.

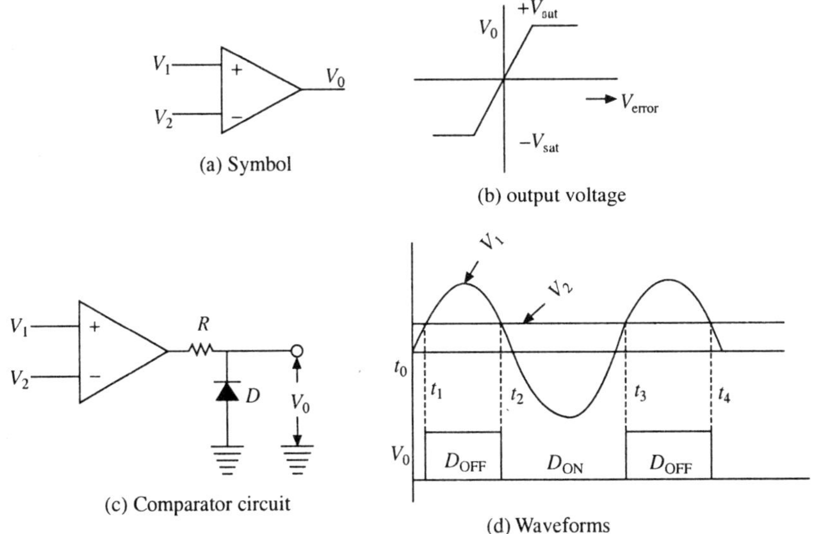

Fig. 13.9 Comparator.

13.10 Voltage Follower

Buffer or voltage follower circuit is shown in Fig. 13.10(a). Non-inverting amplifier circuit is shown in Fig. 13.10(b). The voltage gain of non-inverting amplifier is $A = 1 + R_2/R_1$. In voltage follower circuit, R_2 is made zero and R_1 is made infinity. Hence $V_o/V_{\text{in}} = 1$. This implies $V_o = V_{\text{in}}$. This is called voltage follower since output follows the input signal. It is superior to common collector (emitter follower) configuration and source follower configuration. It is best suited as an impedance matching device between source and a low impedance load.

From (13.1), we know that

$$A = \frac{V_0}{V_{\text{in}}} = 1 + \frac{R_2}{R_1}$$

if
$$R_2 = 0$$

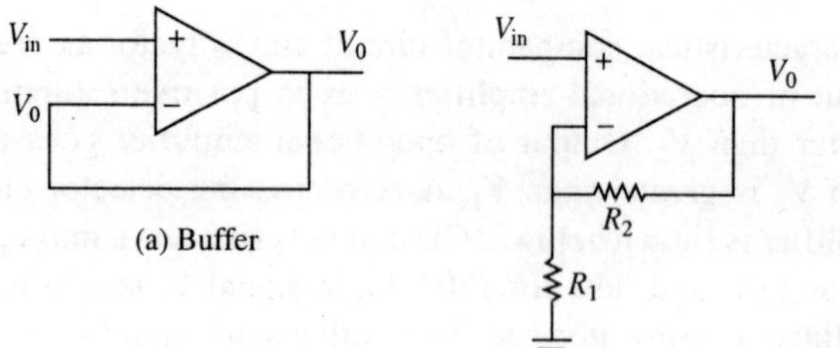

(a) Buffer

(b) Non inverting amplifier circuit

Fig. 13.10 Voltage follower.

and
$$R_1 = \infty,$$

we get
$$\frac{V_0}{V_{\text{in}}} = 1$$

$$\boxed{V_0 = V_{\text{in}}} \tag{13.11}$$

13.11 Phase Shifter

Phase shifter circuit is shown in Fig. 13.11. In the inverting amplifier circuit, R_1 and R_2 are replaced by Z_1 and Z_2. Let the angles of impedances Z_1 and Z_2 be θ_1 and θ_2. θ_1 and θ_2 can be varied from zero to $\pm 90°$ by varying the resistance and capacitance. $V_0 = V_{\text{in}} K \angle\theta$. Output voltage is shifted with respect to input voltage by an angle θ. By changing the variable components in the Z_1 and Z_2, θ can be varied from 0 to 360°. Thus any phase shift between 0 to 360° can be obtained.

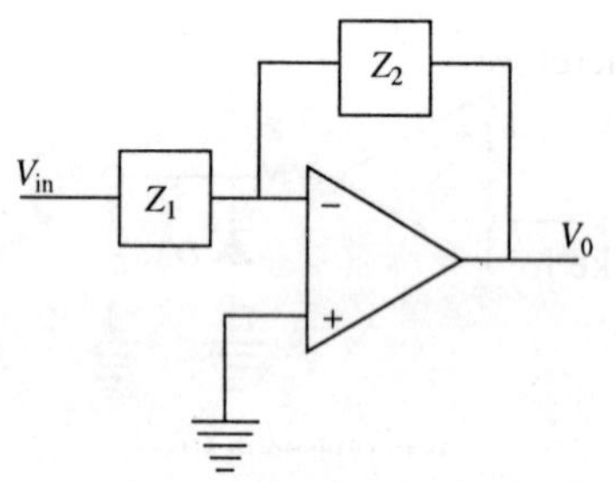

Fig. 13.11 Phase shifter.

Using (13.3) we can write gain as

$$\frac{V_0}{V_{\text{in}}} = \frac{-Z_2}{Z_1}$$

$$\frac{V_0}{V_{in}} = \frac{(-1) \,|\, Z_2 \,|\, \angle\theta_2}{|\, Z_1 \,|\, \angle\theta_1}$$

$$\frac{V_0}{V_{in}} = \frac{1\angle\pi \,|\, Z_2 \,|\, \angle\theta_2}{|\, Z_1 \,|\, \angle\theta_1}$$

$$\frac{V_0}{V_{in}} = \left| \frac{Z_2}{Z_1} \right| \angle\pi + \theta_2 - \theta_1 \quad (13.12)$$

$$\frac{V_0}{V_{in}} = k\angle\theta$$

$$V_0 = V_{in}\, k\angle\theta$$

Z_1 and Z_2 are series combination of R and C. The value of θ can be varied by varying R and C.

13.12 Logarithmic amplifier

Logarithmic amplifier circuit is shown in Fig. 13.12. In place of R_2 of the inverting amplifier, a diode is used. This forms logarithmic amplifier circuit. It can be seen that output voltage is proportional to the log of input voltage. In the place of diode a transistor can also be used since the input characteristic of transistor has a non-linearity similar to that of diode.

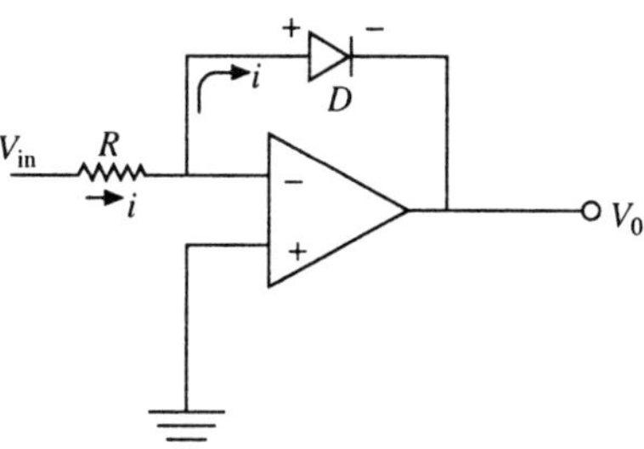

Fig. 13.12 Logarithmic amplifier.

From equation (3.12) current through diode is

$$i = I_0\,(e^{+k_1 V_f} - 1)$$

$$e^{+k_1 V_f} >> 1,$$

Therefore,

$$i \simeq I_0\, e^{+k_1 V_f}$$

$$\frac{i}{I_0} = e^{+k_1 V_f}$$

Take log on both sides

$$\log \frac{i}{I_0} = +k_1 V_f \log_e e$$

$$\log \frac{i}{I_0} = +k_1 V_f$$

$$\frac{1}{k_1} \log \frac{i}{I_0} = V_f \quad (13.13)$$

Applying KVL in output loop

$$V_0 + V_f = 0$$

$$V_0 = -V_f$$

Substitute (13.13) here

$$V_0 = -\frac{1}{k_1}\log\frac{i}{I_0}$$

Substitute $$i = \frac{V_{in}}{R_1}$$

$$\boxed{V_0 = -\frac{1}{k_1}\log\frac{V_{in}}{R_1 I_0}} \qquad (13.14)$$

$$V_0 \propto \log_e V_{in}$$

Therefore output voltage is proportional to log of input voltage.

13.13 Current to Voltage Converter

Current to voltage converter circuit is shown in Fig. 13.13. Photo multiplier tube supplies constant current irrespective of the load resistance connected across it. The device that acts practically as a current source is a photo cell. Current through R_2 is equal to source current since current drawn by operational-amplifier is negligible. Output voltage $V_0 = -I_s R_2$. Therefore output voltage is proportional to source current. Thus this circuit converts current signal into voltage signal.

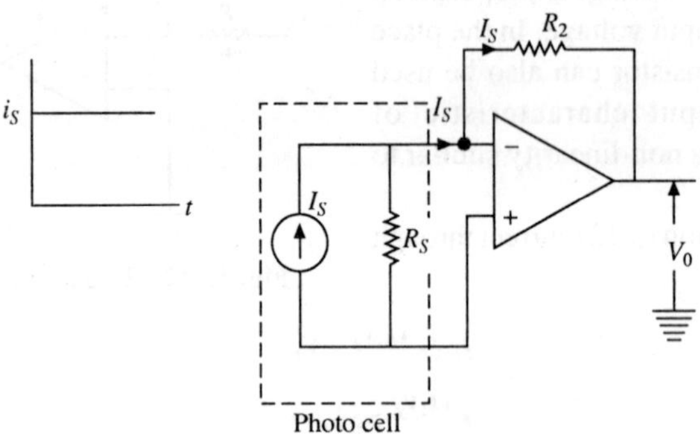

Fig. 13.13 Current to voltage converter.

13.14 Low Pass Filter

Low pass filter circuit is shown in Fig. 13.14(a). Frequency response is shown in Fig. 13.14(b). For low frequencies, the capacitor acts as open circuit. The circuit is equivalent to an inverting amplifier with C_2 open. Low frequency signals pass through the circuit and reach the output. In the high frequency range, capacitor acts as a short circuit. The loading effect due to short circuited capacitor reduces the output voltage. The same can be seen from the graph of output voltage versus frequency. Critical frequency is a frequency where the output voltage falls to 70.7% of normal low frequency output. In db scale, this point is called 3 db point. Voltage gain of above circuit is $A = \frac{-Z_2}{Z_1}$, where Z_2 is parallel combination of R_2 and C_2.

$$Z_1 = R_1$$

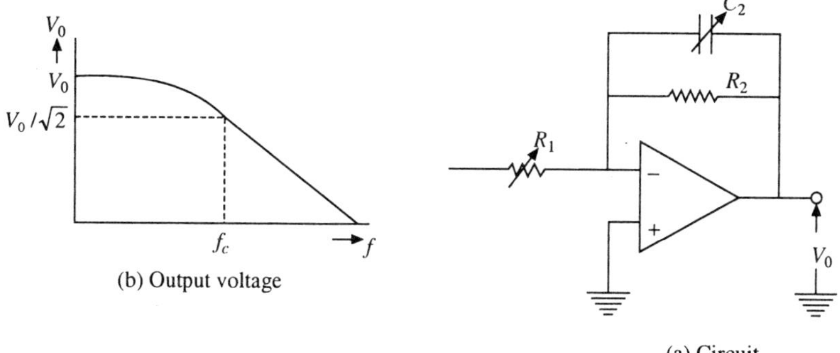

Fig. 13.14 Low Pass Filter.

$$Z_2 = \frac{R_2 \cdot \dfrac{1}{j\omega C_2}}{R_2 + \dfrac{1}{j\omega C_2}}$$

$$Z_2 = \frac{\dfrac{R_2}{j\omega C_2}}{\dfrac{j\omega C_2 R_2 + 1}{j\omega C_2}} = \frac{R_2}{1 + j\omega C_2 R_2}$$

$$\text{Gain}(A) = \frac{-Z_2}{Z_1} = -\frac{1}{R_1} \cdot \frac{R_2}{(1 + j\omega C_2 R_2)}$$

$$|A| = \frac{R_2}{R_1\sqrt{1 + \omega^2 C_2^2 R_2^2}}$$

$$|A| = \frac{|A_m|}{\sqrt{1 + \omega^2 C_2^2 R_2^2}} \quad \therefore A_m = \frac{R_2}{R_1}$$

$$\frac{|A|}{|A_m|} = \frac{1}{\sqrt{1 + \omega^2 C_2^2 R_2^2}}$$

when $\omega = \omega_C, A = \dfrac{1}{\sqrt{2}} A_m$

$$\frac{1}{\sqrt{2}} = \frac{1}{\sqrt{1 + \omega_c^2 C_2^2 R_2^2}}$$

invest and square on both sides

$$2 = 1 + \omega_c^2 C_2^2 R_2^2$$

$$\omega_C^2 C_2^2 R_2^2 = 1$$

Taking root on both sides

$$\omega_c C_2 R_2 = 1$$

$$\omega_c = \frac{1}{C_2 R_2}$$

$$\boxed{f_c = \frac{1}{2\pi C_2 R_2}} \tag{13.15}$$

13.15 High Pass Filter

High pass filter circuit and output response are shown in Fig. 13.15(a) and (b), respectively. For low frequencies the capacitor C_1 acts as open circuit. Applied voltage signal does not reach the input of operational-amplifier since capacitor acts as open. Output voltage is zero, since input voltage to operational-amplifier is zero. As the frequency increases, the reactance offered by the capacitor decreases. Therefore, output voltage increases. In the high frequency range, capacitor acts like short circuit. The above circuit is equivalent to inverting amplifier. Therefore high frequency signals pass to the output. Critical frequency is the frequency at which output voltage is $1/\sqrt{2}$ times the maximum output voltage.

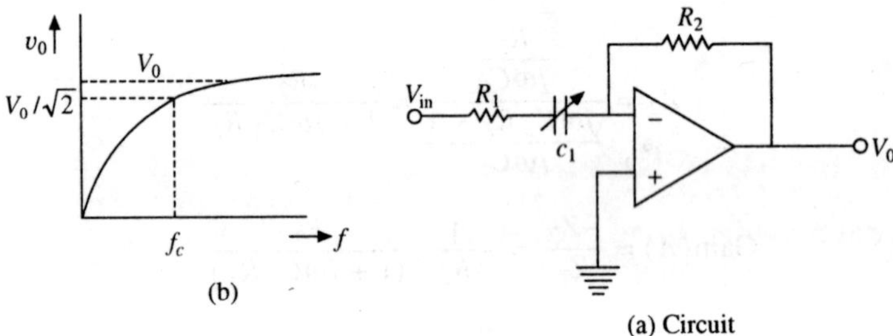

Fig. 13.15 High pass filter.

$$Z_1 = R_1$$

$$Z_2 = \frac{R_2 \cdot \dfrac{1}{jC_1\omega}}{R_2 + \dfrac{1}{j\omega C_1}}$$

$$Z_2 = \frac{\dfrac{R_2}{j\omega C_1}}{\dfrac{j\omega C_1 R_2 + 1}{j\omega C_1}} = \frac{R_2}{1 + j\omega C_1 R_2}$$

$$\text{Gain}(A) = \frac{-Z_2}{Z_1} = -\frac{1}{R_1} \cdot \frac{R_2}{1 + j\omega C_1 R_2}$$

$$|A| = \frac{R_2}{R_1\sqrt{1 + \omega^2 C_1^2 R_2^2}}$$

$$|A| = \frac{|A_m|}{\sqrt{1 + \omega^2 C_1^2 R_2^2}}. \quad \therefore \quad A_m = \frac{R_2}{R_1}$$

$$\frac{|A|}{|A_m|} = \frac{1}{\sqrt{1 + \omega^2 C_1^2 R_2^2}}$$

when

$$\omega = \omega_C, \; A = \frac{1}{\sqrt{2}} A_m$$

$$\frac{1}{\sqrt{2}} = \frac{1}{\sqrt{1 + \omega_c^2 C_1^2 R_2^2}}$$

Invert and square on both sides

$$2 = 1 + \omega_c^2 C_1^2 R_2^2$$

$$\omega_c^2 C_1^2 R_2^2 = 1$$

Take root on both sides

$$\omega_c C_1 R_2 = 1$$

$$\omega_c = \frac{1}{C_1 R_2}$$

$$\boxed{f_c = \frac{1}{2\pi C_1 R_2}} \tag{13.16}$$

13.16 Half Wave Rectifier

Half wave rectifier using operational-amplifier is shown in Fig. 13.16. This circuit is a combination of operational-amplifier and half-wave rectifier. The input A.C. signal is amplified by using an operational-amplifier.

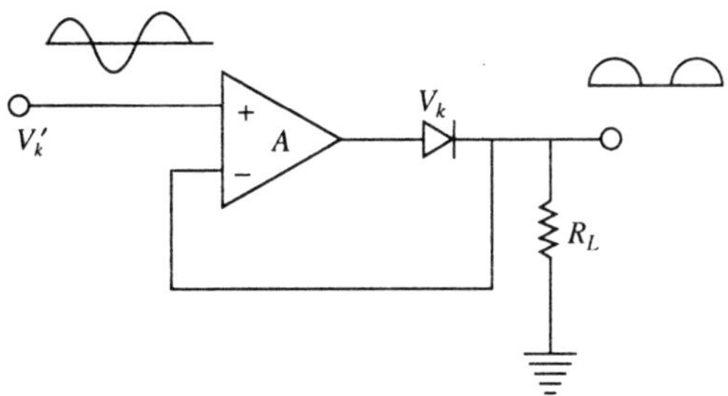

Fig. 13.16 Half wave rectifier.

During the positive half-cycle of the A.C. output diode is forward biased. It conducts and positive half-cycle appears across the load resistance. During the negative half-cycle of A.C., diode is reverse biased, it does not conduct. Thus output across R_L is a half-wave rectified signal.

With this configuration the value of cut-in voltage required reduces to micro volts range. This circuit can perform rectification of weak signals accurately.

$$A = \frac{V_k}{V_k'}$$

13.17 Instrumentation Amplifier

Instrumentation amplifier circuit is shown in Fig. 13.17. This amplifier is nothing but a differential amplifier except that there are two buffers in the two input lines.

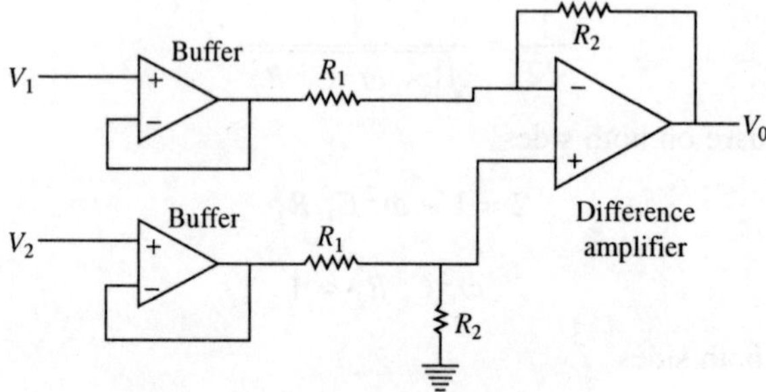

Fig. 13.17 Instrumentation amplifier.

This amplifier has the advantage of high CMRR, high gain, high input impedance and isolation between input and output grounds. Generally this is used as front end amplifier of measuring instruments. In practical systems, power circuit operates at high voltage and high current. Control circuit operate at low voltage and low current. If any extra wire connects the power circuit and control circuit with common ground, a large current flows through control circuit and it gets burnt. To prevent this, grounds of control circuit and power circuit must be isolated.

The above amplifier is ideally suited for providing the isolation between control and power circuits.

13.18 Voltage Controlled Oscillator (VCO)

VCO and its waveforms are shown in Fig. 13.18. Let the initial voltage of the capacitor be zero. When D.C. input voltage is given, capacitor charges exponentially. Voltage across capacitor varies exponentially. Capacitor charges with plate *a* negative since voltage is negative (--v_c). When the capacitor voltage goes slightly above total cut-in voltage V_B, diodes conduct and capacitor discharges

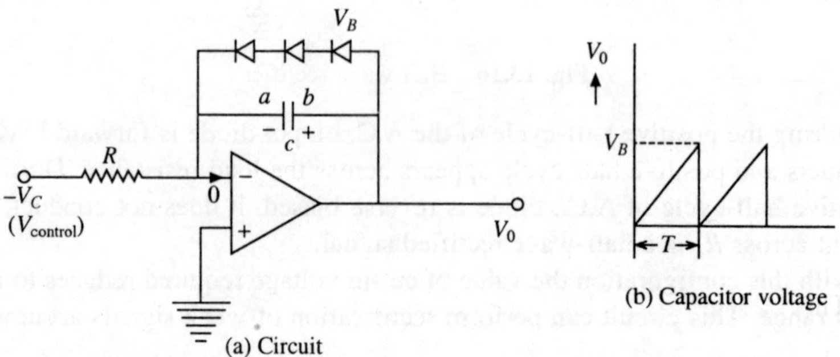

Fig. 13.18 Voltage controlled oscillator (VCO).

quickly to zero, through the diodes. Diodes go off since voltage is reduced to zero. The capacitor again charges exponentially and above process gets repeated. We know that output voltage is equal to capacitor voltage. Output voltage waveform is shown positive since negative control voltage is given to the inverting input.

The output voltage is approximately a saw-tooth wave with negligible flyback period (discharged period). For the sake of analysis capacitor charging is assumed to be linear.

$$\text{Slope of output waveform is } \frac{V_B}{T} = m \tag{13.17}$$

$$0.63\ V_c = m\tau$$

$$m = \frac{0.63 V_c}{\tau} \tag{13.18}$$

Equate (13.17) and (13.18)

$$\frac{0.63 V_c}{\tau} = \frac{V_B}{T}$$

$$\frac{1}{T} = \frac{0.63 V_c}{\tau V_B}$$

$$f = \frac{0.63\ V_c}{\tau\ V_B} \tag{13.19}$$

$$\boxed{f \propto V_c}$$

Therefore frequency of output is proportional to control voltage.

13.19 Phase Locked Loop (PLL)

Block diagram of phase locked loop is shown in Fig. 13.19. Phase detector is basically a mixer circuit. Mixer circuit accepts two signals of different frequencies. Output contains one signal of frequency $(f_{in} + f_o)$ and another signal of frequency $(f_{in} - f_0)$. The output of low pass filter contains only $(f_{in} - f_o)$ since high frequency signal is filtered. This signal is amplified by an amplifier. Output of the amplifier represents error signal. VCO generates a signal V_Y having a frequency f_0. The frequency f_o is proportional to the error voltage v_e. This signal is one of the inputs to the phase detector (v_y).

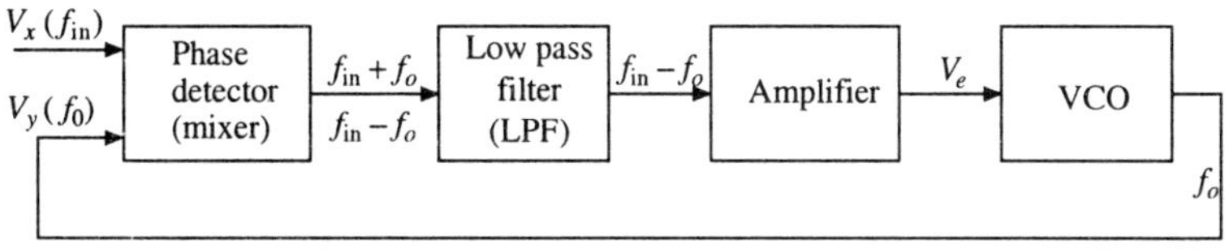

Fig. 13.19 Phase locked loop (PLL).

This system has three different states

1. Free running state,
2. Capture state and
3. Phase locked state.

When no input signal is applied ($v_x = 0$), the circuit is said to operate in free running state. When an input signal is applied, an error voltage is created based on the difference between the frequencies f_{in} and f_o. This error voltage is converted into frequency by the VCO. This process continues until $f_o = f_{in}$. This state is called capture state. When the frequency of VCO is equal to input frequency the system is said to be in phase locked state. In this state output frequency of LPF is zero i.e. a D.C. signal is applied to the amplifier.

Capture range is defined as band of frequencies, those can be locked by VCO when the system has not yet locked. Lock range is a band of frequencies over which VCO can track the incoming signal after it is locked. (Locked condition means $f_{in} = f_o$).

Worked Problems

Ex. 13.1 For the inverting amplifier $R_1 = 1\ k\Omega$, $R_2 = 2\ M\Omega$, Find voltage gain, Z_i and V_o if $V_{in} = 1\mu V$.

$$A_v = \frac{-R_2}{R_1} = \frac{-2M\Omega}{1\ k\Omega} = -2000$$

$$Z_{in} = R_1 = 1k\ \Omega$$

$$V_0 = -A \cdot V_i = -2000 \times 1\ \mu V$$

$$V_0 = -\ 2mV.$$

Ex. 13.2 Calculate output voltage of a non-inverting amplifier with $R_1 = 100$ kΩ, $R_2 = 600$ kΩ and $V_{in} = 2mV$.

$$\text{Output voltage } V_0 = \left(1 + \frac{R_2}{R_1}\right)V_1 = \left(1 + \frac{600}{100}\right)V_1$$

$$V_0 = 7 \times 2 = 14\ mV$$

Ex. 13.3 Find the output voltage of a summer with $R_f = 1$ MΩ, $R_1 = 500$ kΩ, $R_2 = 1$ MΩ, and $R_3 = 1$ MΩ, $V_1 = 1$ V, $V_2 = 2$ V, $V_3 = 3$ V.

$$\text{Output voltage} \quad V_0 = -\left(\frac{R_f}{R_1} \cdot V_1 + \frac{R_f}{V_2} \cdot V_2 + \frac{R_f}{R_3} \cdot V_3\right)$$

$$V_0 = -7\ V$$

Ex. 13.4 A 5 mV (peak), 1 kHz sine wave is applied to the input of Op-Amp integrator for which $R = 100$ kΩ, $C = 1\ \mu$F. Find the expression for output voltage.

$$V_0 = -\frac{1}{RC}\int V\,dt$$

$$V_0 = \frac{-1}{100 \times 10^3 \cdot 1 \times 10^{-6}} \int 5 \times 10^{-3} \sin 2\pi\, 10^3\, t$$

$$V_0 = \frac{-1}{0.1}\int_0^t 5 \times 10^{-3} \sin 2\pi\, 10^3\, t$$

$$V_0 = \frac{1}{40}(\cos 2000\,\pi t - 1)$$

$$V_0 = 7.9 \times 10^{-6}(\cos 2000\,\pi t - 1) \text{ mV}$$

Ex. 13.5 The input to the differentiator is a sine wave with peak value = 5 mV and $f = 1$ kHz. Find the expression for output voltage if $R = 100$ kΩ and $C = 1$ μF.

$$V_0 = -RC\frac{dv_i}{dt}$$

$$V_0 = -\,1000\,\pi \cos 2000\,\pi t \text{ in mV}$$

Ex. 13.6 Calculate the cut off frequency of a first order low pass filter for $R = 1.2$ kΩ and $C = 0.02$ μF

$$f_c = \frac{1}{2\pi CR_2} = \frac{1}{2 \times \pi \times 0.02 \times 10^{-6} \times 1.2 \times 10^3} = 6.63 \text{ kHz}$$

Ex. 13.7 An operational-amplifier adder has $R_f = 12$ k, resistances at input side are $R_{f1} = 12$kΩ, $R_{f2} = 2$kΩ, $R_{f3} = 3$ kΩ. The corresponding inputs are $V_{i1} = 9V$, $V_{i2} = -3V$, $V_{i3} = -1V$. Non- inverting terminal is grounded. Calculate output voltage assuming ± 10V supply.

$$V_0 = -\left(\frac{R_f}{R_{f1}}V_{i1} + \frac{R_f}{R_{f2}}V_{i2} + \frac{R_f}{R_{f3}}V_{i3}\right)$$

$$V_0 = -\left(\frac{12}{12}\cdot 9 - \frac{12}{2}\cdot 3 - \frac{12}{3}\cdot 1\right)$$

$$V_0 = -(9 - 18 - 4)$$

$$V_0 = 13\text{V}$$

In an operational-amplifier output voltage cannot be greater than supply voltage. $V_0 = 10$ volts since output gets clamped to saturation voltage.

Ex. 13.8 An IC 741 Op-Amp has a slew rate of 0.4 $v/\mu s$ and output peak sine wave $V_p = 10$ V. Calculate maximum undistorted frequency for large signal operation.

$$S_R = \frac{dv_c}{dt}$$

$$dt = \frac{d_{vc}}{S_R} = \frac{10}{0.4} = 25\mu s$$

$$f = \frac{1}{t} = \frac{10^6}{25} = 10^4$$

$$f = 10 \text{ kHz}$$

Ex. 13.9 Design a differential amplifier, using operational-amplifier to give $V_0 = 10\,(V_2 - V_1)$, use maximum value of resistance as 10 kΩ.
Let the value of R_1 be 10 kΩ. The value of R_2 is 10 times that of R_1. $R_2 = 10 \times 10$ K = 100 K.

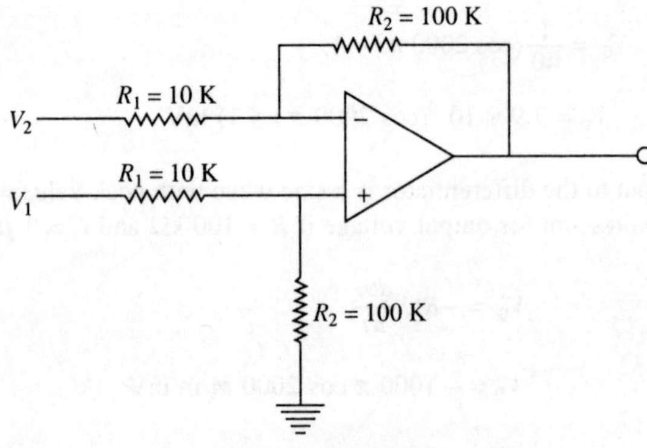

Fig. 13.20

Ex. 13.10 Draw the circuit of a non-inverting amplifier for a gain of 10. Use minimum resistance of 10 kΩ.

$$1 + \frac{R_2}{10\ \text{k}} = 10$$

$$R_2 = 90\ \text{k}$$

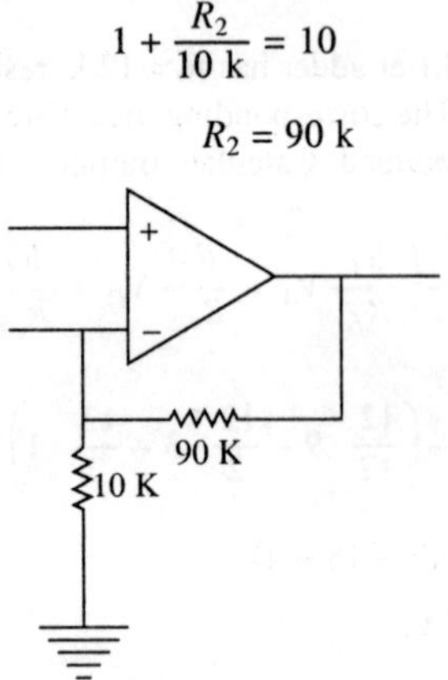

Fig. 13.21 Non-inverting

Ex. 13.11 Figure below shows a non-inverting operational-amplifier summer with $V_1 = 2$ V, $V_2 = -1$ V. What is output voltage?

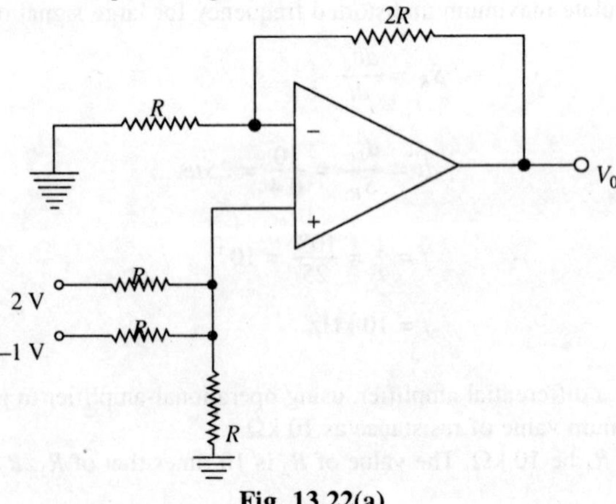

Fig. 13.22(a)

If the inputs are given to the non-inverting terminal, formula given in (13.6) cannot be used. We have to use superposition theorem. +2V is reduced to zero and –1V is retained. This is shown in fig. 13.22(b). Determine the input voltage at '2' by using circuit reduction. Calculate output using the formula

$$V_0 = \left(1 + \frac{R_2}{R_1}\right)V_1; \quad I_1 = \frac{-1}{R + 0.5R}$$

$$I_1 = \frac{-1}{1.5\,R}$$

$$V_2 = I_1 \frac{R}{2} = \frac{-1}{1.5R}\,0.5R$$

$$V_2 = \frac{-1}{3}\,\text{V}$$

$$V_{20}^1 = \frac{-1}{3}\left(1 + \frac{2R}{R}\right)$$

$$V_{20}^1 = -1\ \text{V}$$

Fig. 13.22(b)

Fig. 13.22(c)

–1 V is reduced to zero and 2 V source is retained as shown in fig. 13.22(d). From fig. 13.22(e),

$$I_2 = \frac{2}{1.5R}$$

$$I_2 = 1.33/R$$

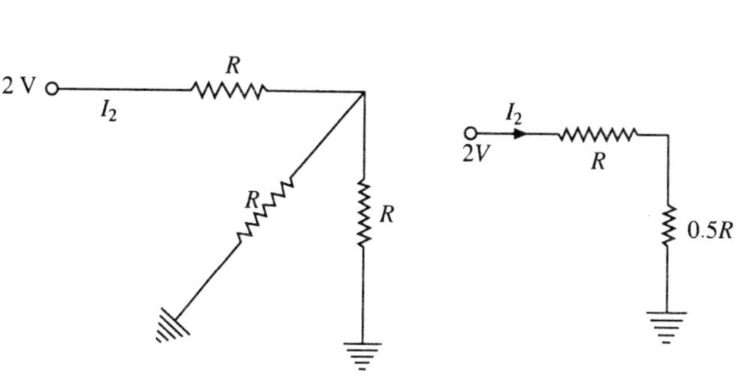

Fig. 13.22(d)

Fig. 13.22(e)

$$V_2 = \frac{1.33}{R} \times \frac{R}{2}$$

$$V_2 = 0.67V$$

$$V''_{20} = +\frac{2}{3}\left(1 + \frac{2R}{R}\right) = +2\text{V}$$

By superposition theorem

$$V_0 = V'_{20} + V''_{20}$$

$$V_0 = -1 + 2$$

$$V_0 = 1\text{V}$$

Ex. 13.12 Figure below shows an operational-amplifier based amplifier. Find steady state output voltage when (i) *S* is open and (ii) *S* is closed.

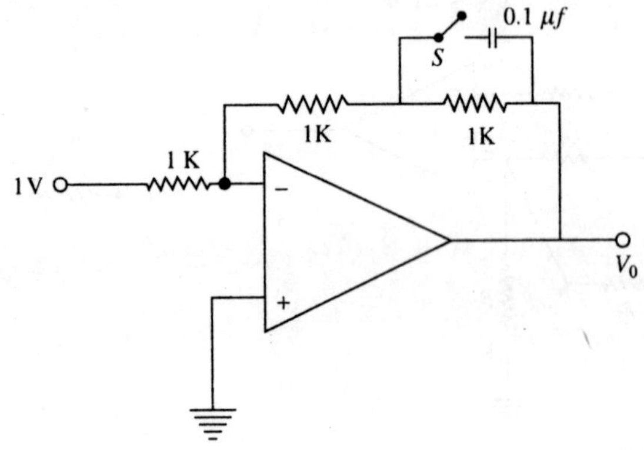

Fig. 13.23

(i) When *S* is open, the circuit acts like an inverting amplifier.

$$V_0 = -V_i \frac{R_2}{R_1}$$

$$V_0 = -1 \times \frac{2\text{k}}{1\text{k}}$$

$$V_0 = -2 \text{ V}$$

(ii) *S* is closed, capacitor acts as open circuit in steady state. Therefore, circuit behaves like an inverting amplifier.

$$A = \frac{-2}{1}$$

$$A = -2$$

$$V_0 = A \cdot V_i$$

$$V_0 = -2 \times 1$$

$$V_0 = -2\text{ V}$$

Ex. 13.13 Prove that transfer function for the circuit shown in fig. 13.24 is $\dfrac{K}{S^2 + (3-k)S + 1}$

$$T(S) = \frac{V_2(S)}{V_1(S)}$$

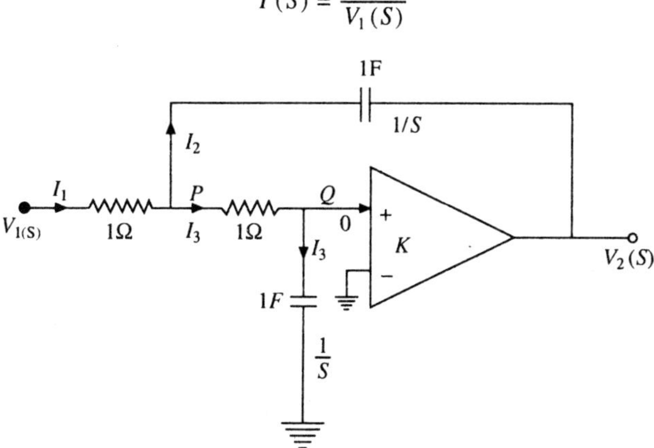

Fig. 13.24

Apply KCL at P

$$I_1 = I_2 + I_3$$

$$\frac{V_1 - V_P}{1} = \frac{V_P - V_2}{1/S} + V_P - V_Q$$

$$V_1 - V_P = (V_P - V_2)\, S + (V_P - V_Q) \tag{13.20}$$

$$k = \frac{V_2}{V_Q} \tag{13.21}$$

$$I_3 = \frac{V_P}{1 + \frac{1}{S}} = \frac{V_P}{\frac{S+1}{S}}$$

$$I_3 = \frac{V_P S}{S + 1}$$

$$V_Q = I_3 \times \frac{1}{S} = \frac{V_P\, S}{S+1} \times \frac{1}{S}$$

$$V_Q = \frac{V_P}{S+1} \tag{13.22}$$

Substitute (13.22) in (13.21). $k = \dfrac{V_2}{V_P}(S+1);\ V_P = V_2\dfrac{(S+1)}{k}$ (13.23)

Substitute (13.21) and (13.22) in (13.20). Eliminate V_p and V_Q so that we can get the ratio V_2/V_1

$$V_1 - V_P = (V_P - V_2)S + \left(V_P - \frac{V_P}{S+1}\right)$$

$$V_1 - V_P = (V_P - V_2)S + V_P\left(\frac{S+1-1}{S+1}\right)$$

$$V_1 - V_P = (V_P - V_2)S + V_P\left(\frac{S}{S+1}\right) \quad (13.24)$$

$$V_1 - V_P = V_P S - V_2 S + V_P\left(\frac{S}{S+1}\right)$$

$$V_1 = V_P + V_P S - V_2 S + V_P\left(\frac{S}{S+1}\right)$$

$$V_1 = V_P\left(1 + S + \frac{S}{S+1}\right) - V_2 S$$

$$V_1 = V_P\left(\frac{(S+1)^2 + S}{S+1}\right) - V_2 S \quad (13.25)$$

Substitute (13.23) in (13.25)

$$V_1 = \frac{V_2(S+1)}{k}\left(\frac{(S+1)^2 + S}{S+1}\right) - V_2 S$$

$$\frac{V_1}{V_2} = \left(\frac{(S+1)^2 + S}{k} - S\right)$$

$$\frac{V_1}{V_2} = \frac{(S+1)^2 + S - kS}{k}$$

$$\frac{V_1}{V_2} = \frac{S^2 + 2S + 1 + S - kS}{k}$$

$$\frac{V_1}{V_2} = \frac{S^2 + 3S + 1 - kS}{k}$$

$$\frac{V_1}{V_2} = \frac{S^2 + S(3-k) + 1}{k}$$

$$\frac{V_2}{V_1} = \frac{k}{S^2 + (3-k)S + 1}$$

Short Questions and Answers

Q. 1. Give typical value of bias current of μA 714 op. amp.
Ans. 2.6 nano amperes.

Q. 2. What are the disadvantages of differentiator circuit?
Ans. 1. High frequency Noise is also amplified.
2. Gain is not stable.

Q. 3. What kind of feed back is present in non-inverting amplifier?
Ans. Negative feedback, since feedback element is connected to negative input terminal.

Q. 4. Define capture range of phase locked loop.
Ans. A band of frequencies those can be locked by VCO when PLL system is in the capture state.

Q. 5. Distinguish between positive and negative feed back.

Ans. In negative feedback, the feedback signal gets subtracted with the source voltage. Over all gain is reduced.

In the positive feedback, the feed back signal gets added to the source voltage. over all gain is increased.

Q. 6. Let the magnitude of the gain of inverting operational-amplifier shown in fig be X with S_1 open. When S_1 is closed the magnitude of gain becomes . . . . . .

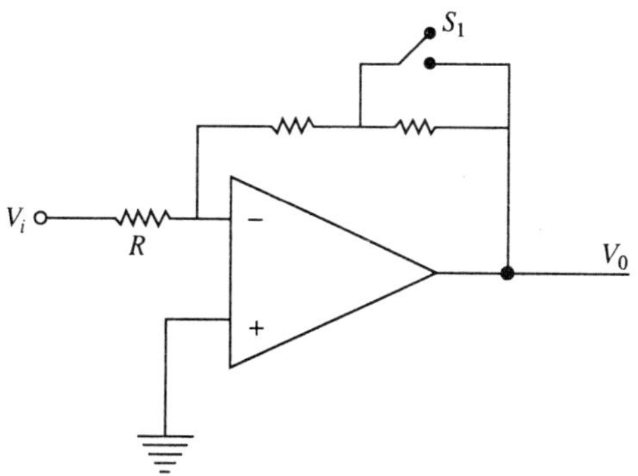

Fig. 13.25

With S_1 open

$$A_1 = \frac{-R_2}{R_1}$$

$$A_1 = \frac{-2R}{R} = X = -2$$

with S_1 closed

$$A_2 = \frac{-R}{R} = -1; \quad A_2 = \frac{X}{2}$$

Q. 7. The circuit shown in fig. 13.26 acts as inverting summer and output is . . .

Ans.

$$V_0 = -R_f\left(\frac{V_1}{R_1} + \frac{V_2}{R_2}\right)$$

$$V_0 = -1\left(\frac{1}{1} + \frac{2}{1}\right); \; V_0 = -3 \text{ V}$$

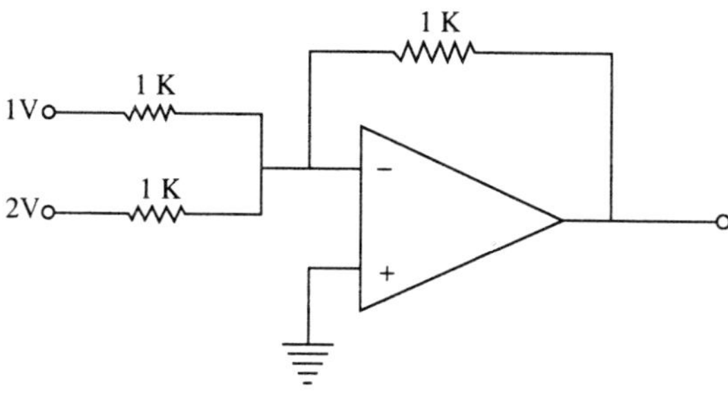

Fig. 13.26

Q. 8. In the figure, the operational-amplifier is ideal. What is the magnitude of the output voltage?

$$A = \frac{-R_2}{R_1}$$

$$A = -10$$

$$V_O = AV_i$$

$$V_O = -10 \times 10$$

$$V_O = -100\ \mu\text{V}$$

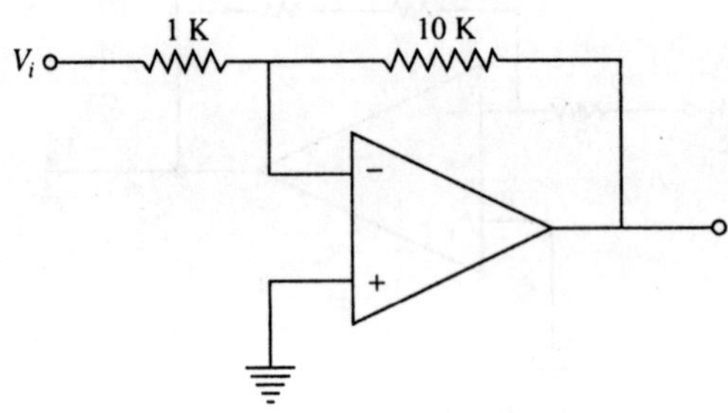

Fig. 13.27

Q. 9. The circuit shown in figure is an *inverting amplifier.*

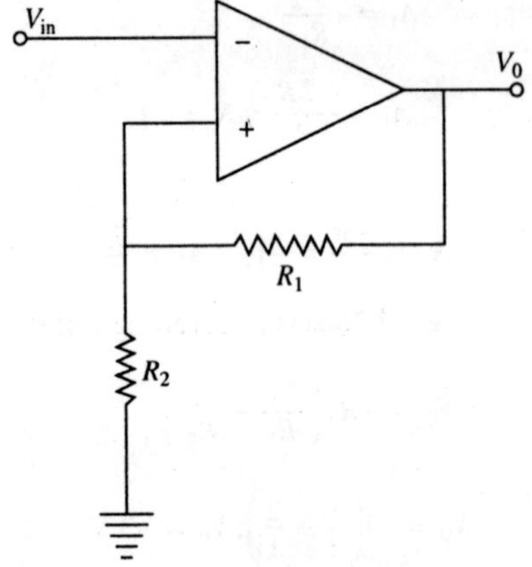

Fig. 13.28

Q. 10. In the given circuit, if the inputs V_1 and V_2 are to be amplified by the same amplification factor, the value of R should be 22 *k*

Ans. $R = 22\ k$

Q. 11. In order that the circuit shown in fig. 13.30 has to work as differentiator, state the modifications.

Ans. Interchange R and C

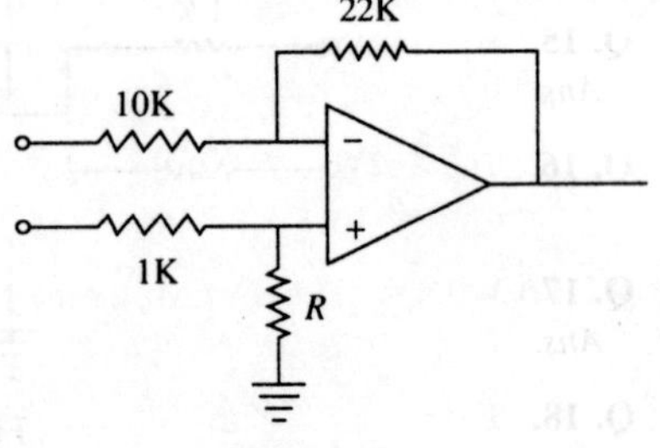

Fig. 13.29

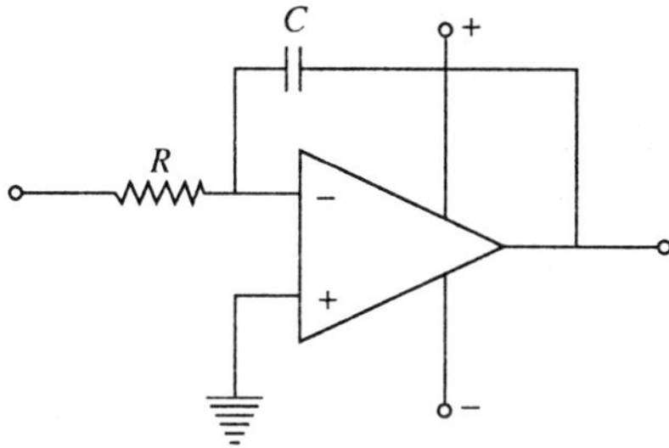

Fig. 13.30

Q. 12. The operational-amplifier circuit shown in Fig. 13.31 can be used as *log amplifier*

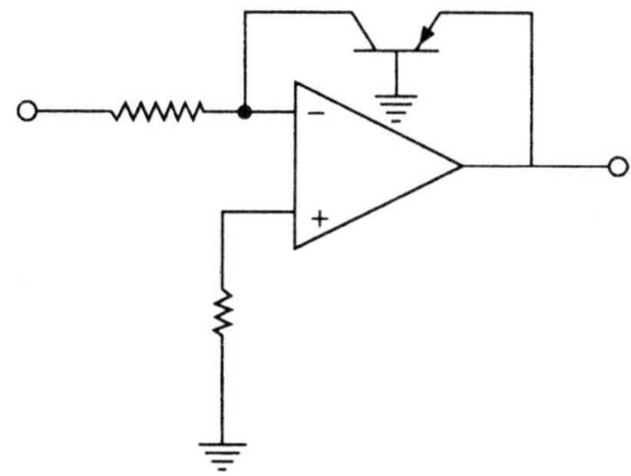

Fig. 13.31

Q. 13. The operational-amplifier circuit shown in figure is *peak detector*

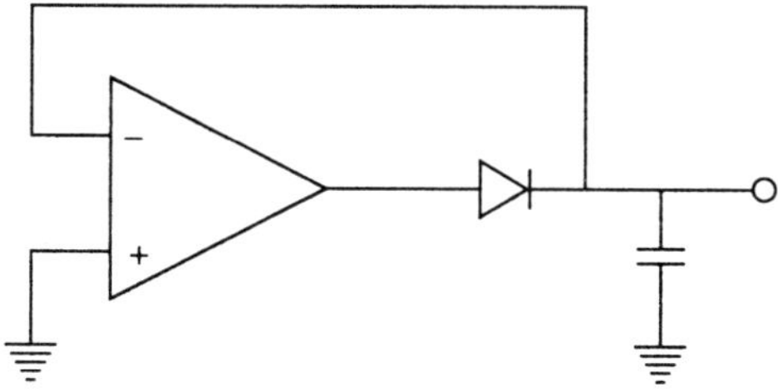

Fig. 13.32

Q. 14. An operational-amplifier used in open-loop is called *comparator* and it produces a *saturated* positive or negative output.

Q. 15. What is the main advantage of inverting amplifier?
Ans. It can handle more than one input at a time.

Q. 16. The features of instrumentation amplifier are *high* Z_i, *high CMRR* and *isolation of grounds*.

Q. 17. What is the use of Miller integrator?
Ans. *To generate ramp voltage*

Q. 18. The essential part of PLL is
Ans. *VCO*

Q. 19. What are the advantages of operational-amplifier based active filters?

Ans. (i) Voltage gain
(ii) Negligible loading effect
(iii) Elimination of inductors

Q. 20. What is the closed loop gain of operational-amplifier inverting amplifier?

$$A = -R_2/R_1$$

Q. 21. What is the voltage gain of operational-amplifier non-inverting amplifier?

$$A = 1 + \frac{R_1}{R_2}$$

Q. 22. High input impedance of operational-amplifier is achieved by
Ans. *Darlington connection*

Q. 23. The following type of amplifier is used in operational-amplifier.
Ans. *Differential amplifier*

Q. 24. What is the reason for error voltages and currents?
Ans. Two sections of differential amplifier are not identical.

Q. 25. Which point of operational-amplifier is used as summing point in an adder?
Ans. *Virtual ground.*

Q. 26. Operational-amplifier is classified as *linear* amplifier.

Q. 27. A differentiator circuit converts a linear ramp into *step voltage*

Q. 28. Adder produces an output which is equal to *algebraic* sum of inputs.

Q. 29. The circuits given in figs. 13.33 and 13.34 represent . . .

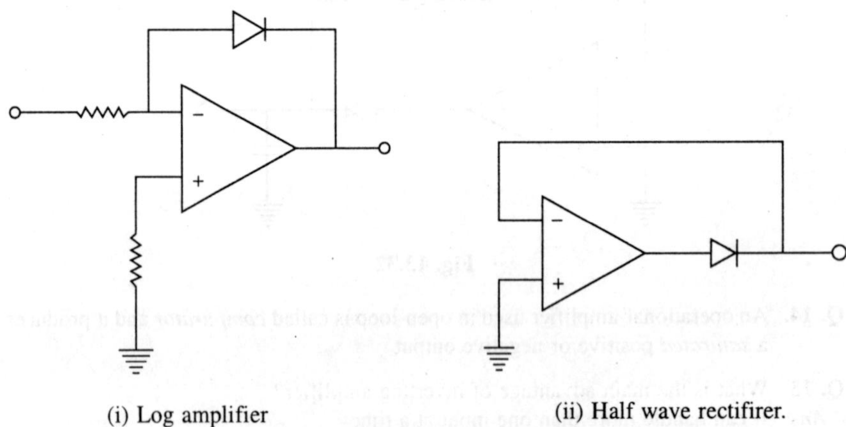

(i) Log amplifier (ii) Half wave rectifirer.

Fig. 13.33 **Fig. 13.34**

30. The function of resistor R_i in differentiator circuit is to reduce the value of high frequency noise.

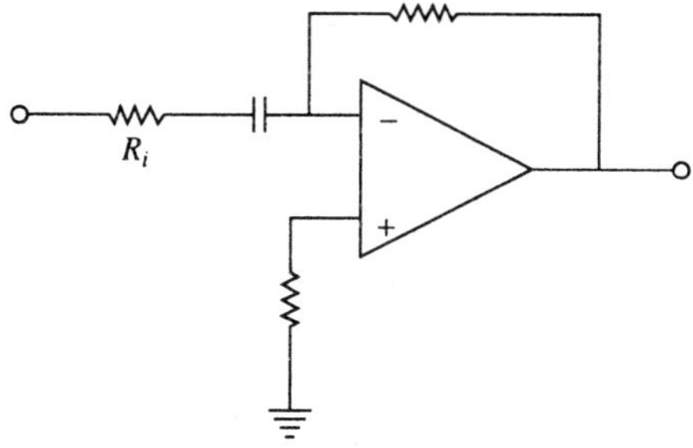

Fig. 13.35

31. The magnitude of peak value of output will be

$$V_0 = -RC\ V_m\ \omega \cos \omega t$$

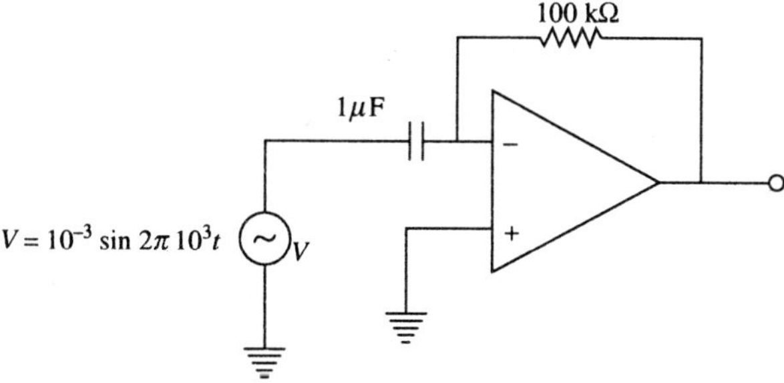

Fig. 13.36

$$\text{peak value} = R_c\ V_m \omega = 10^{-6} \times 10^5 \times 10^{-3} \times \omega$$

$$= 10^{-6} \times 10^5 \times 10^{-3} \times 2\pi \times 10^3$$

Peak value = 0.628 Volts

32. The approximate input impedance of the operational-amplifier circuit shown in figure is

$$R_1 = \frac{V_i}{I}$$

$$Z_{in} = R_1$$

$$Z_{in} = 10\ k\Omega$$

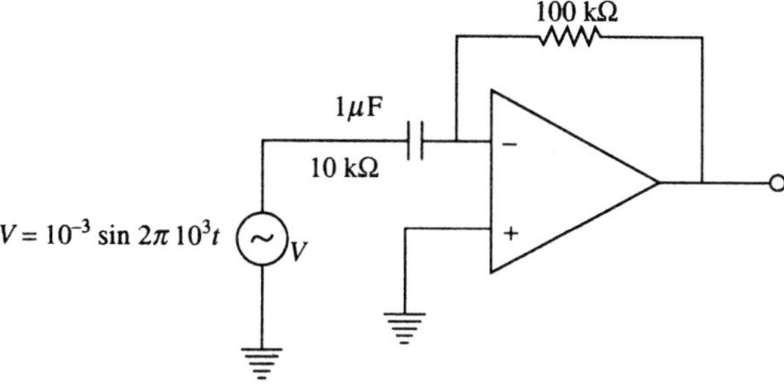

Fig. 13.37

33. When Z >> Z[1], the circuit behaves as . . .

$$A = 1 + \frac{Z'}{Z}$$

If $Z >> Z'$, gain is approximately equal to one. If $A = 1$, output is equal to input. Therefore, the circuit acts as a buffer.

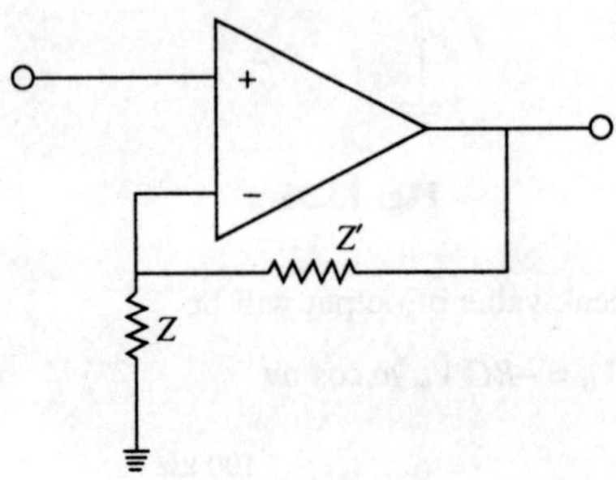

Fig. 13.38

Appendix I

Short Questions and Answers

1. A pure germanium crystal is an——semiconductor and a doped crystal is an——semiconductor
   Ans. intrinsic; extrinsic
2. The overall charge of a doped semiconductor is——
   Ans. positive or negative
3. Donor type impurities must have——valance electrons.
   Ans. five
4. The maximum voltage that can be applied to a diode without destroying it is called——voltage
   Ans. rated
5. In tunnel diode, impurity concentration is of the order of——
   Ans. one part in 10^3 atoms
6. Zener breakdown results from strong——
   Ans. electric field
7. The main function of a rectifier is to convert an A.C. voltage into——D.C. voltage
   Ans. unidirectional
8. When filter capacitor is shorted, output D.C. supply falls to——
   Ans. Zero
9. In a centre-tap full wave rectifier, V_m is the peak voltage between the centre tap and one end of the secondary. The maximum voltage across the reverse biased diode is——
   Ans. 2 V_m.
10. Define regulation.
    Regulation is the ratio of drop to full load voltage.

    Ans. $$\text{Regulation} = \frac{V_{dc(NL)} - V_{dc(FL)}}{V_{dc(FL)}}$$
11. A bridge rectifier is preferable to an ordinary two diode full wave rectifier because it needs much——for the same output.
    Ans. less PIV rated diodes
12. Larger the value of filter capacitor——is the peak current in the rectifying diode.
    Ans. larger
13. The leakage currents in a transistor are due to——carriers
    Ans. minority
14. For a given emitter current, the collector current will be higher if the recombination rate in the——were decreased.
    Ans. base region

15. A transistor is said to be in a quiescent state when——is applied to the input.
Ans. zero signal

16. The value of total collector current I_C in CB circuit is.

Ans. $\propto I_E + I_{CBO}$

17. The intersection of D.C. load line with the given base current curve is the.
Ans. Q-point

18. An improperly biased transistor produces——in the output signal.
Ans. clipping

19. What is the function of bias stabilization circuit?
Ans. Eliminates thermal run away and maintains stable Q-point.

20. MOSFETs gate and channel are——from each other.
Ans. insulated

21. ——electrons are responsible for forming different types of bonds in solids.
Ans. Valence

22. Intrinsic semiconductors have unequal number of electrons and holes. (True/False)
Ans. True

23. Why do we call the h parameters as hybrid parameter?
Ans. They have impedance, admittance, current gain and voltage gain. Hence they are called hybrid parameters.

24. How do you call the negative gate operation of an n channel MOSFET?
Ans. Depletion mode

25. Define stability factor.
It is the rate of change of collector current with respect to leakage current

Ans. $$S = dI_c/dI_{CO}$$

26. A transistor is operating in the active region. Under this condition——is forward biased and——is reverse biased.
Ans. input junction; output junction

27. Why is potential divider bias used in amplifiers?
Ans. It has low stability factor and it eliminates thermal runaway problem.

28. If collector supply voltage is 10 V, what is the collector cut off voltage under D.C. condition?

Ans. $$V_{CE} = V_{CC} - I_C R_C$$

$$\text{if } I_C = 0;\ V_{CE} = V_{CC} = 10 \text{ V}$$

29. If input signal is 1 mV, $R_{AC} = 1$ K ohm, $R_{in} = 0.5$ K ohm and $\beta = 25$. Calculate the output voltage.

Ans. $$A = \frac{\beta R_{AC}}{R_i} = \frac{25 * 1}{0.5} = 50$$

$$V_0 = 50 * 1 = 50 \text{ mV}$$

30. An amplifier has a voltage gain of 400, if the input is 4 mV, then the output voltage is——.

Ans. $$V_0 = AV_i = 400 * 4 = 1600 \text{ mV}$$

31. Give the symbol and basic biasing arrangement for the uni junction transistor.

Ans.

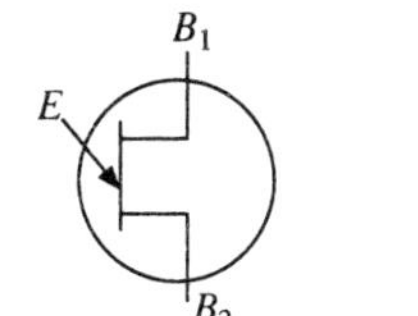

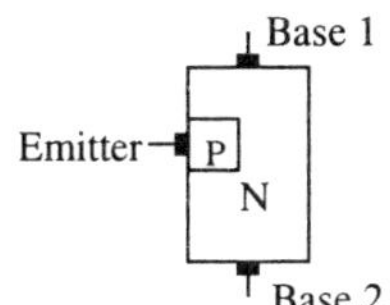

32. The voltage gain of a given common source JFET amplifier depends on
Ans. μ, R_L and r_D

33. Mention the reason for producing distortion in the output signal of a transistor.
Ans. 1. Improper biasing
2. Non linearity

34. Minority carriers in a transistor are responsible for.
Ans. leakage current

35. A particle with zero initial velocity, placed inside a uniform magnetic field.
Ans. deflects along the arc of a circle

36. In a *P*-type semiconductor the——are majority carriers and——are minority carriers.
Ans. holes; electrons

37. By adding donor type impurities in a intrinsic semiconductor we get——type semiconductor.
Ans. N

38. Two types of charge carriers in a semiconductor are——and——.
Ans. electrons; holes

39. Across an open circuited *P-N* junction diode there exists a ——potential.
Ans. barrier

40. When a *P-N* junction is reverse biased, the depletion region width is——.
Ans. increased

41. LEDs largely emit——.
Ans. light energy

42. *PIN* diodes are used as——switches.
Ans. microwave

43. The transition capacitance is mainly due to——
Ans. space charge or depletion layer formed by positive and negative impurity ions.

44. The device which exhibits variable junction capacitance with applied reverse bias voltage is.
Ans. varactor diode

45. On increasing the magnitude of base collector voltage, the effective base width decreases. The effect is called.
Ans. early effect

46. In germanium transistors, the change in i_{co} with temperature poses more problems than a silicon transistors. (True/False)
Ans. True

47. FET has——input impedance.
Ans. high

48. The——region of the UJT characteristics is used to construct oscillators.
Ans. negative resistance

49. DIAC is used as a——device in TRAIC power control systems.
Ans. triggering

50. CB transistor has got——input impedance and——output impedance.
Ans. low; high

51. When FET is operated in the——region of its output characteristics it exhibits a voltage variable resistor characteristics.
Ans. saturation

52. The other name for hybrid model is——model.
Ans. *h*-model or low frequency model

53. When a transistor is used as a switch its operating point swings between—— region and——region.
Ans. cut-off; saturation

54. Heat sinks are used to avoid——problem.
Ans. thermal runaway

55. Define potential.
Ans. Potential is defined as work done per unit charge.

56. Magnetic deflection sensitivity in a CRO is directly proportional to anode voltage. True or False.
Ans. False

57. In an *N*-type semiconductor the——are the majority carriers and——are the minority carriers.
Ans. electrons; holes

58. Semiconductors have negative temperature coefficient of resistance. True False
Ans. True

59. A——is developed across a current carrying metal bar when it is placed in a transverse magnetic field. The effect is called the——.
Ans. voltage; Hall effect

60. The diffusion capacitance is mainly due to——charge carriers.
Ans. temporary

61. Solar cells make use of——effect.
Ans. photovoltaic solar energy conversion

62. For a transistor to operate in active region, the emitter junction is——biased and collector junction is——biased.
Ans. forward; reverse

63. In a fixed bias circuit the——current is maintained constant.
Ans. base.

64. Typical vlaue of h_{fb} for CB transistor is.
Ans. 0.98.

65. The useful region of operation of a Zener diode is——biased region.
Ans. reverse

66. SCR can be called as a——thyristor.
Ans. PNPN

67. CC amplifier has got——input impedance and——output impedance.
Ans. high; low

68. A——device can be used as voltage variable resistor.
Ans. Varistor

69. *h* parameters are called hybrid parameters because.
Ans. they are combination of impedance, admittance, voltage gain and current gain.

70. Cut in voltage of a germanium diode is——volt and silicon diode is——volts.
Ans. 0.3; 0.7

71. If an electron has a constant speed but does not move perpendicularly into a uniform magnetic field, what type of path it will describe?
Ans. Circular arc.

72. What are the applications of CRO?
Ans. Measurement of voltage, current, phase angle and frequency.

73. What it Hall effect?
Ans. When a current carrying semiconductor is kept in a magnetic field, a voltage is induced in it.

74. Why is silicon preferred to germanium in the manufacture of semiconductor devices?
Ans. Silicon has less leakage current.

75. What is potential barrier in *P-N* junction. For silicon *P-N* junction will the potential barrier be about 0.7 V, or 0.1 V or 2.1 V?
Ans. Height of the potential graph for the depletion layer is called potential barrier. 0.7 V

76. Explain the terms knee voltage and break down voltage with respect to diodes.
Ans. Knee voltage is the voltage below which the current through the diode is zero. Breakdown voltage is the voltage at which the diode breaks down. If the voltage of a reverse biased diode is increased beyond breakdown voltage, the current through the diode increases rapidly.

77. What is avalanche breakdown in *P-N* junction diode.?
Ans. When the voltage applied to a reverse biased *P-N* junction is increased, the electric field increases. When minority carrier (electron) enters the electric field, it experiences a force. This electron can break a covalent bond and two electrons come out. These two can knock off two more electrons and this process continues. The current increases abruptly. This is called avalanche breakdown mechanism.

78. What is β of the transistor when alpha is 0.98.

Ans.
$$\beta = \frac{\alpha}{1-\alpha} = \frac{0.98}{1-0.98} = 49$$

79. In a common base connection, the emitter current is 1 mA, $I_{CBO} = 50\ \mu A$, $\alpha = 0.92$, find the total collector current.

Ans.
$$I_C = \alpha I_E + I_{CBO}$$
$$= 0.92 \times 1 + 0.05 = 0.97 \text{ mA}$$

80. Which of the biasing methods provide very good stabilization of the operating point (a) fixed bias (b) potential bias (c) collector feedback bias?
Ans. Collector feedback bias

81. Constructionally what are the basic differences between phototransistor and BJT?
Ans. In phototransistor the energy required at the base is given in the form of light energy. In BJT the source connected at the base supplies the energy required. Phototransistor conducts when light falls on the glass surface. BJT conducts when current flows through the base.

82. For a FET, A.C. resistance is 32 K ohm, the transconductance is 3600 μmho, find the amplification factor.

Ans.
$$\mu = g_m r_d$$
$$= 3600 \times 10^{-6} \times 32 \times 10^3$$
$$= 115.2$$

83. Why is the input impedance of a MOSFET higher than that of a FET?
Ans. Dielectric present between the gate and channel increases the input impedance.

84. Define the terms holding current and latching current with respect to SCRs.
Ans. Holding current is the minimum current required to keep the device in ON state. The current required to take the device from OFF state to ON state is called latching current.

85. At $V_{CE} = 7.5$ volt, the change in collector current is 1.2 mA for a change in base current of 20 μ A, find h_{fe} of the transistor.

Ans. $$h_{fe} = \frac{1.2 \times 10^{-3}}{20 \times 10^{-6}} = 60$$

86. With respect to diode switching, what is the reverse recovery time t_s?
Ans. The time diode takes to turn off after it is reverse biased is called reverse recovery time. This is due to the charge storage.

87. What is gain-bandwidth product?
Ans. Product of gain and bandwidth is called figure of merit.

88. What are various focussing techniques used in CRO?
Ans. 1. Electrostatic focussing 2. Electromagnetic focussing

89. Define ALPHA of a transistor?

Ans. $$\alpha = \Delta I_C / \Delta I_E$$

90. What is thermal runaway?
Ans. When temperature increases, leakage current increases and collector current increases. This continues until the transistor is damaged. It is called thermal runaway.

91. What is pinch-off voltage?
Ans. Drain to source voltage beyond which the drain current remains constant is called pinch-off voltage.

92. Define intrinsic stand off ratio.
It is the ratio to pea.. potential to the applied voltage.

Ans. $$\eta = \frac{V_P}{V_{BB}} = \frac{R_{B2}}{R_{B1} + R_{B2}}$$

93. Define the following terms: (a) Rise time (b) Storage time of a transistor.
Ans. Rise time is the time required by the collector current to rise from 10% to 90% of its maximum value.
Storage time is the time between the instant of applying reverse voltage to the instant of 90% of I_C.

94. What does UJT stand for? Justify the name UJT.
Ans. UJT stands for unijunction transistor. It is called UJT since it has only one junction.

95. What is tunnelling phenomenon?
Ans. The phenomenon of penetration of charge carriers directly through the potential barrier is called tunnelling.

96. Define the following term: Fall time of a diode.
Ans. Fall time is the time taken by the diode current to fall from 90% to 10% of its maximum value.

97. Define and give the expression for deflection sensitivity of a TV picture tube
Ans. Deflection sensitivity is defined as deflection per applied voltage.

$$S = \frac{Ll}{2\,d\,V_a}$$

98. What type of focussing is used in a CRO. Why?
Ans. Electrostatic focussing is used in a CRO. To obtain a sharply focussed beam, electrostatic focussing is used.

99. Give the relationship between mobility and conductivity.

Ans. $$\mu = v\ \sigma/J$$

100. State the expression for drift and diffusion currents.

Ans. $$I_{di} = AqD_n dp/dx$$

$$I_{dr} = A\sigma E$$

101. What is the load line?
Ans. Line drawn on the output characteristic with a slope of $(-1/R_c)$ is called load line.

102. What is the significance of I_{CBO} and I_{CO}?
Ans. I_{CBO} is the leakage current from collector to base with emitter open. I_{CO} is the leakage current from collector to emitter with base open.

103. Which transistor configuration will give more bandwidth?
Ans. Common collector configuration.

104. Define f_B and f_T.

Ans. $$f_b = 1/(2\pi r_{be}C).$$

$$f_T = \beta f_b.$$

105. State two applications of magnetic deflection.
Ans. (1) CRO, 2) TV picture tube.

106. Mention two applications of LDR.
Ans. 1. Burglar alarms 2. Automatic street lighting 3. OFF at dark circuit

107. State two applications of clampers and clippers.
Ans. 1. Radars 2. Television receivers 3. Digital computers

108. What are the ratings of a transistor?
Ans. Voltage rating, continuous current rating and di/dt rating.

109. Which device will give more speed, high input impedance and less noise; BJT or FET?
Ans. FET

110. Define break over current I_K in a Zener diode.
Ans. Minimum value of Zener current required to keep the device in breakdown region is called Zener diode.

111. Define electron—volt.
Ans. It is the kinetic energy gained by an electron moving through a potential difference of 1 V.

112. What is meant by mass—action law?
Ans. The product of number of holes and number of electrons is constant.

$$np = n_i^2$$

113. Define the term diffusion capacitance.

Ans. $C_d = Tg$ where T is life time and g is conductance

Diffusion capacitance is due to the temporary storage of charges near the junction.

114. What are the factors that contribute to the delay time when the transistor is used as a switch?
Ans. Base emitter junction capacitance.

115. How many Joules does an electron volt equal?
Ans. 1.6×10^{-19} Joules

116. What is the force experienced by a charged particle moving parallel to a steady magnetic field?
Ans. $F = 0$

117. What is the trajectory of a charged particle in a uniform magnetic field?
Ans. Circular arc.

118. Is a semiconductor negatively charged when doped with donor atoms?
Ans. Yes

119. What are the two charge carriers and the two conduction mechanisms present in a semiconductor?
Ans. Electrons and holes; 1. Drift current 2. Diffusion current.

120. Why does reverse saturation current vary with temperature?
Ans. As temperature increases, the number of covalent bonds broken increases, free electrons increase and reverse saturation current increases.

121. Write down the expression for diffusion current density due to electrons.

Ans. $$J_n = -qD_n dp/dx.$$

122. Write down the expression for drift current density due to holes.

Ans. $$J_p = -qD_p dp/dx.$$

123. Write down and explain the junction diode equation.

Ans. $$I = I_0(e^{V/KV_t} - 1).$$

I—current through the junction diode
I_o—reverse saturation current
V—voltage applied
K—constant
V_t—volt equivalent of temperature

124. What semiconductor materials are preferred in the manufacture of LEDs?
Ans. Gallium arsenide and gallium arsenide phosphide are preferred.

125. What is varactor diode?
Ans. Varactor diode is basically a reverse biased *P-N* junction. This utilizes the depletion layer capacitance of *P-N* junction. It is used as a variable capacitor. As voltage increases, width of the depletion layer increases and capacitance decreases.

126. What is peak inverse voltage?
Ans. The maximum reverse voltage across the non-conducting diode is called peak inverse voltage.

127. Write down an expression for the efficiency of a single phase full wave rectifier.

Ans. $$\text{Efficiency} = 0.81 \times R_L/(R_f + R_L).$$

138. Is it true that an ideal power supply has zero per cent regulation?
Ans. Yes

129. Sketch the transfer characteristics of DE MOSFET.
Ans.

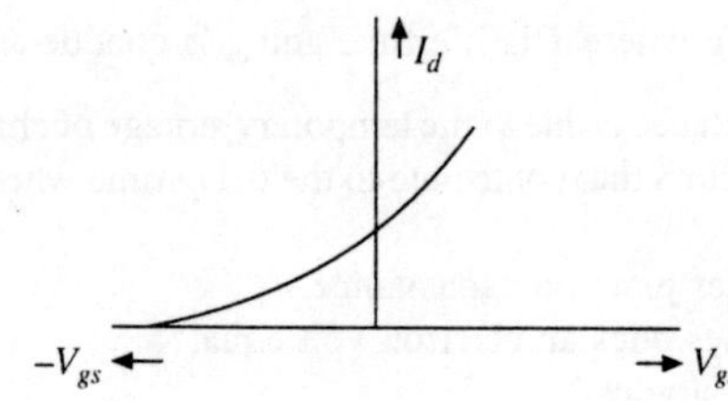

Question and Answer Bank–I

1. What are the regions used when BJT is used as a switch?
   Ans. Cut-off region and saturation region.
2. Gate current of FET is of the order of——.
   Ans. nano amperes
3. Define Latching current of SCR.
   Ans. Minimum current required to take the SCR from OFF state to ON state is called latching current.
4. Most SCRs can be turned OFF by voltage reversal during negative half cycle of A.C. supply up to——.
   Ans. 2 kHz
5. What is thermal resistance of power BJT?
   Ans. Thermal resistance is the resistance to the flow of heat. The heat flows from the junction to the surrounding air. Larger the transistor case, smaller the thermal resistance and vice versa. Thermal resistance is reduced by providing heat sink with the transistor.
6. What is the punch through (or) reach through?
   Ans. Width of the depletion layer depends on the applied voltage. If the voltage applied to the output junction is increased, width of the depletion layer increases. At a particular output voltage both the depletion layers touch. A large current flows through transistor. Such effect is called punch through or reach through.
7. Define holding current of SCR.
   Ans. Minimum current required to keep the SCR in ON state is called holding current.
8. What is the difference between signal device and power device?
   Ans. Signal devices are rated at low voltage and milli ampere of current. Power devices operate at high voltage and high currents. (KV and KA).
9. What is the difference between converter grade and inverter grade SCR?
   Ans. Converter grade SCRs are suitable for low frequency operation. (power frequency—50 Mz). Turn OFF time is higher. Typical value is 100 μs.
   Inverter grade operate at high frequencies. Time period is lesser. Inverter grade SCRs are used in inverter circuits. Typical turn OFF time is 20 μs. They are costlier than converter grade SCRs.
10. What is reverse recovery of SCR?
    Ans. To turn-off the SCR, reverse voltage must be applied after the current is reduced to zero. Electrons in N_2 are attracted by positive plate. Holes in P_1 flows through the negative plate. Hence small current flows through the SCR in reverse direction. Free charges around J_2 are trapped. They can vanish only by recombination. Sum of reverse recovery time and recombination time is called SCR turn off time.
11. Define stability factors S and S' and S''

    Ans. $$S = \frac{dI_c}{dI_{co}};\ S' = \frac{dI_c}{dV_{be}};\ S'' = \frac{dI_c}{dB}$$

12. How do you stabilize D.C. operating point?
    Ans. Voltage divider bias along with compensation can keep the operating point stable.
13. Why does the Q point shift?
    Ans. Q point can shift due to following reasons:
    1. Change in temperature changes leakage current and I_c.

2. Change in V_{be} changes collector current.
3. Change in β changes collector current.

14. A BJT has $I_B = 10\ \mu A$, $\beta = 99$, $I_{CO} = 1\ \mu A$. What is collector current?

Ans.
$$I_c = \beta I_B + I_{CEO};\quad I_c = \beta I_B + (1 + \beta) I_{co}$$
$$= (99 * 10) + (1 + \beta)\, I_{co}$$
$$= 99 * 10 + (1 + 99)\, 1$$
$$I_c = 1090\ \mu A.$$

15. Why common emitter configuration needs biasing?

Ans.
$$I_c = \beta I_B + I_{CEO}$$
$$I_c = \beta I_B + (1 + \beta)\, I_{co}$$
$$\frac{dI_c}{dI_{co}} = 1 + \beta$$

Stability factor is very high. To reduce the stability factor voltage divider bias must be used.

16. Why biasing is not required for common base configuration?

Ans.
$$I_c = \propto I_E + I_{co}$$

$\frac{dI_c}{dI_{co}} = 1$. It has built in stability factor of unity. Therefore biasing is not required.

Question and Answer Bank-II

1. With the increase in gate current the breakdown voltage of an SCR *decreases.*

2. In a UJT what is the relation between V_P and V_{BB}?

$$\eta = \frac{V_P}{V_{BB}} = \frac{V_{EB2}}{V_{BB}}$$
$$V_{EB2} = \eta V_{BB}$$

3. Number of the thyristors connected in parallel provide total *rated* current *less than the sum of individual ratings.*

4. Inverter grade SCRs have t_q *less than* 25 *μs.*

5. Excess *dv/dt* to a thyristor may cause *maloperation*

6. In case of thyristor in series, *both forward and reverse voltages have to be shared.*

7. The collector current is 2.9 mA in a certain transistor. If the base current is 100 μA. What is the value of ∝?

$$\propto = \frac{I_c}{I_E} = \frac{I_c}{I_C + I_B} = \frac{2.9}{3} = 0.97$$

8. An example of solid state device is *FET.*

9. GTO can be turned-OFF by applying *large negative current.*

10. Turn-OFF time of SCR depends on *temperature* and *forward current.*

11. What is a *P-N* junction?

If a pure semiconductor is partly doped with third group and partly with fifth group impurity, a *P-N* junction is formed. It offers low resistance in forward direction and high resistance in reverse direction.

12. What are the members of thyristor family?
SCR, DIAC, TRIAC, SUS, SCS, GTO, LASCR.
13. What is the effect of negative gate current on a normal SCR?
Gate has no control over turn-OFF of the device. Negative gate current produces additional heat at the gate junction.
14. How the forced turn-OFF of an SCR is different from the natural turn-OFF?
In A.C. circuits the current naturally reduces to zero. In D.C. circuit the current is forced to zero by using L and C.
15. The reverse recovery time of a diode is the time required by *majority carriers to recombine and get neutralized.*
16. Compared to that of the thyristor the switching loss of GTO is *much lower since turn-ON and turn OFF times of GTO are less.*
17. A transformer is employed in the triggering circuit for *isolation purpose*
18. A diode is usually connected across the primary of a pulse transformer *to prevent D.C. saturation.*

Bibliography

1. Applied Electronics, R.S. Sedha, S. Chand & Co., 1990.
2. Basic Electronics and Linear Circuits, N.N. Bhargava, TMH, 1984.
3. Basic Electronics, B.L. Theraja, S. Chand & Co, 1991.
4. Electron Devices, S. Rama Reddy, Lecture Notes, University of Madras, 1998.
5. Electronic Devices and Circuits, Millman and Halkias, TMH, 1982.
6. Electronic Devices and Circuits, Sanjeev Gupta, Dhanpat Rai & Sons, 1994.
7. Principles of Electronics, P.A. Malvino, TMH, 1984.
8. Solid State Circuits, S. Rama Reddy, Lecture Notes, University of Madras, 1997.

Index